Systems Engineering Approach to Medical Automation

Artech House Series
Engineering in Medicine & Biology

Series Editors

Martin L. Yarmush, Harvard Medical School
Christopher J. James, University of Southampton

Advanced Methods and Tools for ECG Data Analysis, Gari D. Clifford, Francisco Azuaje, and Patrick E. McSharry, editors

Advances in Photodynamic Therapy: Basic, Translational, and Clinical, Michael Hamblin and Pawel Mroz, editors

Biological Database Modeling, JakeChen and Amandeep S. Sidhu, editors

Biomedical Informaticsin Translational Research, Hai Hu, Michael Liebman, and Richard Mural

Biomedical Surfaces, Jeremy Ramsden

Genome Sequencing Technology and Algorithms, Sun Kim, Haixu Tang, and Elaine R. Mardis, editors

Intelligent Systems Modeling and Decision Support in Bioengineering, Mahdi Mahfouf

Life Science Automation Fundamentals and Applications, Mingjun Zhang, Bradley Nelson, and Robin Felder, editors

Microscopic ImageAnalysis for Life Science Applications, Jens Rittscher, Stephen T. C. Wong, and Raghu Machiraju, editors

Next Generation Artificial Vision Systems: Reverse Engineering the Human Visual System, Maria Petrou and Anil Bharath,editors

Systems Bioinformatics: An Engineering Case-Based Approach, Gil Alterovitz and Marco F. Ramoni, editors

Systems Engineering Approach to Medical Automation, Robin Felder.

Translational Approaches in Tissue Engineering and Regenerative Medicine, Jeremy Mao, Gordana Vunjak-Novakovic, Antonios G. Mikos, and Anthony Atala, editors

Systems Engineering Approach to Medical Automation

Robin Felder
Majd Alwan
Mingjun Zhang

Editors

ARTECH HOUSE
BOSTON | LONDON
artechhouse.com

Library of Congress Cataloging-in-Publication Data
A catalog record of this book is available from the Library of Congress.

British Library Cataloguing in Publication Data
A catalogue record of this book is available from the British Library.

ISBN 13: 978-1-59693-164-0
ISBN 10: 1-59693-164-7

Cover design by Igor Valdman

685 Canton Street
Norwood, MA 02062

10 9 8 7 6 5 4 3 2 1

Contents

Preface

Robin A. Felder and G. Terry Sharrer

Over the past two centuries, automation has transformed industry after industry, making better material goods, delivering them efficiently, and allowing more people to share in a culture called "the good life." Greater choice and lower costs have been the hallmarks of automation's bounty, but strangely—almost dumbfoundlingly—healthcare has followed a different course. It is common knowledge that the United States has the most expensive healthcare system in the world, without corresponding performance in, say, extending life expectancy. Indeed, the main basis of faith in American healthcare is that if it costs so much, it must be good. The truth is that automation and process integration on a scale that powered automobile and computer production is virtually nonexistent in healthcare. It is the industry left behind —incredibly, as it deals with matters of life and death.

"Medical automation" is a term we have coined to define the integrated use of robotics, informatics, process management, and other technologies that streamline healthcare delivery. Like all automation before, greater effectiveness, efficiency, and equitability are the macro-goals of medical automation, with improvements in safety, elimination of fraud, lower costs, and a more personalized approach to healing among many supporting goals. It is not hyperbole to suppose that without medical automation the economic and social security of the United States is threatened, even more so than from the actions of terrorists.

We wrote this book to light the way for automated healthcare. Laymen and professionals will find the basic principles of medical automation in the chapters that follow. Chapters 1 through 3 lay out the concepts of automation, human factors engineering, and mathematical modeling. In-depth coverage of the mathematical principles involved in dynamic simulation, robot swarm modeling, biometrics, and lab-on-a-chip follow in Chapters 4 through 7. Lay readers can find overviews of practical applications in real-world automation for the laboratory, pharmacy, surgery, supply chain, process management, telehealth, in-home geriatric care, connected medicine, and information technology in Chapter 8 through 16. Opportunities for healthcare improvements exist in the way we handle a single living cell or the layout of a laboratory that decreases labor needs by tenfold.

The authors hope these pages inspire as well as inform, as automation is capable of motivating a self-sustaining innovative force. If so, health itself could become a more achievable goal, now and for generations that inherit what we've accomplished.

CHAPTER 1

Introduction to Medical Automation

Robin A. Felder

This chapter introduces the concept of medical automation and describes various applications of this emerging field. Some of the benefits of medical automation will be described, including increased patient safety, reduced costs, reduced burdens on healthcare workers, and ultimately improved patient health.

1.1 Introduction

Medical automation is the application of process automation principles to the practice of medicine. Webster's defines automation as the technique of making an apparatus, a process, or a system that operates using a self-acting or self-regulating mechanism. The principal benefits of automation are not only the reduction in cost to producing an end product, but the improvements in quality of the medical process and increased safety to patients. Measured medical cost reductions through automation can be so dramatic that one wonders why there isn't the equivalent of a "medical gold rush," to rapidly adopt these technologies.

Optimally, medical automation is deployed using a familiar engineering paradigm called a closed loop process. A sensor or data generator begins the process by generating useful data. Data is then sent to a process manager computing element to render a decision based on the data. The decision is then used to effect a change in a process delivered to the patient. A simple example is a glucose sensor in a closed loop with an insulin pump, so that patients do not have to actively participate in deciding what insulin dose to administer. This eliminates the significant elevations in blood glucose that come with manual finger sticking and syringe-based insulin dosing. Health outcomes have been demonstrated to improve based on closed loop systems [1].

1.2 The Need for Improved Process Management in Medicine

Despite major efforts to reduce the increasing cost of medicine, Americans have been unable to cap the rise in medical care to levels that match the rate of inflation. In 2005 (the most recent data), medical care costs increased at a rate (6.9%) that was twice that of inflation and equal to 16% of the gross domestic product in the United States [2]. The two trillion dollar price tag that was measured in 2005 is predicted to increase to four trillion dollars [3] in 2015 as the aging population begins to impact the healthcare system [4]. Annually, Americans spend more than any other country on health insurance premiums, which average $11,500 for a typical family of four and approximately $4,200 for a single person.

Increased spending does not always lead to high quality. Various reports from the Institute of Medicine have estimated that 98,000 Americans die each year due to errors in medicine, including delays in providing appropriate treatment [5].

1.3 Knowledge Empowers Automation

Optimal use of information technology will provide real-time data regarding the facts associated with every medical process, such as location and health of the patient, location and skill level of the human resources, location and operational status of medical equipment, and real-time data regarding the flow of these resources through the medical facility. Armed with knowledge, computer oversight can then schedule routine medical procedures, as well as optimize the process and scheduling to best suit the patient and facility. Administrative tasks can also be facilitated by process control by suggesting process solutions, forecasting trends, and anticipating future needs.

Properly managed processes that have been optimized for lean performance have demonstrated over 50% in productivity gains, greater than 10% reduction in defects, and up to 90% reduction in inventory costs. An unanticipated benefit of lean processes is space gains up to 50% due to better organized floor plans that consume less space resources. Throughput of medical work can improve by up to 80%, one contributing factor being a 75% reduction in distance traveled (data from ValuMetrix, www.valumetrixservices.com). In summary, automated medical care will be self-regulating, which will hopefully have the benefit of preventing inappropriate care or possibly harmful care to patients.

1.4 Personalization of Medicine Enabled by Technology

Outcome-based medicine based on population data is evolving to focus on personalized medicine based on longitudinal data from individuals. For example, many cancer cases present with the same basic disease, but manifest these diseases in unique ways. Thus, no one treatment may help each patient in the same way. Individualized diagnostics coupled with individualized therapies will hopefully yield better outcomes. New biotechnologies and technical innovations are focusing on personalized interfaces that are intimately connected to the patient. Cell phones, wearable computers, and embedded sensors in the home are beginning to gather and interpret real-time physiologic data throughout the day. For example, the Bodybug [6] provides pulse, body position, and skin temperature in order to quantitatively assess daily body activity. Glucose meters linked to cell phones can measure, interpret, and report glucose values for diabetics [7]. Clothing will include sensing and computing elements that measure biochemical substances in the skin and report gradual and otherwise unnoticeable changes in physiology.

1.5 Applications of Automation

The costly health system in the United States presents ideal opportunities for automation [8]. Clinical laboratories generate up to 70% of medically relevant data and yet are plagued by large numbers of medical specimens that need to be processed in a timely fashion. The pharmacy supply chain also presents significant challenges for storage, dispensing, returning, restocking, and billing for the right medication at the right time without burdening the nursing staff. Our aging nursing workforce will soon need to be supplemented with new talent; however, the forecast is for a paucity of new nurses over the next decades. Technology can make up for the nursing gap and also provide a higher degree of safety for patients. Surgery is also evolving through the use of automation. Less invasive robotic technologies are enabling more precise procedures to be performed and allowing patients to return to their normal routine in less time. Other areas that are already benefiting from automation include resource tracking, food service, deliveries by mobile robot, patient check-in and discharge, clinical engineering, sterile supply, and numerous other direct patient contact or service areas.

Automation is even being applied to psychological assessment of patients. Clinical psychologists Jaap Hollander and Jeffrey Wijnberg from Holland developed an automated Internet-based psychological coach called MindMentor [9]. An automated hour-long artificial intelligent session in a question and answer format allows patients to be assessed regarding mental challenges. Laced with humor, intelligence, and personal support, the system apparently demonstrated a direct improvement in patients' sense of well-being by allowing them to quickly focus on their main issues using a goal-directed technique. A clinical trial conducted in 2006 resulted in 47% of the subjects reporting that their problems were solved. While seemingly not impressive, 47% is a higher rate of success than that experienced by direct contact between individual patients and their psychologists.

1.5.1 Bar Coding, RFID, and Wireless Tracking

Bar codes have been the mainstay of labeling in medical systems for only a decade since the medical information systems that were in place were not amenable to systemwide integration. However, most medical systems have installed bar code systems for tracking medications, laboratory specimens, patient charts, and on patient armbands. There is a growing awareness of the benefits of radio frequency identification (RFID) in many industries that need precise inventory control (e.g., automotive manufacturing, cattle breeding, airline baggage handling, and large retail outlets). RFID works by exchanging wireless information between the reader and the tag containing an electronic chip encoded with data, and an associated antenna. There are two main types of RFID tags—an active tag equipped with a battery so it can send out its ID spontaneously, and a passive tag that respond to a nearby reader. Detectors can be placed in doorways, in the ceiling, and in floors, or anywhere that a positive read is required. Thus, items or patients would not be able to be moved from location to location without triggering an update in their location records.

RFID tags have the added benefit of being able to be equipped with temperature measurement sensors. Thus, in an application such as blood banking, the location, length of time in transit, and continuous temperature data could be recorded. Warnings could be issued automatically if the product was not on time, had reduced quality, and was about to be administered to the wrong patient.

1.6 Summary and Conclusion

This brief introduction to medical automation covered the definition of the term medical automation, as well as the closed loop feedback control architecture that maximizes the benefit of automation. Several examples were discussed regarding the use of automation in various medical services. The return on investment that medical facilities can expect from automation was described.

References

[1] Franklin, B. D., et al., "The Impact of a Closed-Loop Electronic Prescribing and Administration System on Prescribing Errors, Administration Errors and Staff Time: A Before-And-After Study," *Quality and Safety in Health Care*, Vol. 16, 2007, pp 279–284.

[2] Catlin, A. C., et al., "National Health Spending in 2005," *Health Affairs,* Vol. 26, No. 1, 2006, pp 142–153.

[3] Borger, C., et al., "Health Spending Projections Through 2015: Changes on the Horizon," *Health Affairs,* Vol. W61, Web Exclusives, www.healthaffairs.org, published online Feb. 22, 2006.

[4] The Henry J. Kaiser Family Foundation, "Employee Health Benefits: 2006 Annual Survey," September 2006.

[5] http://www.iom.edu/?id=12735 (accessed March 1, 2008).

[6] http://www.bodybug.se (accessed April 2, 2008).

[7] http://www.lge.com (accessed April 2, 2008).

[8] 2007 Guide to Patient Safety Technology," *Nursing Management,* December 2006, pp. 18–60.

[9] http://www.mindmentor.com (accessed April 2, 2008).

CHAPTER 2

Human Factors Issues with Automation

Mark W. Scerbo

2.1 Human Factors

Human factors or ergonomics is a discipline concerned with understanding the capacities and limitations of humans and creating technology, tools, and systems that best accommodate the abilities of human users [1, 2]. Human factors is a hybrid of engineering and psychology. Knowledge from a broad range of topics including attention, sensation, perception, human information processing, motivation, physiological processes, teams, anthropometrics, and biomechanics is applied to the design of products and services. Major application areas include computer hardware, software, military systems, aerospace systems, automobiles, consumer products, work environments, and healthcare systems. Some of the many goals of human factors are shown in Table 2.1.

Interest in improving human performance can be traced back the Greek Olympic games held in 776 through 680 B.C. [4]. However, the origins of human factors are often attributed to the efforts of Frederick Talyor to improve worker performance in the early 1900s [4]. The first real studies of worker improvement were conducted by the Gilbreths, who analyzed the motions used by brick layers and were able to document a threefold increase in efficiency by eliminating unnecessary movements [5].

Table 2.1 Fundamental Goals of Human Factors

Reduce errors
Increase safety
Increase reliability of systems
Reduce training requirements
Improve maintainability
Increase efficiency
Increase productivity
Improve the working environment
Reduce fatigue and stress
Increase human comfort
Reduce monotony
Increase convenience of use
Increase user acceptance
Increase job satisfaction
Improve the quality of life

Based on [3].

Human factors efforts to improve technology emerged shortly after WWII. In the United States, the Navy and Army Air Corps created laboratories to study the ability of soldiers and aviators to use military equipment [1]. About the same, the British Royal Air Force began to study the performance of sonar operators [6]. In 1949, Chapanis, Garner, and Morgan published the first book on human factors, *Applied Experimental Psychology: Human Factors in Engineering Design*. Eight years later, the first professional organization, the Human Factors Society, was established followed by the International Ergonomics Association shortly thereafter. The profession grew dramatically from the 1960s through the 1980s largely due to the unique requirements of NASA's space program and the introduction of computer systems. During the same time, Bell Telephone Laboratories began applying human factors principles within their production facilities as well as to their customer products.

At present, human factors professionals work in government and industrial settings. Many work on military applications because most military systems must now be engineered to meet human factors standards. Large numbers of human factors specialists also work in the aerospace industry to ensure the safety and usability of aircraft and equipment used by pilots and air traffic controllers. In addition, most software and e-commerce companies use human factors specialists to ensure the usability of their products and services domestically and in the global marketplace.

2.2 Human Factors in Healthcare

Human factors professionals work on a variety of healthcare issues. Traditionally, much of the work has had a strong focus on occupational safety with researchers studying worker characteristics, job design, tool design, and the work environment. For example, many human factors professionals have examined the effects of lighting, noise, temperature, and even clothing on worker productivity [1]. Effort has also been aimed at redesigning work tools, hand grips, and even keyboards to minimize opportunities for injury including cumulative trauma disorders [7]. Human factors professionals have worked closely with organizations such as the National Institute for Occupational Safety and Health to study and educate the public about worker safety issues and the Occupational Safety and Health Administration to establish safety standards and implement safety programs.

Not all efforts have been directed at the work place, however. Human factors specialists have also addressed many of the challenges with providing effective healthcare in the home and nursing home environments (e.g., [8, 9]). Issues such as telemonitoring, patient-physician communication, redesign of the physical environment, the role of nurses and care givers, as well as patient autonomy and privacy will become increasingly important in the near future as the population continues to age.

Human factors researchers have long been interested in the nature of errors in medicine and their effects on patient safety, beginning with the study of medication errors in the early 1960s [10, 11]. Bogner published the first book on human error in medicine addressing the influences of equipment, team composition, and organizational policies on medical error [12]. The Institute of Medicine would publish its landmark report, *To Err is Human* [13], 5 years later, in which medical errors were estimated to contribute to as many as 98,000 deaths annually in U.S. hospitals. The IOM report generated an unprecedented demand for human factors professionals to help reduce errors in all areas of healthcare.

Another area where human factors has had a significant impact is in identifying weaknesses in the design of medical devices and equipment that contribute to user errors. For example, Andre and Kingsburg recently conducted a human factors analysis of automatic external defibrillators (AEDs) intended for public use [14]. A defibrillator delivers a brief electrical shock to a patient's heart to restore its natural rhythm. Traditionally, only medical personnel with extensive training could use a defibrillator. Today, however, AEDs incorporate intelligent software that can quickly assess cardiac rhythm and deliver the appropriate type of shock. Thus, AEDs have now been designed for the lay person to use and are installed in many public areas such as airports and shopping malls.

Andre and Kingsburg evaluated four different AED systems that were all commercially available [14]. They recruited over 60 adults of different ages and occupations, gave them some fundamental information about AEDs, and led them to a room where they found a fully clothed mannequin lying on the floor and an AED nearby. They were told to use the defibrillator and attempt to deliver a shock to the patient as quickly as they could.

The investigators found important differences among all four AED systems. Perhaps most troubling was the observation that one third of the participants could not complete the task and deliver a shock within the 7-minute time limit with two of the AED systems. For those participants who were successful in delivering a shock with those same two AEDs, they needed 2 to 4 minutes to complete the task. By contrast, all participants using the other two AED systems successfully completed the task in under 2 minutes. None of the four systems, however, resulted in optimal placement of electrode pads, pad separation, or skin contact. Thus, there were design issues with each system.

Problems with the design of medical devices are not limited to those intended for the general public. There is abundant evidence that medical devices contribute to safety problems in hospitals. Vicente describes the history of a patient-controlled analgesia infusion pump that may have been responsible for as many as 667 deaths [15]. More alarming, these deaths occurred over a period of 12 years *after* the Emergency Care Research Institute (ECRI) issued a report stating that this device was susceptible to user programming errors. Vicente and his colleagues analyzed the interface for the device and found that it was cumbersome and error-prone. Using human factors methods, they created an alternative interface that reduced the number of steps needed to program the device by half and completely eliminated the error that was linked to the patient deaths. The alternative interface proposed by Vicente and his colleagues shows how human factors principles can contribute to patient safety through better design, without the need for additional training.

The U.S. Food and Drug Administration maintains a database for reporting adverse events involving medical devices. Al-Tarawneh, Stevens, and Arndt examined entries in the database over a 5-year period from 1997 to 2002 and reported that device-related hospital incidents contributed to over 2,000 fatalities and 40,000 injuries during that interval [16]. In fact, the FDA has issued several reports citing aspects of device design that contribute directly to user error and the need to apply human factors concepts (e.g., [17, 18]). Human factors methods that focus on user characteristics, cognitive demands, error identification, iterative design and testing, managing risk, and evaluating products in the context in which they will be used are paramount for ensuring patient safety [19, 20].

Perhaps there is no better example of the interface between technology and the patient than in the intensive care unit (ICU). All of the potential problems with individual medical devices are compounded in the ICU. However, new and different types of problems can emerge when healthcare providers must interact with a variety of devices, in different configurations, across multiple patients.

According to Donchin [21], most ICUs are equipped with the latest state-of-the-art equipment intended to provide the patient with the best possible care. However, the technology has dramatically altered how patients receive care in the ICU. First, patients are monitored more from the output of the devices than by observing the patients themselves. Second, the data from the devices must be integrated with other sources of data including laboratory results and background information/diagnoses. Third, the physical environment around a patient's bed can be cluttered with devices, wiring, and tubing, thereby limiting access to the patient. Fourth, the assortment of devices that interface with the patient is often produced by different manufacturers. They each convey information about the patient in their own unique way, with their own signaling and alarm systems, none of which have been tested and evaluated in a coherent fashion. Last, the staff often works under very stressful conditions created by the number of patients in the ICU and their unique requirements, the fluctuating urgency of patients' needs, and a physical environment that can be crowded and noisy. Donchin estimates that error rates in the ICU are less than 0.5%. Although that figure is encouraging given the unique demands placed on the ICU staff, he argues that it still translates to a potential for two errors per day per patient. Similar issues have also been described for the operating room [22].

Obviously, there is much more that can be said about the nature of human factors. There are several good introductory books on the subject (e.g., [23, 24]) as well as more advanced treatments (e.g., [1, 2]). There are also more comprehensive professional references including the *Handbook of Human Factors and Ergonomics* [25] and the *International Encyclopedia of Ergonomics and Human Factors* [26]. However, the purpose of this chapter is to address human factors concerns surrounding the use of automation. Automated systems pose an interesting set of challenges for users, and during the past 15 years, a growing number of people in the human factors community have begun to study these particular issues. Thus, the remainder of this chapter will address automation, and its advantages and disadvantages, adaptive automation, and offer some guidelines for designers.

2.3 Automation

Automation has been referred to as the process of allocating activities to a machine or system to perform [27]. Parasuraman and Riley described automation as a machine agent that can execute functions normally carried out by humans [28].

More recently, Sheridan described automation with respect to the mechanization or computerization of sensing, data processing, and action execution functions [29].

Although most people have the impression that automation fully replaces activities once performed by people, there are actually many varieties of automated systems. Sheridan and Verplank described these differences along a continuum beginning with systems that require complete manual interaction to fully automated systems that require no interaction (e.g., the antilock braking system on an automobile) [30]. Between these two extremes are systems that have different levels of automation. Further, the levels are distinguished by both authority and autonomy. For example, on the lower end of automation, systems might offer suggestions to the user about how to proceed. The user has the authority to either veto the suggestion or accept the suggestion and implement the action. At the intermediate levels, the system may have the autonomy to carry out the suggested actions once approved by the user. At the higher end, the system may select a course of action and have the authority and autonomy to implement that action. At this level, the system would merely inform the user of actions taken.

Parasuraman, Sheridan, and Wickens later expanded on these descriptions [31]. Their model addresses different levels of automation within four different function areas that are analogous to the stages of human information processing. *Information acquisition* systems address the detection and registration of data. *Information analysis* systems analyze and summarize data, make predictions and/or inferences, or modify/augment information displays. *Decision selection* systems augment or perform decision-making functions. Last, *action implementation* systems actually execute functions. Thus, systems can be described by their level of automation within and/or across the four function areas.

Rouse and Rouse also described different types of automation; however, their distinction rests with how the technology assists the operator [32]. For example, entire tasks can be *allocated* to the system to perform. Alternatively, tasks can be *partitioned* so that the system and operator are each responsible for controlling some portion of the task. Last, a task can be *transformed* to make it easier for the operator to perform. Automation based on the allocation and partition approaches would have high autonomy because the system has control over all or some portions of the task. The transformation strategy, however, would have lower autonomy because the operator is still responsible for performing the task.

2.3.1 Advantages of Automation

Wickens and Hollands describe several purposes for automation. First, automation can be used to perform functions that are beyond the capabilities of humans [2]. For example, systems that use infrared sensors operate on a range of the electromagnetic spectrum that lies outside of human sensitivity. Second, automated systems can perform functions that humans cannot do well. For instance, the automatic exposure system in a camera compensates for our poor ability to accurately judge lighting conditions by coupling the camera's light meter to its diaphragm/shutter system. Third, automation can provide users with auxiliary aids to assist on difficult tasks. An intelligent tutoring system is an example of automation that analyzes a user's pattern of activities and errors and offers recommendations for improvement. Last, automation is often introduced as a way to reduce costs and/or increase safety.

There are many obvious benefits to automation. First, as more functions are assigned to the system there are fewer for the human to perform. Thus, automation can reduce operator workload. In fact, many automated systems are welcomed when they relieve us from having to perform nuisance or housekeeping activities [33]. For example, a thermostat automatically controls the heating/cooling system in response to the temperature we set, thereby relieving of us from having to continually turn the system on and off. Automation also allows for greater flexibility and can therefore allow humans to control more complex systems and increase total work output [34]. In addition, automation can attenuate variability in human performance, thereby reducing errors. For instance, the daVinci telerobotic surgery system facilitates operations in small areas of the body by automatically filtering and dampening unintended finger movements. Last, automation in aviation has made it possible to reduce flight times, increase fuel efficiency, and navigate more effectively [35].

2.3.2 Disadvantages of Automation

A growing body of research has begun to show that the benefits derived from automation come at a price. Often, there is a good deal of skepticism among operators, particularly when automation is first introduced. Several researchers have argued that automation introduces new and often unexpected problems [34, 36, 37]. Evidence has been gathered from analyses of real automation failures as well as laboratory studies and field experiments. In the following section, several disadvantages of automation will be discussed.

2.3.2.1 Increased Workload

When tasks become automated there are fewer activities for the operator to execute. Thus, in theory automation should lead to a decrease in operator workload. In well-designed systems, the automation should not impose inappropriate demands on operators. However, not all automated systems are well matched for the operator requirements. Wiener indicates that in some instances, automation can actually *increase* workload by requiring operators to interact with automated systems when workload is already high [38]. Thus, some automated systems may operate well under periods of low workload, but may become a burden during periods of high workload.

In addition, as systems become more complex increased automation is often required for humans to manage the complexity. Further, the computer systems and associated software needed to orchestrate the automated systems also increase in complexity [2]. Consequently, the automation in complex systems is configured to operate in different modes and therefore requires users to maintain the system status and functional boundaries of the current mode in memory (see below). Moreover, Woods claims that automation can lead to incongruent goals among subsystems and increase the operator's workload when troubleshooting the system [34].

2.3.2.2 Mode Awareness Problems

Many complex automated systems incorporate multiple modes of automation. For example, the flight management systems on most advanced commercial aircraft can have over a dozen unique modes of operation that address navigation, guidance, system monitoring, and management functions [39]. Complex systems with multiple modes of automation present operators with several challenges. First, they are difficult to learn and operators must invest a significant amount of time to understand the boundaries of each mode of operation. Second, they tend to increase workload because the user must know the operating procedures associated with each mode as well as which mode is active at all times. The complexity of automated systems with multiple modes can lead to errors, particularly when operators find themselves under stressful, high workload conditions.

Degani describes how a mode awareness problem with an automatic blood pressure measurement system contributed to a critical incident in surgery [40]. Normally, blood pressure is measured by wrapping an inflatable cuff around the patient's upper arm, inflating the cuff to restrict arterial blood flow, then deflating the cuff to assess the maximum (systolic) pressure and minimum (diastolic) pressure, respectively. An automated blood pressure measurement system can perform all of the functions needed to obtain both systolic and diastolic measurements on regular intervals as often as desired. This particular device had three primary modes of operation: automatic, manual, and idle. Setting the frequency of the measurement interval takes the device out of manual mode and places it in idle mode. Likewise, while in automatic mode, changing the timing interval requires entering a new value and pressing a start button to place the device back in automatic mode. Thus, changing the timing interval *also* changes the modes.

In this particular instance, the patient undergoing surgery began to bleed and the surgeon asked the anesthesiologist to lower the blood pressure. The anesthesiologist changed the interval on the automated blood pressure measurement system to take more frequent readings and attempted to lower the blood pressure with drugs. After waiting for an interval, the anesthesiologist checked the blood pressure reading and found that it had not changed, so more drugs were administered. This cycle continued for nearly 45 minutes until the surgeon was able to control the bleeding. At this point, they both realized that the blood pressure readings on the device had not changed during the entire time they were distracted trying to control the bleeding. When the anesthesiologist finally obtained a new reading, the patient's blood pressure was found to be dangerously low. Fortunately, the patient survived.

An analysis of the incident revealed that when attempting to increase the frequency of readings, the anesthesiologist failed to press the start button, leaving the device in idle mode. However, examination of the device showed that when changing the interval frequency from within the automatic mode, the display read as if the device were in automatic mode. Thus, there was no distinct information on the display to indicate that the device was in idle mode, waiting for the user to press the start button. The anesthesiologist had probably used that automatic blood pressure measurement system many times before without incident. However, under the stressful emergency conditions that arose on that particular day, a design deficiency kept the true mode of the device hidden from the anesthesiologist precisely when that information was needed most, thereby forcing the anesthesiologist (in this case, unsuccessfully) to keep track of the modes in memory.

2.3.2.3 Monitoring Complacency

There is another consequence of automated systems that perform activities for the operator. This approach to automation often changes the role of the operator from an active participant to a passive monitor of system operations. Unfortunately, this shift in roles creates a new and different set of tasks for operators, tasks for which humans are not well suited.

Over 50 years of research has shown that human operators make poor monitors when asked to spend long periods of time looking for rare events [6]. Further, several studies have shown that monitoring efficiency can also be compromised by automation. For instance, Parasuraman, Molloy, and Singh asked individuals to perform a systems monitoring task requiring them to look for periodic deviations in a set of gauges and manually reset them [41]. The operators had no trouble performing this task by itself or in conjunction with two other tasks. When the systems monitoring task was automated (i.e., the system automatically reset the deviations), however, it created problems for the operators. In the automated condition, the individuals did not need to perform the monitoring task; they could concentrate on the other tasks. In fact, this is precisely what they did. However, on occasion, the automated task failed to reset the gauges and the ability of the operators to notice the gauge deviations declined markedly. Further, their monitoring behavior became increasingly inefficient with longer periods of automation.

Parasuraman, Molloy, and Singh referred to the tendency of operators to put too much trust in automation as *complacency* [41]. Moreover, the impact of complacency is not limited to laboratory research. Parasuraman and Riley argue that automation-induced complacency led to the crash of Eastern Flight 401 in the Florida Everglades in 1972 [28]. The effects of complacency are rarely observed when the operator's sole responsibility is to monitor the system. However, in real-world settings, most operators must perform system monitoring in addition to other tasks. Under these conditions, relying on the automation to execute some tasks so that one can focus efforts on other activities may seem reasonable [29]. However, it is precisely because operators have several tasks competing for their attention that they tend to place too much trust in the automation.

Monitoring complacency is also tied to system reliability. In general, operators are more likely to become complacent when automated systems are highly reliable [31]. For instance, Bailey and Scerbo conducted a study in which operators were asked to perform several tasks including an automated system monitoring task [42]. The operators performed the tasks for 1 hour on three separate occasions. Over the course of the study, there was a single automation failure. Bailey and Scerbo wanted to know how many operators would detect the failure. Their results showed that only three of the nine operators detected the failure, suggesting that highly reliable systems foster complacency.

2.3.2.4 Loss of Situation Awareness and Skills

Several researchers have argued that automation removes operators from the loop, leaving them less aware of the operating state of system [2]. Endsley describes this as a decrease in operator situation awareness, or the ability to understand current conditions and make predictions about future states [43]. This loss of situation awareness can leave operators with a poor understanding of system status and less able to reengage and perform appropriately when they are needed [43].

Along similar lines, dependence on automation can lead to a loss of skills. Many complex skills require continual practice to be executed efficiently. However, manual skills may deteriorate in the presence of long periods of automation [2]. Parasuraman, Mouloua, Molloy, and Hilburn showed that a temporary return to manual operations may serve as a countermeasure to poor monitoring behavior induced by automation [44]. In their study, operators were asked to perform several tasks simultaneously, including an automated monitoring task. During the course of the study, however, the automated task transitioned to manual control for a brief period. The investigators found that overall monitoring performance improved after the period of manual intervention.

2.3.2.5 Ambiguous Alarms

In many automated systems, the operator is required to monitor operations and respond to alarms. The alarms are typically designed to notify the operator of unsafe conditions. Under ideal situations, the system conveys the correct information about the conditions and the operator confirms the information. However, because all systems are designed to operate within limits, they inevitably run up against their boundaries. Further, they do not always function reliably and can even fail. Hence, in less than ideal conditions a system can produce false alarms or indicate all is safe when hazards actually exist.

These two types of system errors have different consequences for how operators choose to respond to the automation. Dixon and Wickens describe these response patterns as reliance and compliance [45]. Specifically, when a system signals safe conditions (i.e., no alarms), the operator becomes reliant on the automation and can focus attention on other activities. Unreliable automation that tends to miss unsafe conditions eventually degrades reliance, by forcing the operator to shift attention away from other activities to reassess the true conditions. By contrast, highly reliable automated systems that tend to miss unsafe conditions enable the operator to focus on other activities, but create the conditions for complacency described above. On the other hand, a compliant response pattern describes operators who switch attention from other activities to address alarms (whether they are true or false). Automation that tends to generate false alarms will degrade compliance, resulting in slower responses or an unwillingness to respond to true alarms. Further, Dixon and Wickens showed that these patterns are moderated by reliability and workload. In general, the effects are more pronounced when automation is more reliable and concurrent workload demands are high.

2.4 Creating More Effective Automation

Parasuraman and Riley discussed human performance issues surrounding automation with respect to its use, misuse, disuse, and abuse [28]. Regarding its *use*, operators make decisions about when to activate or disengage automation based on a complex interaction of factors, including their level of mental workload, perceived risk, trust in the system, and confidence in their own abilities. Moreover, these factors are mediated by individual differences among operators. Consequently, it is difficult to predict how a given operator will use automation under different circumstances.

As noted above, although automated systems tend to be highly reliable they can still fail. Operators tend to *misuse* automation when they become complacent. Overreliance on automation can result from an inability to detect system failures or biases in decision heuristics (i.e., taking mental shortcuts to reduce effort). Thus, there is a need for system designers to understand that operators may place too much trust and reliance on automation.

Automated systems can also lead to underutilization or *disuse*. Users need time to become familiar with new technology and to establish a mental record of its reliability and operational boundaries. Thus, new technology may not be fully utilized at first. Further, automated systems that fail to meets the user's expectations regarding reliability may be rejected. Automation disuse is often observed with warning and alerting systems that generate high levels of false alarms. It is not uncommon under these conditions for operators to ignore the alarms or disengage the alarm system.

Parasuraman and Riley reserve the term *abuse* for automation that is created by designers or implemented by management without regard for its impact on operator performance [28]. They argue that operators cannot be held solely responsible for their interactions with automation. The designers of automated systems and the organizational climate under which the technology is created and used also contribute to its effectiveness or its consequences. Automation is often developed for economic reasons or when the cause of an accident points to human error. However, human error is not eliminated by removing the human operator. In fact, doing so merely replaces the human operator with the human designer. Also, as noted above, building systems that change the role of an active system operator to that of a passive monitor of automation creates jobs for which humans are poorly suited.

Parasuraman and Riley offer several suggestions for designing more effective automation [28]. Many of these have been summarized in Table 2.2. Moreover, they stress that problems associated with automation often stem from differing expectations among operators, designers, and managers. They argue that overall improvements in the effectiveness of automated systems will be much more likely when all constituents share a better appreciation of these issues.

2.5 Adaptive Automation

Several researchers have argued that one way to address some of the problems associated with automation is to give the user more flexibility over how the automation

Table 2.2 Guidelines for Creating More Effective Automated Systems

Use	*Misuse*	*Disuse*	*Abuse*
Increase operator knowledge of how automation works.	Design for overreliance by reducing concurrent workload or adding more training.	Designers of alerting systems must balance the sensitivity of the system with the base rate of hazardous conditions.	Automation should be implemented so that it maintains operator involvement.
Educate operators on how to make rational decisions about using automation.	Avoid jobs where operators must monitor highly reliable automation under high workload.	Designers should consider presenting operators with likelihood alarms when hazardous conditions are possible.	Operator error must be considered along with designer error in automated systems.
Reduce cognitive workload by allowing operators to easily turn automation on or off.	Provide operators with information about the status of automation.		Management must be educated about automation-related issues.

Based on [28].

operates [28, 46]. As such, researchers and developers have begun to consider the merits of *adaptive automation.* In adaptive automation the level of automation or type of automation can be modified in real time. More important, changes in the state of automation can be implemented by *either* the human or the system [46, 47]. Parasuraman et al. argue that adaptive automation allows the level or modes of automation to be coupled more closely to operator needs at any given moment [44].

Adaptive automation systems can be implemented in either an *adaptable* or *adaptive* manner [48]. In adaptable systems, changes among modes or levels are initiated by the user. In adaptive systems, the changes can be initiated by both the user and the system.

Adaptable and adaptive technology can also be described with respect to authority and autonomy. As noted above, Sheridan and Verplank describe several levels of automation running along a continuum from manual to fully automatic [30]. According to their description, as the level of automation increases, systems become more autonomous and take on greater authority. With respect to Scerbo's description [48], the operator maintains authority over invoking changes among levels of automation in adaptable systems. Thus, one could say that adaptable systems reflect a superordinate-subordinate relationship between the operator and the system. On the other hand, the authority to initiate changes is shared in adaptive systems.

Just who should have control over changes among modes of operation has been the subject of much debate. On the one hand, there are those who argue that operators should always have authority over the system because (1) operators are ultimately responsible for the behavior and safety of the system, and (2) operators may be in the best position to manage resources when they can control changes in the state of automation [49, 50]. On the other hand, there is some evidence to suggest that operators are not necessarily the best judges of when automation is needed. Often, changes in levels of automation are needed at the precise moment when the operator is too busy to make those changes [38]. Others have shown that humans may be incapable of making good decisions in time critical situations (e.g., whether to drive through a yellow light or abort a take-off in an aircraft [40]). Instead, the optimal decisions may be made when the operator and the automation share control [51]. Last, Scerbo has argued that in some hazardous situations where the operator or system is vulnerable, it would be extremely important for the system to have authority to invoke automation [46]. For example, the Ground Collision-Avoidance System (GCAS) developed and tested on the F-16D is designed to prevent the aircraft from breaking through a pilot-determined minimum altitude. The automation can usurp control of the aircraft to avoid a collision with the terrain, and then return control to the pilot quicker than any human pilot can respond to the hazard [52].

The development of adaptive automation systems is still fairly new; however, there have been some interesting examples. For example, in the 1990s, the U.S. Army sponsored the development of the Rotorcraft Pilot's Associate (RPA). The goal was to create an intelligent system associate that could serve as a "junior crew member" [53]. The RPA had a Cognitive Decision Aiding System (CDAS) that detected and organized incoming data, assessed internal information about the aircraft and external information about targets and mission status, and fed the information into a series of planning and decision-making modules. The Cockpit Information Manager served as the adaptive automation portion of the system and was designed to make inferences about current and impending activities for the crew, allocate tasks among crew members (and the aircraft), and reconfigure cockpit displays to help the crew execute those activities. Initial evaluations from a sample of pilots indicated that the RPA did indeed provide the right information at the right time.

A different approach to adaptive automation was reported by Pope and his colleagues [54]. They developed the first brain-based adaptive system that used an operator's own EEG signals to estimate his or her level of mental workload and then adjusted task requirements to maintain a moderate level of engagement. Subsequent research with the system showed that the brain-based approach had beneficial effects on performance [55].

On the commercial side, several automobile manufacturers including Infinity, Lexus, and Cadillac now incorporate adaptive cruise control systems in their vehicles [46]. Adaptive technology is also beginning to find its way into smart homes that manage heating and lighting based upon living patterns in the occupants [56, 57].

2.6 Conclusion

Human factors is a discipline concerned with improving the fit between humans and technology. It has its roots in military applications, but now impacts almost every product or system with which users interact. Human factors professionals have made significant contributions to safety in the areas of aviation, automobile design, power plant operations, space exploration, as well as healthcare systems.

Recently, members of the human factors community have focused their attention on some of the unique characteristics of automated systems. Although many automated systems decrease workload and increase efficiency and safety, they often create unintended consequences for their operators. Automated systems can lead to skill degradation, loss of situation awareness, over or underreliance, and even increases in workload. The unintended consequences of automation can create little annoyances or lead to huge catastrophes.

It is important to understand that automation is neither inherently good nor bad. It does, however, change the nature of work [37, 46]. Woods suggests that it may be inappropriate to view automation in terms of costs and benefits [34]. Instead, he argues that automation *transforms* the nature of work. Delegating tasks to an automated system does not leave a gap in the operator's responsibilities. That gap is quickly filled with other activities. Thus, automation often results in a redistribution of resources and it becomes incumbent upon designers and management to understand the impact of that shift in resources throughout the system and even the organization [28].

Vicente has argued there are two driving forces responsible for shaping the evolution of technology [15]. On one side are the efforts of developers to demonstrate what technology can do, and on the other side are the desires and needs of the consumer. Often, the zeal to apply new and more sophisticated automation to solve real problems is not balanced with the effort needed to comprehend the full extent of its impact. The healthcare industry is not immune to these two forces. There is an almost urgent need to improve efficiency and contain costs throughout the healthcare system. Hospitals and providers are implementing automated pharmaceutical delivery systems, automated patient monitoring systems, automated inventory tracking systems, and even more tightly coupled electronic health record and service systems at a brisk pace. One can only hope that the proponents of healthcare automation give careful consideration to the lessons learned about automation failures from other high-risk domains.

References

[1] Sanders, M. S., and E. J. McCormick, *Human Factors in Engineering and Design, 7th. Edition,* New York: McGraw-Hill, 1993.

[2] Wickens, C. D. and J. G. Hollands, *Engineering Psychology and Human Performance, 3rd Edition,* Upper Saddle River, NJ: Prentice Hall, 2000.

[3] Chapanis, A., *Human Factors in System Engineering*, New York: Wiley, 1995.

[4] Bailey, R. W., *Human Performance Engineering: A Guide for System Designers,* Englewood Cliffs, NJ: Prentice-Hall, Inc., 1982.

[5] Gilbreth, F. B., *Brick Laying System*, New York: Clark Publishing, 1911.

[6] Warm, J.S. (ed.), *Sustained Attention in Human Performance,* Chichester: Wiley,1984.

[7] Marras, W. S., "Biomechanics of the Human Body," in *Handbook of Human Factors and Ergonomics, 2nd Edition,* pp. 233–267, G. Salvendy (ed.), New York: Wiley, 1997.

[8] Lathan, C. E., A. Kinsella, and M. J. Rosen, et al., "Aspects of Human Factors Engineering in Home Telemedicine and Telerehabilitation Systems," *Telemedicine Journal,* Vol. 5, 1999, 169–175.

[9] Zimmerman, D. R., "Human Factors and Ergonomics in Nursing Home Care," in P. Carayon (ed.), *Handbook of Human Factors and Ergonomics in Health Care and Patient Safety* (pp. 899–920). Mahwah, NJ: Erlbaum, 2007.

[10] Safren, M. A., and A. Chapanis, "A Critical Incident Study of Hospital Medication Errors—Part I, *Hospitals,* Vol. 34, 1960a, 32–34, 57–66.

[11] Safren, M. A., and A. Chapanis, "A Critical Incident Study of Hospital Medication Errors—Part II," *Hospitals,* Vol. 34, No. 53, 1960b, pp. 66–68.

[12] Bogner, M. S. (ed.), *Human Error in Medicine*, Hillsdale, NJ: Erlbaum, 1994.

[13] Kohn, L., J. Corrigan, and M. Donaldson (eds.), *To Err is Human: Building a Safer Health System,* Washington, DC: Institute of Medicine, National Academy Press, 1999.

[14] Andre, A. D., and J. R. Kingsburg, "When Every Minute Counts, All Automatic External Defibrillators Are Not Created Equal," in *Human Performance, Situation Awareness and Automation: Current Research and Trends, Vol. II,* pp. 76–81. Mahwah, NJ: Erlabaum, 2005.

[15] Vicente, K., *The Human Factor*, New York: Routledge, 2004.

[16] Al-Tarawneh, I. S., W. J. Stevens, and S. R. Arndt, "An Analysis of Home and Hospital Medical Device Incidents in the MAUDE Database," *Proceedings of the 48th Annual Human Factors and Ergonomics Society Annual Meeting,* HFES, 2004, Santa Monica, CA, pp. 1718-1722.

[17] "Human Factors Implications of the New GMP Rule Overall Requirements of the New Quality System Regulation," Center for Devices and Radiological Health, Food and Drug Administration, Rockville, MD, 1998; retrieved January 25, 2008, from www.fda.gov/cdrh/humfac/hufacimp.html.

[18] "Medical Device Use-Safety: Incorporating Human Factors Engineering into Risk Management," Center for Devices and Radiological Health, Food and Drug Administration, Rockville, MD, 2000; retrieved January 25, 2008, from www.fda.gov/cdrh/humfac/1497.html.

[19] Stanton, N.A, P. M. Salmon, and G. H. Walker., et al., *Human Factors Methods: A Practical Guide for Engineering and Design,* Burlington, VT: Ashgate, 2005.

[20] Ward, J., and, J. Clarkson, "Human Factors and Engineering and the Design of Medical Devices," in P. Carayon (ed.), *Handbook of Human Factors and Ergonomics in Health Care and Patient Safety*, Mahwah, NJ: Erlbaum, 2007, pp. 367–382.

[21] Donchin, Y., "The Intensive Care Unit May Be Harmful to Your Health," in *Misadventures in Health Care: Inside Stories,* pp. 127–144, M. S. Bogner (ed.), Mahwah, NJ: Erlbaum, 2004.

[22] Seagull, F. J., and, P. M. Sanderson, "Anesthesia Alarms in Surgical Context: An Observational Study," *Human Factors,* Vol. 43, 2001, pp. 66–78.

[23] Adams, J. A., *Human Factors Engineering*, New York: Macmillan, 1989.

[24] Wickens, C. D., J. D. Lee, and Y. Liu, et al., *An Introduction to Human Factors Engineering, 2nd. Ed.,* Upper Saddle River, NJ: Pearson, Prentice Hall, 2004.

[25] Salvendy, G. (ed.), *Handbook of Human Factors and Ergonomics, 3rd Ed.*, New York: Wiley, 2006.

[26] Karwowski, W. (ed.), *International Encyclopedia of Ergonomics and Human Factors,* Boca Raton, FL: CRC, Taylor & Francis, 2006.

[27] Parsons, H. M., "Automation and the Individual: Comprehensive and Comparative Views," *Human Factors,* Vol. 27, 1985, pp. 99–111.

[28] Parasuraman, R., and V. Riley, "Humans and Automation: Use, Misuse, Disuse, and Abuse," *Human Factors,* Vol. 39, 1997, pp. 230–253.

[29] Sheridan, T. B., *Humans and Automation: System Design and Research Issues,* New York: Wiley, 2002.

[30] Sheridan, T. B., and W. L. Verplank, *Human And Computer Control of Undersea Teleoperators,* Man-Machine Systems Laboratory Report, MIT, Cambridge, 1978.

[31] Parasuraman, R., T. B. Sheridan, and C. D. Wickens, "A Model for Types and Levels of Human Interaction with Automation," *IEEE Transactions on Systems, Man, and Cybernetics—Part A: Systems and Humans,* Vol. 30, pp. 286–297, 2000.

[32] Rouse, W. B., and S. H. Rouse, "A Framework for Research on Adaptive Decision Aids," Technical Report AFAMRL-TR-83-082, Wright-Patterson Air Force Base, OH: Air Force Aerospace Medical Research Laboratory, 1983.

[33] Rouse, W. B., *Design for Success: A Human Centered Approach to Designing Successful Products and Systems,* New York: Wiley, 1991.

[34] Woods, D. D., "Decomposing Automation: Apparent Simplicity, Real Complexity," in *Automation and Human Performance: Theory and Applications,* pp. 3–17, R. Parasuraman and M. Mouloua (eds.), Mahwah, NJ: Erlbaum. 1996.

[35] Wiener, E. L, "Cockpit Automation," in *Human Factors in Aviation,* pp. 433–461, E. L. Wiener and D. C. Nagel (eds.), San Diego: Academic Press, 1988

[36] Parasuraman, R., and M. Mouloua, *Automation and Human Performance: Theory and Applications*, Mahwah, NJ: Lawrence, 1996.

[37] Wiener, E. L., and R. E. Curry, "Flight-Deck Automation: Promises and Problems," *Ergonomics,* Vol. 23, pp. 995–1011, 1980.

[38] Wiener, E. L., "Human Factors of Advanced Technology ('Glass Cockpit') Transport Aircraft," Technical Report 117528, Moffett Field, CA:NASA Ames Research Center, 1989.

[39] Billings, C. E., "Human-Centered Aircraft Automation Philosophy: A Concept and Guidelines," Moffett Field, CA: NASA Technical Memorandum No. 103885, 1991.

[40] Degani, A., *Taming HAL: Designing Interfaces Beyond 2001,* New York: Palgrave Macillan, 2004.

[41] Parasuraman, R., R. Molloy, and I. L. Singh, "Performance Consequences of Automation-Induced 'Complacency'," *International Journal of Aviation Psychology,* Vol. 3, 1993, pp. 1–23.

[42] Bailey, N. R., and M. W. Scerbo, "Automation-Induced Complacency for Monitoring Highly Reliable Systems: The Role of Operator Trust an Task Complexity," *Theoretical Issues in Ergonomics Science,* Vol. 8, 2007, pp. 321–348.

[43] Endsley, M. R., "Automation and Situation Awareness," in *Automation and Human Performance: Theory and Applications,* pp. 163–181, R. Parasuraman and M. Mouloua (Eds.), Mahwah, NJ: Lawrence Erlbaum, 1996.

[44] Parasuraman, R., M. Mouloua, and R. Molloy, et al., "Training and Adaptive Automation II: Adaptive Manual Training," Technical Report CSL-N92-2. Washington, DC: Cognitive Science Laboratory, Catholic University of America, 1992.

[45] Dixon, S. R., and C. D. Wickens, "Automation Reliability in Unmanned Aerial Vehicle Control: A Reliance-Compliance Model of Automation Dependence in High Workload," *Human Factors,* Vol. 48, pp. 474–486, 2006.

[46] Scerbo, M. W., "Theoretical Perspectives on Adaptive Automation," in *Automation and Human Performance: Theory and Applications*, pp. 37–63, R. Parasuraman and M. Mouloua (eds.), Mahwah, NJ: Erlbaum,1996.

[47] Rouse, W. B., "Adaptive Allocation of Decision Making Responsibility Between Supervisor and Computer, in *Monitoring Behavior and Supervisory Control*, pp. 295–306, T.B. Sheridan and G. Johannsen (eds.), New York: Plenum Press, 1976.

[48] Scerbo, M. W, "Adaptive Automation, in *International Encyclopedia of Ergonomics and Human Factors, 2nd Edition,* pp. 1893–1896, W. Karwowski (ed.), Boca Raton, FL: CRC Press, 2006.

[49] Billings, C. E., and D. D. Woods, "Concerns About Adaptive Automation in Aviation Systems," in *Human Performance in Automated Systems: Current Research and Trends,* pp. 264–269, M. Mouloua and R. Parasuraman (eds.), Hillsdale, NJ: Lawrence Erlbaum Assoc., 1994.

[50] Malin, J. T., and D. L. Schreckenghost, "Making Intelligent Systems Team Players: Overview for Designers," NASA Technical Memorandum 104751, Johnson Space Center, Houston, TX, 1992.

[51] Inagaki, T., Takae, Y., and N. Moray, "Automation and Human-Interface for Takeoff Safety," *Proceedings of the 10th International Symposium on Aviation Psychology*, pp. 402–407, 1999.

[52] Scott, W. B., "Automatic GCAS: 'You Can't Fly Any Lower'," *Aviation Week & Space Technology, February*, 1999, pp. 76–79.

[53] Miller, C. A., and M. D. Hannen, "The Rotorcraft Pilot's Associate: Design and Evaluation of an Intelligent User Interface for Cockpit Information Management," *Knowledge-Based Systems,* Vol. 12, 1999, pp. 443–456.

[54] Pope, A. T., E. H. Bogart, and D. Bartolome, "Biocybernetic System Evaluates Indices of Operator Engagement," *Biological Psychology,* Vol. 40, 1995, pp. 187–196.

[55] Scerbo, M. W., F. G. Freeman, and P. J. Mikulka, "A Brain-Based System for Adaptive Automation," *Theoretical Issues in Ergonomic Science,* Vol. 4, 2003, pp. 200–219.

[56] Mozer, M. C., "Lessons from an Adaptive House," in *Smart Environments: Technologies, Protocols, and Applications,* pp. 273–294, D. Cook and R. Das (eds.), New York: Wiley, 2004.

[57] Scerbo, M. W., "Adaptive Automation," in *Neuroergonomics: The Brain at Work,* pp. 239–252, R. Parasuraman and M. Rizzo (eds.), Oxford: Oxford University Press, 2007.

Selected Bibliography

Billings, C. E., "Human-Centered Aircraft Automation Philosophy: A Concept and Guidelines," NASA Technical Memorandum No. 103885, Moffett Field, CA, 1991.

Freeman, F. G., P. J. Mikulka, and L. J. Prinzel, "Evaluation of an Adaptive Automation System Using Three EEG Indices with a Visual Tracking Task," *Biological Psychology,* Vol. 50, 1999, pp. 61–76.

Sarter, N. B., and D. D. Woods, "How in the World Did We Ever Get into That Mode? Mode Error Awareness in Supervisory Control," *Human Factors,* Vol. 37, 1995, pp. 5–19.

CHAPTER 3

Mathematical Modeling for Medical Automation

Weimin Tao and Mingjun Zhang

3.1 Introduction

This chapter introduces various mathematical modeling techniques for medical automation. A couple of application examples are also described for the two major mathematical modeling fields in medical automation: medical process and medical tool modeling. Conclusions are presented at the end.

3.1.1 Medical Automation

Medical automation usually involves automation for clinical applications, which include medical robotics, medical laboratory automation, medical process automation, and drug delivery. Goals of medical automation are to improve the efficiency, reliability, and quality of clinical activities.

During the past 30 years, automation has crossed over from industrial applications to the medical fields. Automation starts from hard-tooled machines that conduct simple repetitive work to advanced robotics that can conduct various types of work through software programming. Medical robotics is one of the important fields for medical automation, which is being applied to computer-integrated surgery [1], radiology cancer treatment [2], and clinical and pharmaceutical laboratory automation and services [3], to name a few. In addition to the use of mechanical robotics, advanced control technologies are also being applied to medical automation, such as bioreactor control [4] and model-based diagnosis system for monitoring and control of the human cardiovascular system [5].

Robot-assisted medical automation is one of the most extensively applied technologies. Many medical robot applications have passed FDA approval and have been put into routine clinical operations. The medical robots can assist with various surgical procedures including radiosurgery (or telesurgery) to cure brain, liver, kidney, and spine [1] tumors in the human body.

With precision control, image guidance, and other advanced automation technologies, surgical robots perform better than human surgeons in terms of (1) ability to treat otherwise unreachable targets, (2) reduced error rate and operation time, (3) high accuracy to minimize the damage to healthy tissue, and (4) ability to perform long or delicate surgeries without fatigue. Besides surgical robots for medical automation, robots are also used for clinical or pharmaceutical lab automation, hospital conveyance system, instrument storage and retrieval system, and delivery of medical supplies. Robot-assisted medical automation normally involves physical motion in medical activities. For such applications, in addition to modeling the medical process itself, the modeling of the robotic dynamics is also critical to the success of medical automation.

There are other medical activities that do not need medical robots to perform physical operation. In such cases, control technologies may be used for medical automations. A bioreactor is a kind of apparatus (e.g., a large fermentation chamber) for growing organisms such as bacteria or yeast that are used in the biotechnological production of substances such as pharmaceuticals, antibodies, or vaccines. They have been widely used in industry to produce chemical compounds or medical material (e.g., proteins or therapeutic drugs) synthesized by microorganism. The objective of a bioreactor's control is normally to maximize or minimize certain performance indexes (e.g., cell production or protein concentration). Proper control could result in high yield of pure product at low cost. However, in order to construct an appropriate control scheme, it is important to have a good understanding of bioreactor modeling issues, which, at present, are still based on rough approximations. Current bioreactor modeling approaches can be classified as fundamental models and empirical models [6].

Besides the above-mentioned control tools used for implementing medical automation, there are other technologies to automate the medical processes in different levels of abstraction. Here are just a few examples:

- *Medical data management automation.* For example, RALS-plus for medical data analysis, which automatically manages, reports, and electronically transfers patient urinalysis data to the RALS-Plus database and the hospital's laboratory information system.
- *Drug discovery automation.* For example, sample handling automation for drug discovery analysis [7], "magnetic" separation technology, using magnetic particles, is a quick and easy method for sensitive and reliable capture of specific proteins, genetic material, and other bimolecules [8].

3.1.2 Modeling and Automation

Similar to the automation in other science and engineering fields, reliable, efficient, and precise control of the medical activities relies on a good mathematical model. A mathematical model is an abstract model that uses mathematical language to describe the behavior of a system. Mathematical models are used widely in engineering disciplines, such as physics, biology, and electrical engineering, and in the social sciences, such as economics, sociology, and political science, to better understand the nature of complex systems and to quantitatively study dynamics problem. Physicists, engineers, computer scientists, and economists use mathematical models extensively to solve complex problems in dynamic systems. A mathematical model is also a representation of the essential aspects of an existing system (or a system to be constructed) that captures the relationship between inputs and outputs of the system. More mathematical models have been recently applied to solve medical automation problems. Unfortunately, systematic study and understanding of the approaches has not been well explored in the open literature for medical automation. This chapter introduces various mathematical modeling approaches for medical automation.

A general automation system consists of two major parts: the process/plant and the control mechanism. The process/plant is subject to the control actions from the control mechanism. The control action will lead the process/plant to the desired status or evolution. A distinguished feature of an automation system is that there is no human intervention in the control mechanism or human intervention is at a high level (e.g., supervisory level). The relationship between the process/plant and control mechanism can be graphically categorized into two classes:

1. *Open loop control system.* In the open loop system, the control action is generated by the control mechanism without taking the process/plant's information into account. Open loop automation is simple and easy to implement, but its control performance is limited depending on the application (Figure 3.1).
2. *Closed loop control system.* In the closed loop control system, the control mechanism takes the process/plant's feedback information into account to generate the control action. The control loop control system is complex, but its control performance will normally outperform the open loop control system (Figure 3.2).

In automation industries such as electrical engineering and robotics engineering, if the mathematical model of the process/plant is given, there is a fairly mature control theory to analyze the system performance and to design/develop the control mechanism to achieve the automation.

Similar to automation in other fields, one important factor for medical automation is the modeling of the medical processes and mechanisms (e.g., process/plant), which include the following two major modeling tasks:

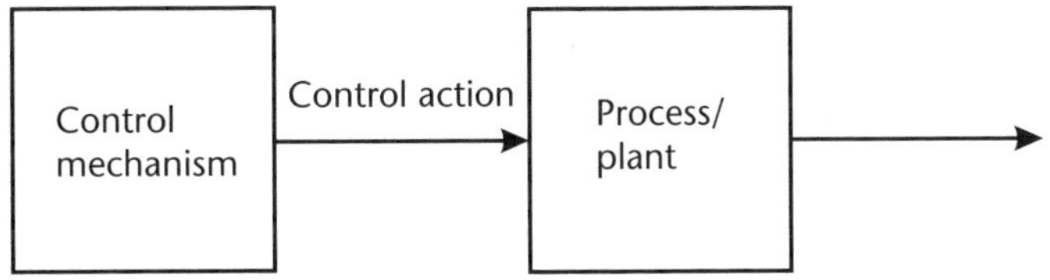

Figure 3.1 Open loop control system.

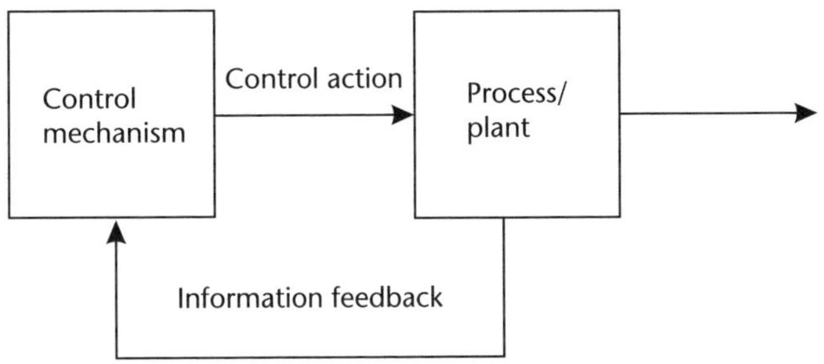

Figure 3.2 Closed loop control system.

1. *Modeling the tools for executing a medical process.* There are many tools such as medical robots, medical equipment, and medical instruments for implementing medical automation. These tools should be modeled as well as the dynamic actions of the tools in order to better understand how they will work in real life.
2. *Modeling the medical process/phenomenon.* This includes the process of white blood cell production, epidemic transmission process, and disease behavior for medical diagnosis. By modeling these medical processes and phenomena, the key elements affecting the medical behavior can be identified, and related control methods may be developed for medical automation.

How mathematical modeling affects medical automation depends on the application domain. Theoretically speaking, if the model of a medical process is obtainable and there is a requirement to automate the process, it is possible to apply general control theory to develop the control mechanism to control that process for medical automation. However, the implementation of the control techniques in medical domain may be limited by physical equipment and process. In light of this fact, our main focus in this chapter is on the modeling itself without going too far into modeling the relationship with the medical automation technologies except for certain specific medical robotic systems. In these cases the mathematical modeling is systematically related to the automation and its control.

3.2 Mathematical Modeling

Mathematical models use mathematical languages such as algebraic, logical, and differential equations to describe the behavior of natural processes. The type of equations used depends on the nature of the system to be modeled. Sometimes, structure diagrams may be used to assist in understanding the relationship of individual modules of the process/system. Mathematical modeling approaches in medical automation share similarities of the modeling in other industrial processes. It is also unique in certain aspects, such as accuracy and sensitivity of medical model, and approaches used for model validations. This section summarizes the general approaches often used in medical automation.

3.2.1 Basic Approaches for Model Creation

There are three basic model creation approaches: analytical modeling, empirical modeling, and hybrid modeling.

3.2.1.1 Analytical Modeling

Analytical modeling is mainly based on the fundamental laws of conservation (physical or chemical laws such as mass and energy balances) to describe the system/process dynamics and their input-output relationship. The analytical modeling may provide a good insight of what happens in the system, which is also imperative for system control in automation.

There are a number of conservation laws such as mass, energy, and momentum conservation laws that dictate the underlying relationship of the state variables and independent variables. The conservation laws are expressed in their respective balance equations such as heat balance, force balance (Newton's law, LaGrange equation), and continuity equations that can be summarized below:

- *Mass conservation:* The law of mass conservation states that matter or mass of a closed system cannot be created/destroyed but can be rearranged. This principle applies well in many fields such as chemical, mechanics, and fluid fields although it doesn't apply to fields described by special relativity and quantum mechanics where energy and mass can be converted among each other.
- *Energy conservation:* The law of the conservation of energy states that the total amount of energy in any closed system is constant, but its form can be changed (e.g., from electrical energy to heat to kinetic energy). The first law of thermodynamics is a precise statement of the law of the conservation of energy in thermodynamics. This law is widely used in mechanics and thermal dynamics.
- *Momentum conservation:* The law of conservation of momentum states that the total momentum of a closed system of objects is constant. The law of momentum conservation can be expressed in force balance, Newton's law, and in other mathematical relationships such as the Lagrange equation, which is most often used in modeling the dynamics of robotics including medical robots.

Besides the fundamental conservation law discussed above, there are also other conservation laws to describe the phenomenon occurred in various engineering disciplines, such as conservation of charge in electrical engineering and the conservation of moment in mechanical engineering.

Depending on the nature of the problem, the state variables may have various relationships with other state variables or independent variables. These relationships can be expressed as various auxiliary laws. Below are two examples:

- Fick's law describes the diffusion of a material from high concentration area to low concentration area, and can be used with the mass conservation law.

$$N = -D\frac{dC}{dX} \tag{3.1}$$

Where N is the mass of diffused material, D is the diffusion coefficient, C is the concentration, and X is the diffusing distance.

Fick's law has been widely use used as the basis for modeling the transport processes in biopolymers, pharmaceuticals, and in cells (e.g., neurons) [9] For example, Fick's law is used as a basis for a diffusion model to calculate the dynamics of the change in fusion pore diameter (a basketlike structures at the cell plasma membrane) that are "important, as any modulation of release processes could have impacts on specific physiological functions."[9].

- Thermodynamic laws describe the heat change in terms of energy with respect to temperature change under constant pressure, and can be used with energy conservation law.

$$dU = \delta Q - \delta w \tag{3.2}$$

where dU is the heat change, δQ is the temperature change, and δw is the coefficient.

Thermodynamics deals with the concepts of heat and temperature and their connection to properties of matter and to processes in natural and constructed systems. Thermodynamics in biological sciences is called biological thermodynamics, which study the "energy transductions that occur in and between living organisms, structures, and cells and of the nature and function of the chemical processes underlying these transductions." [10]. Biological thermodynamics is important since "the transformation of energy from a more to a less concentrated form is the driving force of all biological processes or chemical processes that are responsible for the life of a biological organism."[10]. Since the biological and chemical processes are tightly connected to medical processes, biological thermodynamics is also an important tool for modeling medical processes. The "convenience kinetics" approach used for modeling biochemical systems in a simple and standardized way was introduced in [11]. This approach is based on thermodynamics law.

There are also many other auxiliary laws to describe the relationship of state variables like pressure, speed, temperature, and energy with independent variables like time and space.

3.2.1.2 Empirical Modeling

Empirical modeling employs some mathematical relations to approximate the process based on empirical knowledge. The model parameters are identified through process input and output data. The empirical model can be used where the underlying physical or auxiliary laws are not very well understood. An example of empirical modeling is modeling the performance of fluids in microscopic nanoscale structures. It has the disadvantage of limited model operating range within which the model parameters are created. There are a lot of medical processes that can be described using the existing equations (e.g., hyperbolic saturation can be expressed as the Michaelis-Menten equation). Selecting the best fit equation with matching parameters is the job of empirical modeling. Some popular functions listed below may be used for empirical modeling:

- Linear function:$y = ax + b$, which shows a straight-line relationship between state variable y and independent variable x;
- Exponential function: $y = e^{ax}$, if a>0, y will increase quickly with x. If a<0, x>0, y will decrease steeply to 0.
- Saturation function: $y = a(1 - e^{bx})$, if b<0, y will be saturated to a as x increases. If b>0 and x>0, the function declines from 0.

- Triangular function: $y = \begin{cases} a + bx, if\ x < c \\ e - fx, if\ x > c \end{cases}$: where $a + bc = e - fc$. This function consists of two linear functions.

Besides these popular functions, there are other functions such as the power function, Richards function, Blumberg function, and Hill function.

3.2.1.3 Hybrid Modeling

Sometimes the two modeling approaches may be combined for process modeling to make use of their respective advantages. For example, the preliminary process modeling may be based on conservation laws but some of their parameters may be obtained by empirical modeling.

In the pharmaceutical industry, the fundamental and auxiliary laws in chemical, molecule, and mechanics disciplines can provide good knowledge about the structure and properties of various candidate drug substances. However, some medical effects/behavior cannot be modeled by the analysis approach and need to be assisted with the empirical model due to the complexity of the biological mechanism. A hybrid modeling approach (also called the semiempirical approach) for general treatment of solute-solvent interactions (GSSI) in computer-assisted drug design (CADD) is introduced in [12]. In this application, the analytical model based on molecule dynamics (MD) is not sufficient for modeling the entropic effects, so the empirical model is used. "Our semiempirical general treatment of solute-solvent interactions (GSSI) represents a useful compromise between the insufficient rigor of empirical approaches and the excessive rigor of MD methods. Our approach is similar to molecular dynamics with respect to modeling the enthalpic aspects of solute-solvent interaction and attempts to model the entropic contributions in a more practical manner" [12].

3.2.2 Mathematical Modeling Techniques for Automation

A mathematical model consists of a number of variables, representing the properties of the process/system and a number of equations, describing the relationship of the variables. Some of the variables (state variables) are dependent on other variables (independent variables). Each equation consists of variables and operators applied on the variables. The operators can be algebraic, functional, and differential operators, which give rise to algebraic, general function, and differential equations. The solutions of the equations show the evolution of the state variables versus the independent variables (e.g., time and space).

3.2.2.1 Structure (Schematic) Diagram for Modeling

For the mathematical modeling of a big process/system, its structure diagram will be very helpful in creating the models. The structure diagram should outline the elements in the models and how they are connected. There is no formal format for the structure diagram. Any diagram that helps in understanding the system structure and the interrelationship of the system elements may serve the purpose. One example is the block diagram (or signal flow diagram) used in control system modeling and design. The block diagram is an abstraction of the system structure, which shows the functional relationship of system elements and the flowpaths of system information (e.g., control signals and sensor data). The block diagram is very helpful in deriving the system model in terms of differential equations or Laplace transfer functions.

Figure 3.3 shows a structure diagram for modeling the white blood cell production from stem cells. Where S is the number of hematopoietic stem cells, M is the number of matured WBC. τ_S and τ_M are proliferative cycle time and mature delay time, respectively. *F*(*M*) and *K*(*S*) are the differentiate rate and proliferative rate, respectively. A is the division factor that amplifies the differentiation. α and γ_S are the disappearance rate of M and the apoptosis rate of proliferative cells.

From Figure 3.3, it is easy to see how the WBC is produced and the relationship between stem cell and mature WBC. The mathematical model from this diagram is described in Section 5.4.2.

Figure 3.4 shows a typical block diagram of a feedback control system, which may be used for controlling medical robots.

Figure 3.4 shows the main system structure and each module's input/output relationship. Each module is expressed in its Laplace transfer function. From this figure, it is easy to find the output Laplace transform:

$$Y(S) = H(S)U(S) \tag{3.3}$$

where *H(S)* is the transfer function and *U(S)* is the input Laplace transform. The transfer function *H(S)* can also be obtained based on the block diagram.

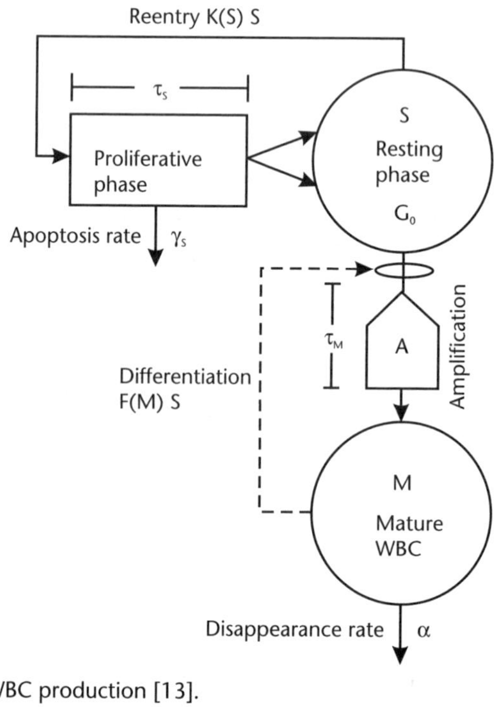

Figure 3.3 Model of WBC production [13].

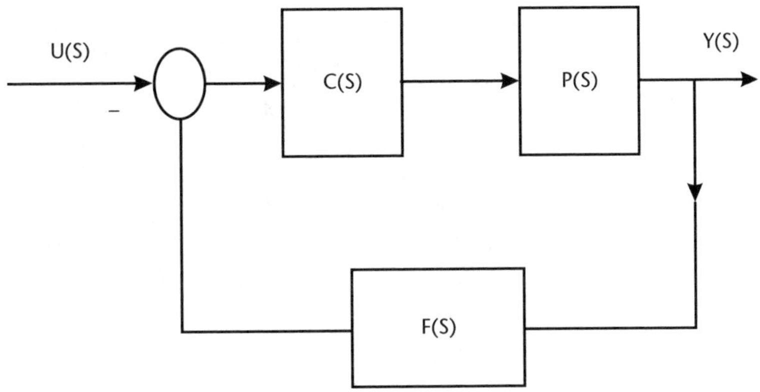

Figure 3.4 Block diagram of a feedback control system.

$$H(S) = \frac{C(S)P(S)}{1 + F(S)C(S)P(S)} \tag{3.4}$$

where $P(S)$ is the plant Laplace transform, $F(S)$ is the Laplace transform of feedback function, and $C(S)$ is the controller's Laplace transform.

Applying reverse Laplace transform to $Y(S)$, we may get output solution $y(t)$, which shows the evolution of output versus time. The block diagram is widely used in industrial automation including medical robotics control for control design and analysis. Refer to Section 3.4.1.4 for further details.

3.2.2.2 Algebraic Equation Modeling

In algebraic equation (AE) modeling, the state variables and the independent variables are combined with algebraic operators. For example: $y = ax^2 + x$.

The solution for state variables in AE modeling is straightforward or easy to derive. The ultrasound image of the human body for medical examination can be taken as an example [14]. The ultrasound images record different time of tumor cross section (the egg-shaped image in Figure 3.5). The radiologist can use these images to analyze quantitatively the tumor evolution. The contour representing the intersection between the ultrasound observation plane and the ultrasound image of the egg-shaped object can be defined as a third-order polynomial as below:

$$\sum_{i,j\geq 0, i+j\leq 3} a_{ij} x^i y^i = 0 \tag{3.5}$$

where x, y is the contour coordinate variables in the observation plane. Equation (3.5) is an algebraic equation modeling the egg-shaped ultrasound image.

3.2.2.3 Ordinary Differential Equation Models for Continuous Time System

Most natural phenomena involves change of its variables (e.g., motion of an object, chemical reaction). Some variables (or functions) depend on some independent variables. A differential equation describes the relationship between the variables, its derivatives (the change of the variable versus the change of another variable) and

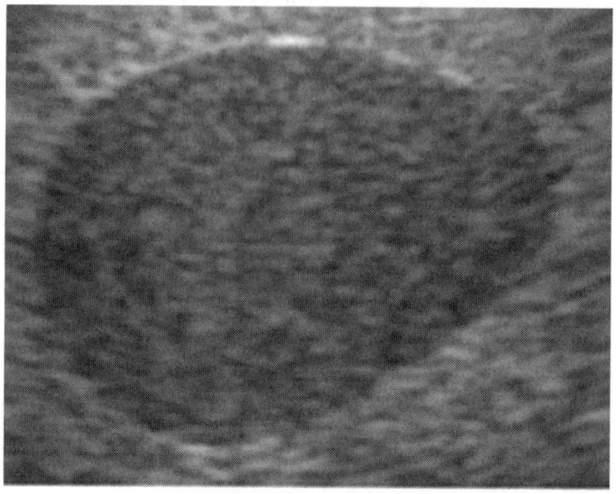

Figure 3.5 Egg-shaped ultrasound image.

the independent variables. The ordinary differential equation (ODE) contains only one independent variable. The partial differential equation (PDE) contains more than one independent variable that requires the description for the partial derivatives.

General Description

An ordinary differential equation has the form of:

$$F(x, y, y, \ldots, y^{(n)}) = 0 \tag{3.6}$$

where x is the independent variable, y is the function of $x, y^{(n)} = \frac{d^n y}{dx^n}$ is the n-th derivative of x. The n is the order of the ODE equation. If F can be written as the linear combination of derivatives of y, (3.6) is called the linear ODE, which can be expressed as:

$$\sum_{i=0}^{N} a_i(x) \frac{d^i y}{dx^i} = f(x) \tag{3.7}$$

The above equation (3.7) is an inhomogeneous equation. If $f(x)=0$, (3.7) is the associated homogeneous equation. If the coefficients a_i (x) are constant, (3.7) becomes the constant coefficients linear equation.

Linear ODE is widely applied to modeling applications. One reason is that many nature phenomena can be modeled by a linear ODE or a set of linear ODEs with sufficient accuracy. Another reason is that general analytical approach is available for solving the linear ODE (though the analytical solution does not always exist). Unlike the linear ODE, there is no general technique to solve an ODE if it is nonlinear. In such cases and the linear ODE cases where analytical solutions are not available, the numerical approach with aid of computer computation may be used to obtain an approximated solution. In this chapter, our discussion is restricted to linear system models.

For nth order linear ODE of (3.7), where $f(x)=0$, its complete solution is:

$$Y(x) = y_g(x) + y_p(x)$$

where $y_g(x)$ is its general solution of the corresponding homogeneous equation and $y_p(x)$ is its particular solution.

For constant coefficient linear ODE, we have:

$$y_g(x) = \sum_{i=1}^{N} B_i u_i(x) \tag{3.8}$$

where $u_i(x)$ ($i=1,\ldots, N$) are N linearly independent solutions of homogeneous equation of (3.7). B_i are the coefficients that may be computed based on initial or boundary conditions.

For constant coefficient linear ODE, $u_i(x)$ depends on the roots of the corresponding characteristic equation:

$$\sum_{i=0}^{N} a_i r^i = 0 \tag{3.9}$$

If r_i denotes the i-th root of the characteristic equation, we have:

$u_i = e^{r_i x}$; if r_i is a single real root.

$u_i = e^{r_i x}; u_{i+1} = xe^{r_i x}; \ldots; u_{i+m-1} = x^{m-1} e^{r_i x}$; if r_i is the real root with m multiplicity.

$u_i = e^{r_i x}\cos(v_i x); u_{i+1} = xe^{r_i x}\cos(v_i x); \ldots; u_{i+m-1} = x^{m-1} e^{r_i x}\cos(v_i x)$; if the root is: $r_i + v_i i$ with multiplicity m.

For constant coefficient linear ODE, $y_p(x)$ may be obtained by examining the form of $f(x)$ and assuming $y_p(x)$ is a linear combination of all independent functions from $f(x)$, then applying the $y_p(x)$ to (3.7) to compute the coefficients of the linear combination functions. Two examples to model the medical process using ODE are given below:

Pharmacokinetic model

A simple two-one compartment combined pharmacokinetic model is given here to show the process of drug absorption and distribution by a constant coefficient linear ODE. The structure diagram of the two-compartment model is shown in Figure 3.6.

Assume the drug is tetracycline and is taken orally. Use X_1 and X_1 to denote the amount of tetracycline in the two compartments (gastrointestinal tract and plasma), respectively, and C_1 and C_2 to denote the transfer rate of them ($C_1 \neq C_2$). Based on mass conservation law, we have two first-order homogeneous equations to describe the drug diffusion process:

$$\begin{cases} \dot{X}_1 = -C_1 X_1 \\ \dot{X}_2 = C_1 X_1 - C_2 X_2 \end{cases} \tag{3.10}$$

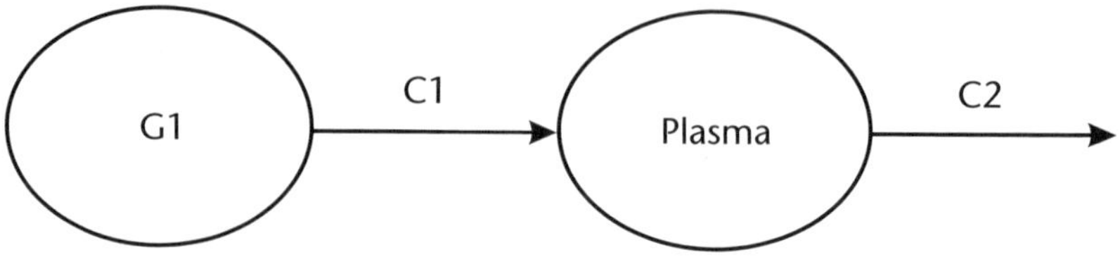

Figure 3.6 The structure diagram of pharmacokinetic model.

Given the initial condition $X_1(0)$=D, and $X_2(0)$=0, and solve the above equations, we have:

$$\begin{aligned} X_1 &= De^{-c_1 t} \\ X_2 &= D\frac{C_1}{C_2 - C_1}(e^{-c_1 t} - e^{-c_2 t}) \end{aligned} \tag{3.11}$$

The above solution X_1 and X_2 shows the evolution of tetracycline in GI and plasma over time.

Epidemic model

The epidemic model describes the transmission behavior of some diseases. A classical model called Kermack-Mckendrick model is given below:

$$\begin{aligned} dx(t) &= -\beta x(t)y(t)dt, & x(0) &= x_0 > 0 \\ dx(t) &= \beta x(t)y(t)dt - \gamma y(t)dt, & y(0) &= y_0 > 0 \\ dz(t) &= \gamma y(t)dt, & z(0) &= 0 \end{aligned} \tag{3.12}$$

where $x(t)$ denotes the size of the susceptible (can be infected) subpopulation, $y(t)$ denotes the size of the infective subpopulation, and $z(t)$ denotes the size of the removals (restore to health and won't get infected again) subpopulation at time t. β is the infection rate and γ is the removal rate. Equation (3.12) is a set of nonlinear ODE, which describes the behavior of three groups of population under the disease transmission. Its analytic solution is available for the process analysis and further progression control.

By resolving the ODE, the function dynamics with respect to the independent variable can be examined, which will help to understand and further control the dynamic process for medical automation. However, solving a general ODE described in (3.7) requires a number of steps from finding a general solution and a particular solution to computing the coefficients based on the initial or boundary conditions. To simplify this process, Laplace transforms may be used for solving the linear ODE, which converts the differential equation to an algebraic problem and gets the complete solution in a simple way.

Laplace Transform

The Laplace transform of a function $f(t), t \in [0, \infty]$ is defined as

$$L(f(t)) = F(S) = \int_0^\infty f(t)e^{-St}\,dt$$

where S is the Laplace operator. One of distinguished features for Laplace transform L is that it can convert the differentiation of $f(t)$ to multiplication of $F(s)$ by S:

$$L\left(\frac{d(f(t))}{dt}\right) = SF(S) - f(0) \tag{3.13}$$

This feature allows the Laplace transform to turn differentiation to algebraic expression. With Laplace transform, we can convert the linear constant coefficients ODE (3.7) into the following S domain algebraic equation:

$$\sum_{i=0}^{n}\left[a_i\left(S^i y(S) - \sum_{k=0}^{i-1} S^{i-1-k} y_0^k\right)\right] = F(S)$$

$$Y(S) = \frac{F(S)}{\sum_{i=0}^{n} a_i S^i} + \frac{\sum_{i=0}^{n}\sum_{k=0}^{i-1} a_i S^{i-1-k} y_0^k}{\sum_{i=0}^{n} a_i S^i} \tag{3.14}$$

where, $Y(S)$ is Laplace transform of $y(t)$, y_0 is the initial value of $y(t)$.

Apparently the denominator of (3.14), called *characteristic polynomial* here, is the same as the characteristic equation of constant coefficient ODE in (3.7).

From (3.14), we can get $y(x)$ by inversing $Y(s)$, which can be done by looking up the inverse of Laplace transform table.

In the pharmacokinetic model of (3.10), applying the Laplace transform, we have:

$$\begin{cases} SX_1(S) = -C_1 X_1(S) + X_1(0) \\ SX_2(S) = C_1 X_1(S) - C_2 X_2(S) + X_2(0) \end{cases} \tag{3.15}$$

This is an algebraic equation, and its solution is:

$$\begin{cases} X_1(S) = \dfrac{X_1(0)}{S + C_1} \\ X_2(S) = \dfrac{X_2(0) + \dfrac{C_1 X_1(0)}{C_1 - C_2}}{S + C_2} - \dfrac{\dfrac{C_1 X_1(0)}{C_1 - C_2}}{S + C_1} \end{cases} \tag{3.16}$$

Applying the inverse Laplace transform, we can get the $X_1(t)$ and $X_3(t)$, the same as solution (3.11).

Vector Equation

To simplify the notation of linear ODE (3.7), a vector-based first-order system is often used to describe the model:

Define: $y_0 = y, y^1 = \dfrac{dy}{dt}, \ldots y^{n-1} = \dfrac{d^{n-1}y}{dt}$ and $Y = \left[y^0, y^1, \ldots, y^{n-1}\right]'$

Equation (3.7) can be expressed as:

$$\dot{Y} = AY + B \tag{3.17}$$

where:

$$A = \begin{bmatrix} 0 & 1 & 0 & \dots & 0 \\ 0 & 0 & 1 & \dots & 0 \\ \dots & & & \dots & \\ 0 & 0 & 0 & \dots & 1 \\ -\frac{a_{N-1}}{a_N} & -\frac{a_{N-2}}{a_N} & -\frac{a_{N-3}}{a_N} & \dots & -\frac{a_0}{a_N} \end{bmatrix}; B = \left[0 \ 0 \ 0 \dots f(t)/a_N\right]'$$

A different definition of $y^0, y^1, \dots, y^{n-1}$ can lead to different expression of A.

State Space Equation for Control System Modeling

The control system modeling is similar to general linear ODE modeling. However, due to its unique features and importance in medical automation, its state space equation is described in this section.

For multiple input multiple output (MIMO) control system, its state space equation of a linear control system can be expressed as follows:

$$\begin{cases} \dot{X} = A(t)X(t) + B(t)U(t) \\ Y = C(t)X(t) + D(t)U(t) \end{cases} \tag{3.18}$$

where $X \in R^n$ is the $n \times 1$ state variable, $Y \in R^p$ is output variable and $U \in R^m$ is system input. A is an $n \times n$ system matrix, B is an $n \times n$ input matrix, C is an $n \times n$ output matrix, and D is a $p \times m$ feedforward matrix.

The state space equation shows the relationship between system state, output, and input. The state variables are user defined and there are numerous ways to define them, as such, the state space equation may be expressed differently. The state space equation is a way to describe the system of linear ODE.

For linear time-invariant (LTI) system (or the linear constant coefficient system), the A,B,C,D matrix in (3.18) is constant. The solution of (3.18) is:

$$\begin{aligned} X(t) &= e^{-A(t-t_0)}X(t_0) + \int_0^t e^{A(t-\tau_0)}Bu(\tau)d\tau \\ Y(t) &= CX(t)Du(t) \end{aligned} \tag{3.19}$$

where t_0 is the initial time.

Take the pharmacokinetic model of (3.10) as an example. Since there is no input *U(t)* so *B* and *D* are zero, and the output is defined as $Y = X_2$, we have (3.10) in state space form:

$$\begin{cases} \dot{X} = AX(t) \\ Y = C(t)X(t) \end{cases} \tag{3.20}$$

where $A = \begin{bmatrix} -C_1 & 0 \\ C_1 & -C_2 \end{bmatrix}, C = [0 \quad 1]$

According to the LTI system solution (3.19), we have:

$$\begin{aligned} X(t) &= e^{-A(t-t_0)} X(t_0) \\ Y &= CX(t) \end{aligned} \tag{3.21}$$

where $t_0 = 0$. The above solution is the same as solution (3.11).

3.2.2.4 Difference Equation Models for Discrete-Time System

General Description

A discrete-time system describes the model in a time sequence with a fixed interval. The function values are presented in a series of discontinuous points known as time series, which are usually discretized in time. When a computer control system involved is in medical automation such as medical robotics, the systems to be modeled may be expressed as discrete-time system. The difference equation for a discrete-time system is similar to the continuous time ODE.

Assuming $y(kT)$ is a series function value at time kT, where k is from 0 to any integer number and T is the discrete time interval, the nth order linear difference equation corresponding to (3.7) can be written as:

$$\sum_{i-0}^{N} a_i(kT)y(kT - iT) = f(kT) \tag{3.22}$$

which has the same characteristic equation as the linear constant coefficient homogeneous equation (3.7).

$$\sum_{i=0}^{N} a_i r^i = 0$$

The general solution for the homogeneous equation (3.7) is:

$$y_g(kT) = \sum_{i=1}^{N} B_i r_i^k$$

The B_i is the coefficients determined by the boundary conditions.

The complete solution of (3.19) is the general solution plus a particular solution. As with the continuous system, the discrete time system can also be expressed in state space form:

$$\begin{aligned} X[(k+1)T] &= G(T)X(kT) + H(T)u(kT) \\ Y[(k+1)T] &= C(T)X(kT) + D(T)u(kT) \end{aligned} \tag{3.23}$$

If the discrete-time system (3.23) is derived from continuous time system (3.18), we have

$$G(T) = e^{AT} \ and \ H(T) = \int_0^T e^{A(T-\tau)} B d\tau$$

In medical automation, the epidemic model (e.g., bird flu epidemic mode) can be described as a difference equation (e.g., logistic growth model):

$$P_{n+1} = P_n + kP_n(L - P_n)$$

where L is the maximum possible size or carrying capacity of the population, k is a nonzero constant. P_n is the number of people infected at the beginning of discrete time n (e.g., day n). The above equation models the evolution of bird flu infection.

Z-Transform

For convenience and simplification, the Z-transform is used for dealing with difference equation. The one-sided Z-transform is defined as:

$$Z\{y(k)\} \equiv Y(z) = \sum_{k=0}^{\infty} y(k) z^{-k} \tag{3.24}$$

where T is set to 1 without losing generality.

Similar to the Laplace transform, which converts differentiation to algebraic expression, the Z-transform converts the difference equation to algebraic equation by dealing with time shift in the difference equation. For example:

$$Z\{y(k-1)\} \equiv z^{-1} Y(z) + y(-1)$$

The Z-transform is used in a similar way to resolve the difference equation.

3.2.2.5 Partial Difference Equation Model

The ODE and AE approaches serve well in most modeling cases. However, there are cases that PDE cannot be avoided or will provide better modeling results. PDE has to be used if the dependent variables depend on more than one independent variable. Taking the waves propagation in a medium as an example, the wave propagation rate depends not only on time but also on space variables. To describe such a propagation process, PDE will have to be used. Similarly, PDE may be used to model the recovering human granulopoiesis after high-dose chemotherapy with stem cell support, since evolution of the white blood cell density depends not only on time but also on the cell maturation rate [15].

Besides wave propagation, PDE is also used for other propagation of certain properties of objects such as sound, heat, electrostatics, electrodynamics, fluid flow, and elasticity, which are distributed in space or in both time and space. A PDE is represented as an equation involving functions and their partial derivatives. The equation's order is the highest order of all partial derivatives. The following PDE is a first-order PDE, which describes the evolution process of a property (e.g., temperature) carried by a flow along a one-dimensional duct with a velocity a:

$$\frac{\partial u}{\partial t} + a \frac{\partial u}{\partial x} = 0 \tag{3.25}$$

Same as ODE, a PDE can also be categorized as linear and nonlinear equation.

A PDE sometimes may be solved using a Backlund transformation, characteristics, integral transform, separation of variables, or Fourier transform approaches but most of the time it can only be solved by numerical methods.

In medical automation, PDE has been widely used for biomedical image analysis [16], white blood cell production [15], and many other applications. To model the white blood cell production after chemotherapy, the cell density may be expressed by a convection-reaction PDE as follows:

$$\frac{\partial u}{\partial t} + \frac{\partial (\beta u)}{\partial x} = \alpha u$$

where t is time, x is the cell maturity level, $u(x, t)$ is the cell density that depends on both time and cell maturity level. α is the cell proliferation rate and β is the maturation rate.

3.2.2.6 Stochastic Modeling

Uncertainty and fluctuations usually exist in every real system. Many biological processes (such as growth process, molecule motion process, etc.) involved in medical automation are full of uncertainty. In the deterministic modeling we have discussed before this section, the uncertainty effect is negligible so the signals and the mathematical model of a system are created without uncertainty and the time behavior can be reproduced by repeated experimentation. In the stochastic systems, the uncertainty effect from system model parameters or its signals is significant, and thus has to be taken into account. In such cases, the values of the signals or the variables occurring in the system can only be estimated with the help of the methods of probability and statistics. The results are presented as expected values together with the bounds of error.

A stochastic process/system has one or more random variables that represent uncertainty and fluctuation. For a random variable x with a distribution $p(x)$, the probability to observe x between a and b is:

$$\int_a^b p(x)dx$$

The stochastic process refers to the process the stochastic system evolves in time. There are normally two modeling approaches to study the stochastic process: distribution evolution equation and the stochastic differential equation.

The *distribution evolution equations* describe the evolution of probability distribution for variables correlated in time. For example, given Markov process with a conditional distribution, $p(x_t | x_{t-1})$ we can use Chapman-Kolmogrov equation:

$$p(x_{t+\varepsilon} | x_{t-\tau}) = \int_{-\infty}^{\infty} p(x_{t+\varepsilon} | x_t) p(x_t | x_{t-\tau}) dx_t \tag{3.26}$$

to find the distribution two time steps ahead , and then get from the equation below:

$$p(x_{t+\varepsilon}) = \int_{-\infty}^{\infty} p(x_{t+\varepsilon}, x_t)dx_t$$

The *stochastic differential equation* (SDE) describes the relationship and evolution of function variables with random variables. The SDE is a differential equation with one or more random variables that follow the stochastic process. The random variables also lead the function variables to a stochastic process. A typical first-order SDE is of the form:

$$dX_t \,/\, dt = \mu(X_t,t) + \sigma(X_t,t)\xi(t), \quad X(0) = X_0 \tag{3.27}$$

or

$$dX_t = \mu(X_t,t)dt + \sigma(X_t,t)dW(t), X(0) = X_0$$

where $\xi(t)$ is the random variable, μ is a deterministic term called drift coefficient and $\sigma(X_t,t)$ is the diffusion coefficient. The solution of the above SDE is the X that satisfies the following integral form of SDE:

$$X(t) = x_0 + \int_0^t \mu(X, s)ds + \int_0^t \sigma(X, s)ds \quad \text{for all time t>0}$$

If the drift coefficient has the form: $\mu(X_t,t) = \mu X$, and the diffusion coefficient has the form: $\sigma(X_t,t) = \sigma X_t$, where σ and μ are constants, we have the equation for geometric Brownian motion:

$$dX_t = \mu(X_t,t)dt + \sigma X_t dW(t) \tag{3.28}$$

It has an analytic solution:

$$X_t = X_0 e^{((\mu - \delta^2/2)t + \sigma W_t)}$$

For the deterministic epidemic model (3.12), after introducing the random variable $W(t)$as a standard Brownian motion, its corresponding stochastic differential equation can be defined as follows [17]:

$$\begin{aligned} dX_t &= -\beta(X_t,Y_t,Z_t)X_tY_tdt + X_t\sigma(X_t+Y_t+Z_t)dW_t, & X_0 &= x_0 > 0 \\ dY_t &= -\beta(X_t,Y_t,Z_t)X_tY_tdt - \gamma Y_tdt + Y_t\sigma(X_t+Y_t+Z_t)dW_t, & Y_0 &= y_0 > 0 \\ dZ_t &= \gamma Y_tdt + Z_t\sigma(X_t+Y_t+Z_t)dW_t & Z_0 &= 0 \end{aligned} \tag{3.29}$$

The existence of the solution of above SDE can also be proved available [17].

3.2.2.7 Neural Network Modeling

The neural network modeling approach constructs the computer model of biological neurons of the human brain to resolve the problems that can only be solved by a human being. Applications such as machine learning, cognitive science (problem solving, speech generation and recognition), neurobiology (modeling how the brain works), mathematics (nonparametric statistical analysis and regression), and philosophy, are the fields where neural network exhibits good performance. In medical automation, the neural network is mainly used for disease diagnosis and drug discovery.

The neural network emulates brain neurons mainly in terms of structure and general properties without too much concern about the way neurons work to realize its functions. In this way, a much simpler neuron model can be constructed without emulating the way biological neurons work. From this point of view, the neural network approach is an empirical-based modeling approach. The basic neural network element to model a neuron is called the node or unit, which has the following mathematical model (see Figure 3.7(a)):

$$y_i = f\left(\sum_j w_{ij} y_j\right) \tag{3.30}$$

where y_i is the output of node i, f is the node's activation function, w_{ij} is the weight from node j to node i, y_i is the input of node i from node j. y_i can also be the input of other nodes. Apparently, the output of a node is the weighted sum called net input times its activation function. All the node's input and output can be interconnected to form a network called the neural network. Figure 3.7(b) shows an example of a

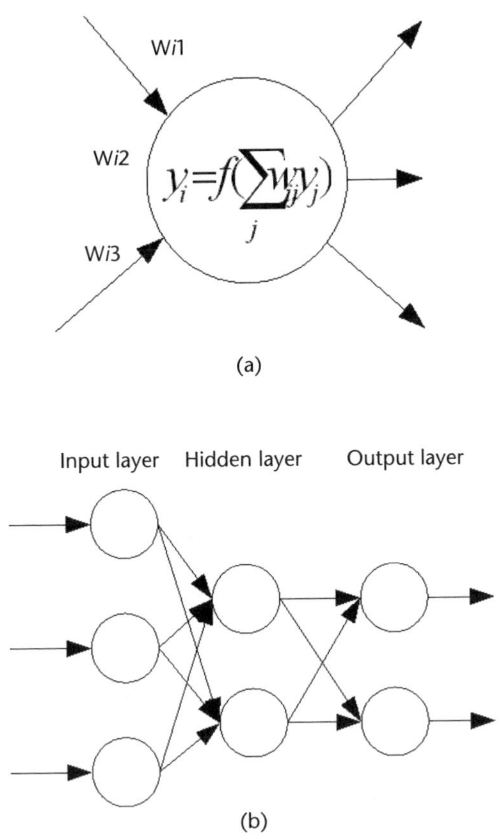

Figure 3.7 A neural network structure diagram. (a) Node structure, and (b) network structure.

neural network with three input nodes, two hidden nodes (middle layer), and two output nodes. In real applications, there may be more nodes and hidden layers.

To use the neural network to model a medical process for medical diagnosis and prediction involves defining the network structure, selecting the activation function, and determining the weights. This modeling procedure can be outlined as follows:

- Choose a few network structures (how many layers, how many inputs/outputs, and how they are connected);
- Use a significant amount of actual data to train the network, which means doing experiments with each network structure and retaining the best network based on selection error;
- In the above training process, the network structure may be adjusted based on training results (avoid underlearning and overlearning);
- In the end, the network structure, activation function, and parameters are determined, and the model is created.

Once the model is created, we may feed it with new input data and get the model output. For example, we may input a patient(s) medical information and get the model output as diagnosis or prediction results. Two things must be noted here:

- *Pre- and postprocessing.* The neural network deals with numerical data for input and output. Any symbolic data has to be converted to numerical form for processing.
- *Training data.* The training data has been representative and has a good coverage of the actual situation, which will ensure the model can generate meaningful and realistic results.

The nature of the neural network is its nonlinearity and ability to model very complex processes/systems.

In medical automation, the neural network has been employed for decision support by automating the process in clinical diagnosis and data analysis and interpretation, and for drug discovery.

- *Clinical diagnosis.* The neural network application ranges from commercial products (e.g., C.Net2000 for cardiac examination through electrocardiogram analysis and BioSleep for sleep analysis) for routine clinical use to research prototype [18]. The commercial products are proven to be equivalent or better than the average human specialist but with much better efficiency.
- *Data analysis.* The neural network has been used for image/signal analysis and interpretation. Applications include segmentation and classification of MRI, CT, and eye image for intelligent medical diagnosis; differentiation of contingent negative variation (CNV) evoked response waveform for patients with Huntington's disease, Parkinson's disease, schizophrenia, and so forth [19].
- *Drug discovery.* The drug discovery process is complex and lacks a general approach. The neural network has been used to predict a drug's mechanism of action from its pattern of activity against malignant cell lines [20].

3.2.2.8 System Identification for Modeling

In order to obtain the mathematical model of a system in a tractable and sufficiently accurate form, both the system structure and parameters should be identified. Our previous sections presented various ways to describe the model structure in terms of equations with the assumption that the system mathematical models have been derived from physical or chemical laws. System identification deals with how to build mathematical models for dynamical systems from experimental data. System identification can be applied to and accomplished by both fundamental approach (white box) and empirical approach (black box). For fundamental approach, the system identification is based on the physical laws that govern the process to be modeled. The phenomenological behavior and variables relationship are represented as equations (e.g., ODE or PDE) derived from the known physical laws. The system identification is to estimate the model parameters directly from input and output data, which is called parametric fitting. For empirical approach, the physical laws underlying the phenomenological behavior of a process is unknown, the model structure/equation and their parameters have to be inferred from measured data. Figure 3.8 shows the system identification diagram.

System identification approaches can be categorized based on its objectives:

- *Function fitting approaches.* This is mainly used for identifying the system parameters assuming the system model structure is known. Methods used include least square fitting for linear system and nonlinear systems.
- *Model selection approaches.* The most commonly used model is polynomials. However, there are some functions that are not suitable for polynomials to match (e.g., a model structure with poles). Other model alternatives may be applied such as the Pade approximant, which is suitable for model with poles, the spline model, which may provide superior local behavior to polynomials, and the orthogonal model, which allows successive corrections to be added without changing the coefficients already been found [2].
- *Filtering and time variant parameter estimation.* This is mainly used for smoothing, noise reduction, or signal separation. Approaches used include matched filters, Wiener filters, and Kalman filters.

3.3 Applications of Mathematical Modeling in Medical Automation

Many conventional modeling approaches used in automation have been explored in and applied to medical automation fields. This section gives a brief overview of the mathematical modeling approaches applied in medical robotics, neutrophil production, and the immune system.

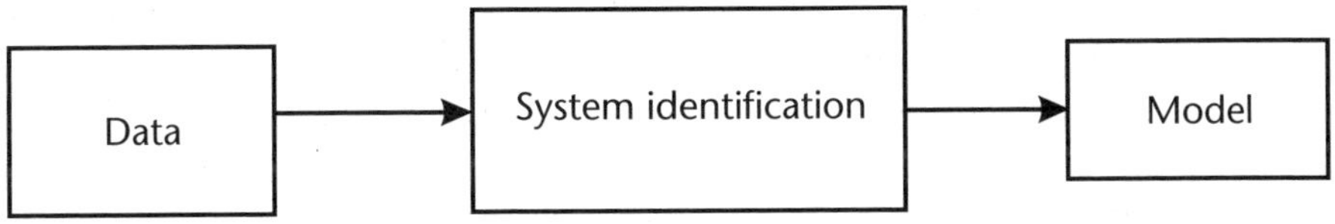

Figure 3.8 System identification.

3.3.1 Modeling of Medical Robotics

3.3.1.1 Overview of Medical Robotics

Medical robots have been used in medical fields such as surgery (remote surgery, minimally invasive surgery, and unmanned surgery), clinical and hospital service, as well as pharmacy and laboratory automation. The PUMA 560 (Programmable Universal Machine for Assembly) was the first robot used for performing neurosurgical biopsies with great precision in 1985. It was used for performing a transurethral resection of the prostate 3 years later. The PUMA 560 also led to the development of other surgical robots (e.g., ROBODOC, which was the first surgical robot approved by FDA). From the first surgical application done by the PUMA 560 robot in 1985, to the first unmanned robotic surgery done in 2006, and up to the time of this writing, robotic surgical technology has emerged as a new approach for surgical therapy with the development in image processing, computer, and network technologies. Besides surgical robots, medical robots are also used in other areas such as medical material handling, mobile service robots, and laboratory automation robots. In this section, as an example, we start with a generic robotics equation and follow with modeling a SCARA robot for medical material handling.

3.3.1.2 Generic Robotics Equation

The modeling techniques for medical robots are similar to other industrial robot (such as articulated robot modeling and mobile robot modeling).

For an n-joints robot, through Newton-Euler or Lagrangian approaches, the dynamics equation can be written following general state space form in joint space:

$$M(\Theta)\ddot{\Theta} + V\left(\Theta,\dot{\Theta}\right) + G(\Theta) = \tau \tag{3.31}$$

where Θ is n-joint angle variables, $M(\Theta)$ is $n{\times}n$ mass matrix, $V(\Theta,\dot{\Theta})$ is $n{\times}1$ centrifugal and Coriolis terms, $G(\Theta)$ is $n{\times}1$ gravity terms, and τ is $n{\times}1$ vector of force or torque terms.

Equation (3.31) can also be formulated in Cartesian state space:

$$M_x(\Theta)\ddot{X} + V_x\left(\Theta,\dot{\Theta}\right) + G_x(\Theta) = F \tag{3.32}$$

where, M_x is $n{\times}n$ Cartesian mass matrix, and V_x is a vector of velocity terms in Cartesian space. G_x is a vector of gravity terms in Cartesian space. F is a force-torque vector acting on the robot end-effector and is a Cartesian position vector of the robot end-effector. The terms in joint space and Cartesian space are related through a Jacobian term.

The general model of (3.31) and (3.32) are nonlinear equations. The control of such nonlinear system can be either a primitive linear control approach or more advanced nonlinear control approach (e.g., computed torque method).

The modeling of medical robot is a typical example of fundamental modeling approach since its dynamics equation is obtained based on physical laws, which are clear and mature enough to describe the relationship of all involved variables (e.g., torque/force, position, velocity, and acceleration).

3.3.1.3 The Modeling of a SCARA Robot

A SCARA robot, as shown in Figure 3.9, can be used for medical material handling. The left figure is the robot photo. The right figure is the diagram of the robot arm set. The robot arm may move along the *xyz* axis in Cartesian coordinates. The *x*, *y* axes are in horizontal plane and *z* axis is vertical to *x*,*y* plane. The robot end-effector cannot rotate around *x* and *y* axes, so the *z* motion is decoupled from *xy* motion. The *z* motion is only translational through a lead screw with fixed load, which is relatively easy to describe. For the motion in the *xy* plane, we can first express its model in joint coordinates and then convert it to Cartesian coordinates through forward kinematics. Based on the Lagrange equation:

$$\frac{d}{dt}\left(\frac{\partial K}{\partial \dot{\Theta}_i}\right) - \frac{\partial K}{\partial \Theta_i} = \tau_i, \quad i - 1{,}2$$

where K is the kinetic energy of the robot arm. Θ_1, Θ_2 are the motor rotation angle and generalized coordinates.

We have:

$$\begin{bmatrix} M_{11} & M_{12} \\ M_{21} & M_{22} \end{bmatrix}\begin{bmatrix} \ddot{\Theta}_1 \\ \ddot{\Theta}_2 \end{bmatrix} + \begin{bmatrix} H_{11} & H_{12} \\ H_{21} & H_{22} \end{bmatrix}\begin{bmatrix} \dot{\Theta}_1^2 \\ \dot{\Theta}_2^2 \end{bmatrix} + \begin{bmatrix} C_1 \\ C_2 \end{bmatrix}\dot{\Theta}_1\dot{\Theta}_2 = \begin{bmatrix} \tau_1 \\ \tau_2 \end{bmatrix} \tag{3.33}$$

where M is mass matrix, H is centrifugal term, and C is Coriolis term. The robot arm is designed to have $l_1 = l_2 = l$ and $r1 = r2 = r$ to ensure the end-effector axis going through the origin of the robot center O, which means $\alpha = \Theta_2 - \Theta_2$ and $\Theta_t = \Theta_1 + 0.5\alpha$. hence we have $\Theta_t = 0.5(\Theta_1 + \Theta_2)$. Only Θ_1 and Θ_2 are independent variables. τ_1 and τ_2 are joint torques of joint 1 and joint 2, which are mainly the respective motor torques.

In Figure 3.9(b), are mass, joint length, the distance between the center of mass and origin O, rotational inertia of joint i.

$$\begin{aligned}
M_{11} &= I_1 + 0.25I_t + m_2 l^2 + m_t l r_t \cos(\alpha/2) \\
M_{12} &= 0.251I_t + (m_2 r_2 + m_t l)l\cos(\alpha) + m_t l r_t \cos(\alpha) \\
M_{21} &= M_{12} \\
M_{22} &= I_2 + 0.251I_t + m_t l + m_t l r_t \cos(\alpha/2) \\
H_{11} &= 0.25 m_t l r_t \sin(\alpha/2) \\
H_{12} &= -(m_2 r_2 + m_t l)l\sin(\alpha) - 0.75 m_t l r_t \sin(\alpha) \\
H_{21} &= (m_2 r_2 + m_t l)l\sin(\alpha) - 0.75 m_t l r_t \sin(\alpha/2) \\
H_{22} &= -0.25 m_t l r_t \sin(\alpha/2) \\
C_1 &= -0.5 m_t l r_t \sin(\alpha/2) \\
C_2 &= 0.5 m_t l r_t \sin(\alpha/2)
\end{aligned}$$

The dynamics equation (3.33) can be used for robot performance analysis and control design. The parameters in the above model may be obtained either through direct measurement or through system identification. Normally in industrial practice, the results from direct measurement are used as the starting point for system identification. The mathematical model (3.33) is a nonlinear ODE.

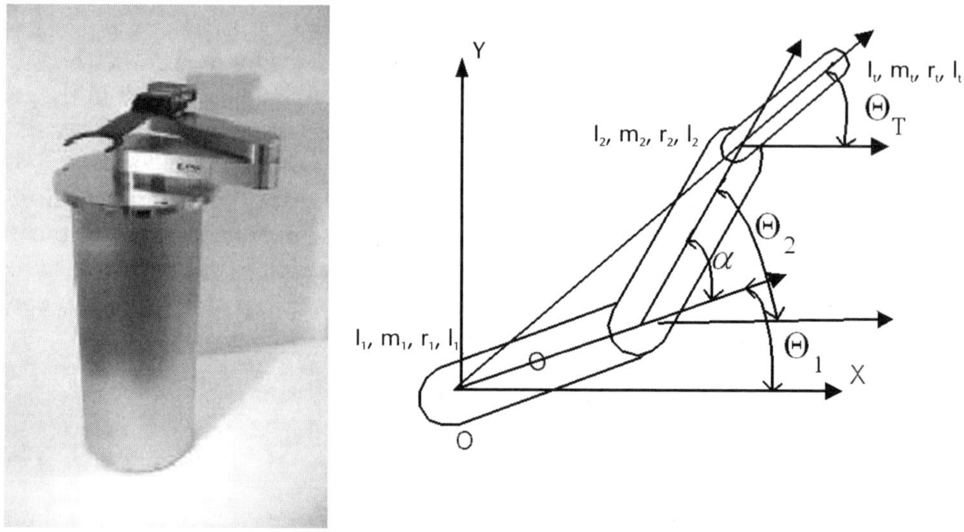

Figure 3.9 A SCARA robot for medical automation. (a) Overview of the robot, and (b) top view of the robot arm diagram.

3.3.1.4 Modeling, Analysis, and Control of a Track Robot for Medical Automation

In this section, a track robot that may be used for medical material handling is described. The modeling of the track robot plant is done through the law of force conservation. Its parameters are obtained through system identification. A block diagram with a Laplace transform was created for the whole system modeling including the control mechanism. The model is then analyzed using traditional control theory such as root locus approach. The control mechanism is incorporated for robot motion automation.

Figure 3.10 shows a track robot that may be used for medical material handling.

Modeling with the Law of Momentum Conservation

The robot track plant includes a SCARA robot, a platform for mounting the robot, and a track-driving system. The track-driving system consists of three pulleys, two belts, and the track motor. A track belt connects the robot platform and pulley 2. Pulley 2 and pulley 1 are directly connected by a solid-steel shaft. Pulley 1 and pulley 0 (motor pulley) are connected by a timing belt. Pulley 0 and an encoder are mounted on motor shaft.

By applying Lagrange's equation, the robot track plant equation can be written as follows:

$$a_1\ddot{\theta} + a_2\dot{\theta} = \mathbf{M} \tag{3.34}$$

where

$$a_1 = I_0 = (I_1 + I_2)\left(\frac{r_0}{r_1}\right)^2 + mr_2^2\left(\frac{r_0}{r_1}\right)^2 \tag{3.35}$$

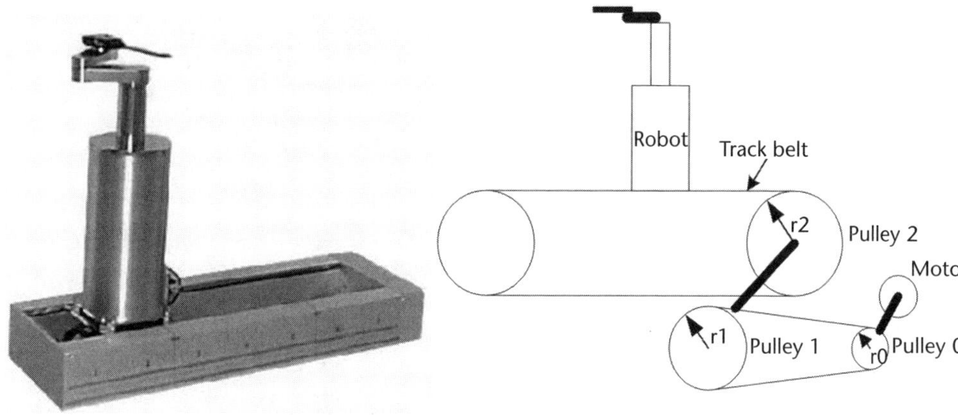

Figure 3.10 A track robot: (a) robot, and (b) schematic drawing.

$$a_2 = c_0 = (c_1 + c_2)\left(\frac{r_0}{r_1}\right)^2 + c_m r_2^2 \left(\frac{r_0}{r_1}\right)^2 \tag{3.36}$$

$$\begin{aligned} M &= M_m + M_f \\ M_f &= -m_1 Sign(\theta) \end{aligned}$$

- I_0, I_1, I_2 is the rotational inertia of motor pulley 0, pulley 1 and pulley 2;
- c_0, c_1, c_2, c_m are the dynamic coefficients of motor pulley 0, pulley 1/pulley 2, and track trail;
- r_0, r_1, r_2 is the radius of motor pulley 0, pulley 1 and pulley 2;
- M_m is the motor torque;
- M_f is the static friction torque;
- m is the weight of robot plus robot platform;
- m_f is the average static friction value;
- θ is the rotational angles of motor pulley 0.

From (3.34), the transfer function of θ/M can be written as:

$$G(s) = \theta / M = \frac{1}{S(a_1 S + a_2)} \tag{3.37}$$

Block Diagram

The above transfer function is only for a track robot plant. To model the whole system, we need to consider both the robot plant and the control mechanism. The whole system is a closed-loop control system (refer to Figure 3.2), where the plant information is the encoder data with feedback to the control mechanism (the controller). The control mechanism consists of a DAC/amplifier module and PID controller (a very popular control system in industries). Figure 3.11 shows the track system block diagram with the control mechanism.

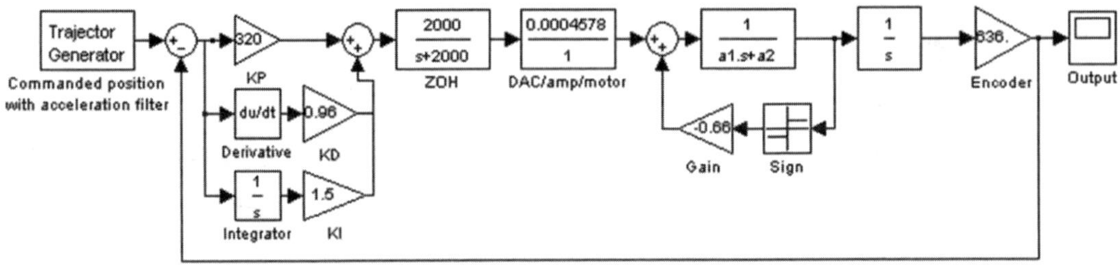

Figure 3.11 Block diagram of a robot track system.

System Identification

The track robot plant parameter a_1 and a_2 is first obtained from the direct mechanical data (mass, pulley radius, etc.) measurement. However, due to the difficulty in getting the precise mechanical data (rotational inertial, static friction, etc.), the track robot plant parameter a_1, a_2 can't be obtained precisely from direct mechanical data measurement. A system identification approach is used to get a better result of a_1, a_2.

The parameters to be estimated are a_1, a_2 in (3.34).

First, the space state equation is written for (3.34).

For the model of (3.34), defined $X = \dot{\theta}, u = M \, and \, Y = X$, we have:

$$\begin{cases} \dot{X} = AX + Bu \\ Y = CX \end{cases}$$

where $A = -\dfrac{a_2}{a_1}; B = -\dfrac{1}{a_1}; C = 1.$

By measuring the motor torque data (to compute M) and track encoder data (to compute $\dot{\theta}$), we may estimate the model parameters A and B (hence a_1, a_2). A number of track torque and motion data (forward, backward, short/long distance) were recorded for system identification and a set of suitable average parameters were computed as model parameters.

MatLab utilities are used for system identification. First, motor torque (M_m) and motor angle (θ) are recorded from actual robot motions, then the actual input data M and actual output data $\dot{\theta}$ are computed from the motor torque and angle. Applying the actual input/output data and initial parameters to the system identification model generates the best-estimated model parameters.

Model Validation

After parameter estimation, the model output data is compared with the actual output data for validation. The validation utilizes both system identification input data and some random selected input data to generate model outputs for comparison. In both validations, $\dot{\theta}$ data is used for comparison. The left plots of Figure 3.12 show the model output and the actual output of $\dot{\theta}$ (Y axis in rad/s) versus time (x axis in second) driven by the input data recorded for system identification.

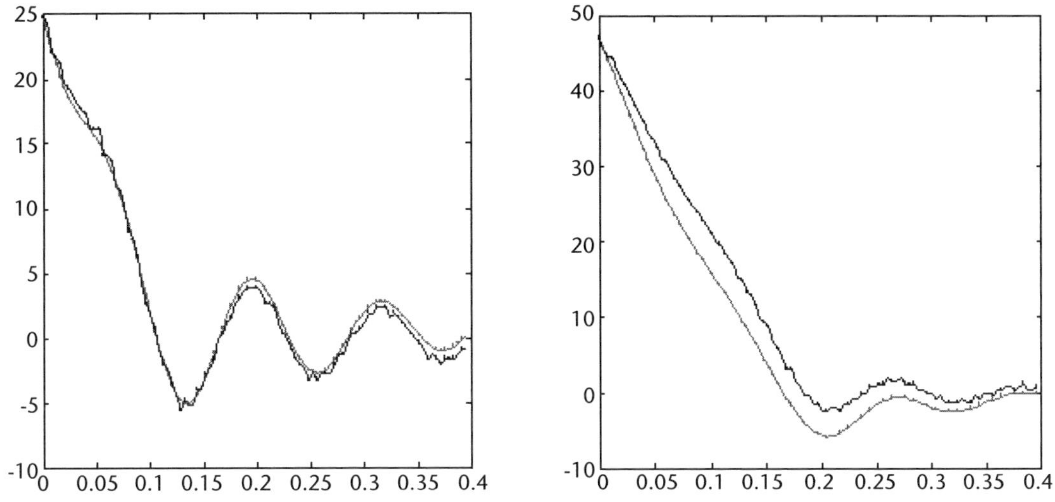

Figure 3.12 Model output and actual output comparison.

System Analysis Based on the Model

The conventional root locus approach in control theory is utilized for performance analysis. Figure 3.13 shows the root locus with the given model. The curve shows how close loop poles change with KP and KD. The square shows close loop poles with the current PID parameters.

By using root locus utility, we can easily investigate the impact of PID parameter (KP and KD) on damping ratio and natural frequency, which are related to track vibration amplitude and tracking error. Some findings using root locus approach are summarized below:

- The track system is extremely underdamped with the current parameters. The damping ratio is about 0.16. The two dominant poles are far from the critical damping ratio region (damping ratio: 0.707). This is the root-cause of the vibration.
- Though the damping ratio can be improved by increasing the KD or decreasing the KP (either or both), the tuning subjects to certain constraints (e.g., decreasing KP subjects to the tracking error limit and increasing KD subjects to electrical noise). Though it's impossible to tune the PID parameter to critical damping ratio under current system constraints, the PID parameter tuning can significantly improve the system performance.

This section shows how the modeling of the robot plant helps in analyzing the system performance. This analysis helps in adjusting the control mechanism (PID tuning), which is critical to the robot's motion automation. The Laplace transform, block diagram, and system identification are also demonstrated to be useful in modeling the robot plant and system performance analysis.

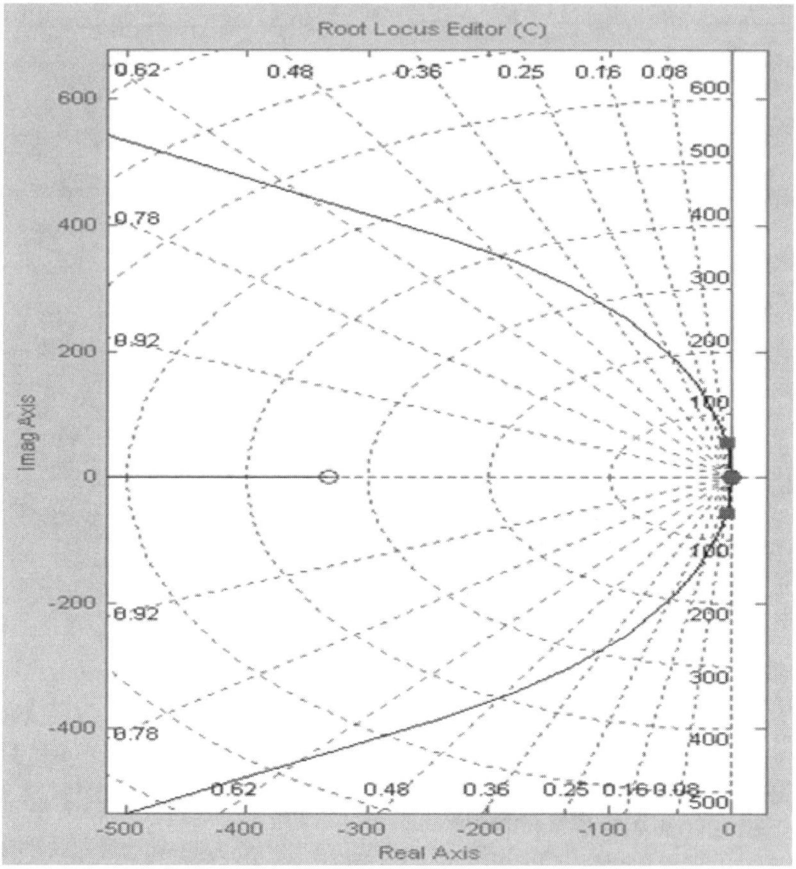

Figure 3.13 Analysis using root locus approach.

3.3.2 Dynamics Modeling of Neutrophils Production from Stem Cells

All blood cells come from a unique source, the stem cells, however, the process regulating the production of neutrophils is still not fully understood. Dynamics modeling of this production process is hence critical in understanding and regulating this process.

Though still in its early stage, stem cell-based therapies are advancing rapidly. Stem cell-based therapies are a major area of investigation in cancer research because the activity of genes that are essential for the process of cell division and differentiation is abnormal in some medical conditions such as cancer. Control of stem cell growth has always been one of the most challenging issues for culturing stem cells. Stem cell-based production is a complicated dynamic process, where multiple factors (precursors, growth factors, temperature, pressure, etc.) contribute to the final outputs. Some factors are controllable and observable. Many others are not. To better understand and control the stem cell-based production, a mathematical model is needed. A good mathematical model may not only help researchers to better understand the production dynamics, but also can be used for culturing neutrophils effectively.

The chemical or biological laws underlying neutrophil production are still not clear, so the modeling of its production process is mainly based on an empirical modeling approach.

Saiyed et al. [8] proposed a two compartmental differential equations model with two discrete delays to describe the white blood cell (WBC) production from hematopoietic stem cell (HSE). The two differential equations are as follows:

$$\frac{dM}{dt} = -\alpha M + AS_{\tau M} F(M_{\tau M})$$

and

$$\frac{dS}{dt} = -SF(M) - SK(S) + 2e^{-\gamma_S \tau_S} S_{\tau S} K(S_{\tau S})$$

where S is the number of hematopoietic stem cells and M is the number of matured WBC. τ_s and τ_M are proliferative cycle time and mature delay time, respectively. $F(M)$ and $K(S)$ are the differentiate rate and proliferative rate. A is the division factor that amplifies the differentiation. α and γ_S are the disappearance rate of M and apoptosis rate of proliferative cells.

Based on the above model, the oscillations in WBC count observed in hematological disease such as cyclical neutropenia (CN) can be explained as the changes in stem cell apoptosis rate, which has been observed in neutrophil precursors in CN patients. However, this model cannot explain the oscillation behavior when the apoptosis rate is decreased below normal in hematological disease such as periodic myelogenous leukemia (PCML). Since the above model was obtained based on an empirical modeling approach, its application range could be restricted and further investigation is required to improve the model.

3.3.3 Modeling of Biological Immune System

The biological immune system consists of a group of cooperating cells with specialized functions to defense against infections. The immune system is able to identify the foreign pathogen agent by detecting its associated molecular patterns called antigens, and respond the pathogen agents in two different ways to fight against it: innate response and acquire response. The innate response takes effect immediately after the pathogen agent is detected. The phagocytic cells and some protein/cytokines are then involved to fight the infection. The acquired response will take effect after repeated exposures to a given infection. The B and T cells are then used to fight the infection. The modeling of immune system is normally for the immune competition. It deals with the dynamics of immune cells, therapy subsistence (e.g., cytokines and vaccine), abnormal cells, and carriers of pathology. The current immune system modeling is mainly based on an empirical modeling approach. Researchers proposed some models and used the clinical data to fit the models. The detailed immune system modeling is very complex and difficult, however, "the immune system presents some overriding principles that allow mathematicians to access the biological realities with mathematical tools", and some "developed concepts like the Michaelis-Menton interactions can be used effectively to model the biological realities of the immune system" [21].

Below is a mathematical model describing the tumor immune dynamics for cancer treatment.

Usman et al. [21] proposed a mathematical model for adoptive immunotherapy. This specific branch of the immune system deals with cancer through acquired response. The model is based on the well-known Kirschner model [22] and is formulated below:

$$\begin{cases} \dfrac{dx}{dt} = cy - \mu_2 x + \dfrac{p_1 xz}{g_1 + z} + s_1 \\ \dfrac{dy}{dt} = r_2 y(1 - by) - \dfrac{axy}{g_2 + y} \\ \dfrac{dz}{dt} = \dfrac{p_2 xy}{g_3 + y} - \mu_3 z + treatment(x,t) \end{cases}$$

where x is the number of effector cells, which are T-lyphocytes cells (can be considered as immune cells), y is the number of tumor cells, and z is the number of cytokines (refers to the interleukin IL-2 in this model).

For the first equation, there is the following explanation:

- C is the antigenicity. The effector cells grow at a rate proportional to the size of the tumor cell and its antigenicity;
- μ_2 is the death coefficient of effector cells;
- $\dfrac{p_1 xz}{g_1 + z}$ indicates the effector cells are activated by the number of cytokine IL-2 (Michaelis-Menton kinetics);
- s_1 is the injection of effector cells.

For the second equation, there is the following explanation:

- $r_2 y(1 - by)$ indicates the tumor cells grow logistically to a fixed carrying capacity;
- When $by > 1$, tumor growing rate becomes negative;
- $\dfrac{axy}{g_2 + y}$ indicates the tumor killing rate by effector cells (based on Michaelis-Menton kinetics).

For the third equation, there is the following explanation:

- $\dfrac{p_2 xy}{g_3 + y}$ indicates that the cytokine IL-2 is created by the effector cells, at a rate that approaches a maximum as the tumor grows indefinitely;
- $\mu_3 z$ indicates the cytokine IL-2 decays at a constant rate;
- *treatment* (x,t) indicates the injection of IL-2.

The above model predicted a variety of phenomena that occurred in a real-world cancer situation. The objective of this model is to "provide clinicians and mathematical model builders a unique way to combine clinical data with the existing mathematical system. This methodology will help predict whether the treatment is working or is leading to toxicity which is vital I tumor suppression" [21]. Apparently, if the model is accurate and we take the *treatment* (x, t) as the control action and min(y) as the control goal, we may use the control theory to develop a control mechanism (treatment) to automate the treatment process for medical automation.

3.4 Discussion and Conclusion

In this chapter, we summarized the basic technologies utilized for the mathematical modeling in medical automation. Most of the technologies are described with an example in medical automation modeling either for medical process or for the tools to realize the medical process. Three mathematical modeling applications were described to exhibit the usage of the modeling technologies. Medical automation is still at its beginning stage and is still undergoing rapid development. Recent advances in technologies have enabled medical automation to do things that were not perceivable a few years ago. The mathematical modeling technology is one of the most important enabling technologies for medical automation. Its advances in technology development will reveal the nature of the medical process and help in implementing medical automation.

References

[1] Taylor, R. H, and D. Stoianovici, "Medical Robotics in Computer-Integrated Surgery," *IEEE Transactions on Robotics and Automation,* Vol. 19, No. 5, 2003, pp.765–781.

[2] Adler, J. R., F. Colombo, M. and P. Heilbrun, et al, "Toward an Expanded View of Radiosurgery," *Neurosurgery,* Vol. 55, No. 6, 2004, pp.1374–1376.

[3] "Robotics and Laboratory Automation in Pharmaceuticals Analysis," Standard Article, Nigel North, SmithKline Beecham Pharmaceuticals, Harlow, U.K.

[4] Mailleret, L., et al., "Nonlinear Adaptive Control for Bioreactors with Unknown Kinetics," *Automatica*, Vol. 40, 2004, pp. 1379–1385.

[5] Tanha, A. R. and D. Popovic, "An Intelligent Monitoring System for Patient Health Care," *Proceedings of the 19th Annual International Conference of the IEEE*, Vol. 3, No. 30, 1997, pp. 1086–1087.

[6] Soni, A. S., "A Multi-Scale Approach to Fed-Batch Bioreactor Control," Master's Thesis, School of Engineering, University of Pittsburgh, 2002.

[7] Rossi, D. T., "Integrating Automation and LC/MS for Drug Discovery Bioanalysis," *Journal of Automated Methods and Management in Chemistry*, Vol. 24, No. 1, 2002, pp. 1–7.

[8] Saiyed, Z. M, S. D. Telang, and C. N. Ramchand, "Application of Magnetic Techniques in the Field of Drug Discovery and Biomedicine", *Biomagn. Res. Technol.*, Vol. 1, No. 2, 2003.

[9] Haller, T., P. Dietl, and K. Pfaller, et al., "Fusion Pore Expansion is a Slow, Discontinuous, and CA2+-Dependent Process Regulating Secretion from Alveolar Type II Cells," *Journal of Cell Biol.*, Vol. 155, 2001, pp. 279–289.

[10] http://en.wikipedia.org/wiki/Biological_thermodynamics.

[11] Liebermeister, W., and E. Klipp, "Bringing Metabolic Networks to Life: Convenience Rate Law and Thermodynamic Constraints," *Theoretical Biology and Medical Modelling,* 2006, pp. 3–41.

[12] Deanda, F., K. M. Smith, and J. Liu, et al., "GSSI, a General Model for Solute-Solvent Interactions. 1. Description of the Model," *Mol. Pharmaceutics*, Vol. 1, No. 1, 2004, pp.23–39.

[13] Bernard, S., J. Belair, and M. C. Mackey "Bifurcations in a White Blood Cell Production Model," *Comptes Rendus Biologies*, Vol. 327, 2004, pp.201–210.

[14] Bachta, W., and A. Krupa, "Towards Ultrasound Image-Based Visual Servoing," *Proceeding of IEEE Int. Conf. on Robotics and Automation*, ICRA 2006, May 2006, pp. 4112–4117.

[15] Ostby, I., and R. Winther, "Stability of a Model of Human Granulopoiesis Using Continuous Maturation," *Journal of Mathematical Biology*, Vol. 49, No. 5, November 2004, pp. 501–536.

[16] McInerney, T., and D. Terzopoulos, "Deformable Models in Medical Image Analysis: A Survey," *Medical Image Analysis*, Vol. 1, No. 2, 1996, pp. 91–108.

[17] Stanek, J., "Stochastic Epidemic Models," *Proc. 15th Annual Conference–WDS 2006*, Prague, 2006, pp. 82–87.

[18] Lisboa, P. J. G., "Neural Networks in Medical Journals: Current Trends and Implications for BioPattern,." *Proc. 1st European Workshop on Assessment of Diagnostic Performance (EWADP)*, Milan, July 7–9, 2004, pp.99–112.

[19] Sordo, M., "Introduction to Neural Networks in Healthcare," OpenClinical: Knowledge Management for Medical Care, 2002.

[20] Weinstein, J., et al., "Neural Computing in Cancer Drug Development: Predicting Mechanism of Action," *Science*, Vol. 258, pp. 447–451.

[21] Usman, A., and T. Jackson, "Application of the Mathematical Model of Tumor-Immune Interactions for IL-2 Adoptive Immunotherapy to Studies on Patients with Metastatic Melanoma or Renal Cell Cancer," *The Rose-Hulman Undergraduate Mathematics Journal*, Vol. 6, No. 2, 2005.

[22] Kirschner, D., and J.C. Panetta, "Modeling Immunotherapy of the Tumor-Immune Interaction," *Journal of Mathematical Biology*, Vol. 37, No. 3, 1998, pp. 235–252.

CHAPTER 4

Contact Dynamic Simulation of Micro-/Nanoscale Manipulation for Medical Applications

Xiumin Diao, Ou Ma, and Mingjun Zhang

This chapter describes dynamics modeling and simulation of an atomic force microscope (AFM)-based manipulation of a micro-/nanoscale object using the compliance-based contact dynamics modeling technique (also referred to as the penalty method) for potential medical applications. The modeling technique has been well developed and applied in macroscale applications. Its application to microscale and nanoscale cases is, however, relatively new. The dynamic model developed in this chapter includes van der Waals forces, electrostatic forces, contact forces (for modeling repulsion), and friction forces with consideration of contact geometry, stiffness, damping, and friction properties of all the physical interactions when manipulating a micro-/nanoscale object such as the inner mass of a stem cell. The model can simulate dynamic behavior of interactions between a micro-/nanoscale object and its surrounding environment. The dynamic simulation of an AFM-based nanomanipulation of the inner mass of a stem cell is presented. To demonstrate the validity of the modeling approach, a nanomanipulation simulation example that matches some published data is included.

4.1 Introduction

Nanotechnology aims to precisely assemble and form new materials or devices that have desirable properties or functionalities that are otherwise very difficult to obtain using traditional manufacturing technologies. Such an assembly task can be done either stochastically by a self-assembly process [1, 2], or deterministically by a nanoscale manipulator [3–5]. The latter approach is similar to the dominant engineering approaches in the macro world and thus may take advantage of many mature engineering technologies and experiences. The approach is also attractive to many engineers and researchers because of its deterministic nature. However, the approach faces several significant challenges due to the limitations of the currently available technologies. The main limiting factors are the difficulties to accurately sense the nanoscale information in real time, to accurately control manipulator's three-dimensional (3-D) motion (i.e., six degrees of freedom (6-DOF) motion) in nanosecond level, and to accurately model and predict the dynamic response of physical contacts occurring in the nanomanipulation. This chapter proposes a modeling and simulation approach for nanomanipulation using a compliance-based contact dynamics modeling technique (also referred to as the penalty method in the literature). The method can be used to help study possible manipulation of the inner mass of a stem cell.

Research of nanomanipulation has been actively conducted since the early 1990s [3–13]. A force-controlled nanoparticle pushing system using an AFM as a manipulator was proposed in [3, 4]. In these works, modeling of the van der Waals forces, capillary forces, electrostatic forces, contact forces, and friction forces, during the 2-D pushing task in ambient conditions, were analyzed. However, the authors did not further explain how to calculate the parameters of the force models and how the models work. Furthermore, no verification of the force models was presented. A teleoperated nanoscale touching system was proposed in [14], where a continuum nanoscale contact mechanics model of van der Waals forces, capillary forces, electrostatic forces, and contact forces was introduced. A simple model of the interaction forces between the manipulator tip, the substrate, and the nanoscale objects was presented in [15] for both qualitative and quantitative analyses. The free body diagram method was used to analyze the three main types of forces, namely, adhesive, repulsive, and frictional forces. The authors considered only the overall effect of the interaction forces. No detailed sources of the interaction forces were investigated as the authors thought it was impossible to obtain all the parameters of the force models. Li and Xu [8] developed a nanomanipulator for 2-D assembly of nanoscale objects via nanomanipualtion. Zhang et al. [9] improved the efficiency and accuracy of AFM-based nanomanipulation by actively controlling the rigidity of the AFM beam. Thermal drift is considered as the main external uncertainty. Various compensation techniques were discussed in [10, 11] to overcome the thermal drift. In fact, the AFM-based nanomanipulation examples discussed in the open literature are mainly focusing on 1-D and 2-D cases [12, 13, 16]. To the best of our knowledge, no AFM-based nanomanipulation example in full 3-D space (i.e., simultaneous operation in all three axes) has been reported in the open literature. Nain and Sitti [17] recently developed a novel 3-D polymer fiber manipulation method by controlling the pulling of a liquid polymer using an AFM probe.

Instead of limiting to simple operations such as pushing, pulling, and tapping, an assembly job of practical value needs more sophisticated manipulations such as full 3-D positioning/posing, twisting, rotating, grabbing, and inserting. Recently, the idea of human-in-the-loop (HIL) operation using an AFM has been proposed [3, 4]. As illustrated in Figure 4.1, the AFM in such an HIL nanomanipulation system is used as both a sensing device and a manipulating device. It was expected that the human operator can directly see interactions on the nanoworksite from the AFM image on a computer screen and operate the nanomanipulator through a haptic human-machine interface.

The HIL nanomanipulation system shown in Figure 4.1 has two worksites: a real worksite and a virtual worksite, described as follows:

Real nano worksite. The real nanoscale worksite consists of the tip of an AFM-based nanomanipulator, the nanoscale objects to be manipulated, and the surrounding environment. This real nanoworksite can be imaged from AFM scan-

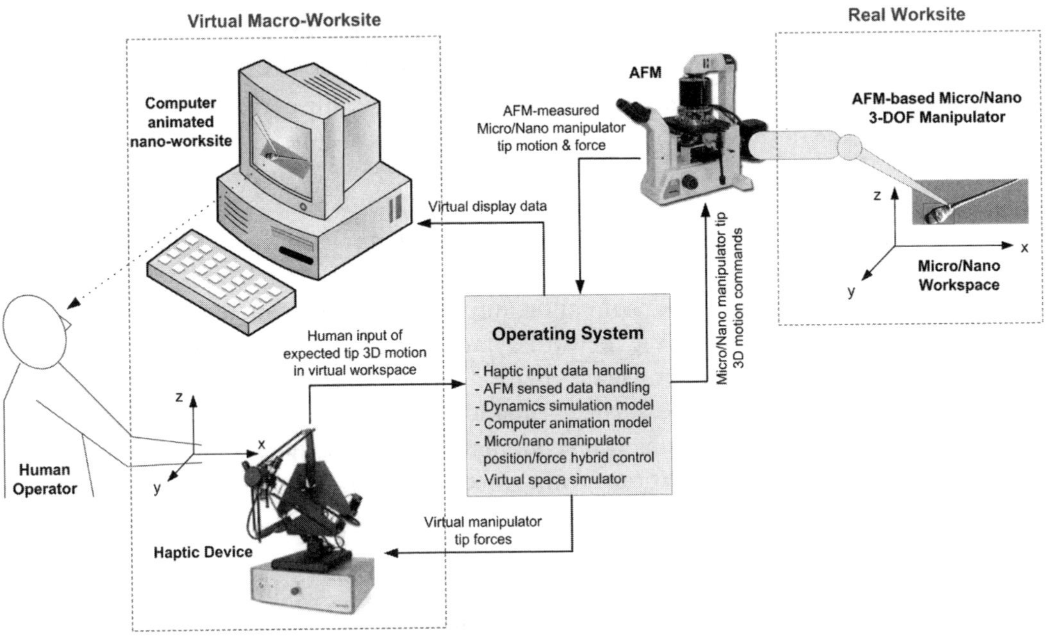

Figure 4.1 Conceptual illustration of the 3-D HIL micro-/nanoscale nanomanipulation.

ning and operated by the nanomanipulator. It should be pointed out that the images generated by AFM are still images rather than sequential video images.

Virtual macroworksite: The simulated macroscale virtual worksite consists of a computer-generated video display of the real nanoworksite and a haptic device for a human operator. The virtual display is graphically generated based on the dynamic simulation of the operation of the real nanoworksite. The dynamic simulation, which has inevitable drifting errors, will be periodically corrected whenever an up-to-date image of the real nanoworksite becomes available from the AFM scanner. Since this worksite is in macroscale, it can be visualized and operated directly by a human operator.

The main advantage of such an HIL nanomanipulation system is to allow human intelligence, learning or adapting abilities, and practical experience to enhance the manipulation control in the nanoworld. Unfortunately, such an AFM-based HIL nanomanipulation system still remains in concept. The main difficulty is that an AFM cannot scan the nanoworksite fast enough to provide the human operator with a real-time view of the worksite. Solutions have been proposed to resolve this problem, such as to predict the dynamic motion of the nanomanipulation using a quasi-static model (for predicting the 2-D motion of the nanoscale object) [7] or an augmented reality (for virtual-reality viewing) [18, 19]. As pointed out in [20, 21], in order to make the HIL nanomanipulation practical, several key problems need to be solved, which include the following [16]:

1. To obtain images of the nanoworksite in real time;
2. To feel and respond to high-frequency dynamics in the nanoworld;
3. To reflect correct scaling of feedback force;

4. To resolve time-scale issue between the macroworld (the haptic device and human operator) and the nanoworld (the real worksite);
5. To apply advanced control strategies for desirable performance. Solutions to some of these problems require a good dynamic model to accurately predict the dynamic response of the system.

Since the virtual display of such an HIL nanomanipulation relies on the dynamic simulation of the real nanoworksite, a key requirement of such a system is to have a high-fidelity dynamic model of nanomanipulation in the real nanoworksite as well as a capability of solving the forward dynamics problem in real time. The research described in this chapter is aimed at developing a dynamic model of nanomanipulation that can be used to enhance the capability of the HIL nanomanipulation technology. Using the compliance-based contact dynamics modeling technique, nanomanipulation of the inner mass of stem cells is modeled and simulated.

4.2 Modeling of Nanoscale System Dynamics

Since nanoscale mechanics have yet to be fully understood, various types of modeling approaches have been proposed and discussed in the field. Quantum electromechanical models enable detailed and precise modeling of specific nanoscale interactions. However, they are computationally intensive, limited only to special materials and cases, and difficult to scale up [2, 21]. The second modeling approach is to use the molecular dynamics and Monte Carlo type of intermolecular models. These models also provide relatively accurate results although they are very slow and cannot model multitime and multilength scale phenomena effectively. Another approach is to approximate the nanoscale dynamics using continuum physical models. The main advantages of this approach are its good computational speed and scaling capability. However, for a large variety of materials, realistic continuum nanosimulator tools do not exist [21]. The proposed modeling approach focuses on the continuum modeling approach.

We have made the assumption that both the AFM tip apex and the micro-/nanoscale object are assumed to be spherical; the substrate is assumed to be a plane. As common practice, the gravitational force is relatively small and thus, neglected [3]. As pointed by [3, 15], there are two basic types of interaction forces in AFM-based nanomanipulation: adhesive force and repulsive force. Each type of these forces comes from many sources. The main components of the adhesive force are van der Waals force, electrostatic force, and capillary force. The repulsive force mainly comes from contact force and friction force. Since capillary forces can be greatly reduced or even eliminated in experiment in a liquid or completely dry (high vacuum) environment [22], they are also ignored. The forces modeled in this preliminary study are listed in Table 4.1. Detailed modeling of each of the individual forces is provided in the next section.

The motions and forces between the AFM tip, the micro-/nanoscale object, and the substrate are illustrated in Figure 4.2 for a noncontact situation and in Figure 4.3 for a contact situation. The AFM tip is controlled to move towards the

Table 4.1 Forces Modeled in This Study

Name of the Force	*Tip—Inner Mass*	*Tip—Substrate*	*Inner Mass—Substrate*
Van der Waals force	Yes	Yes	Yes
Electrostatic force	Yes	No	No
Contact force	Yes	Yes	Yes
Friction force	No	Yes	Yes
Rotational friction moment	No	Yes	Yes

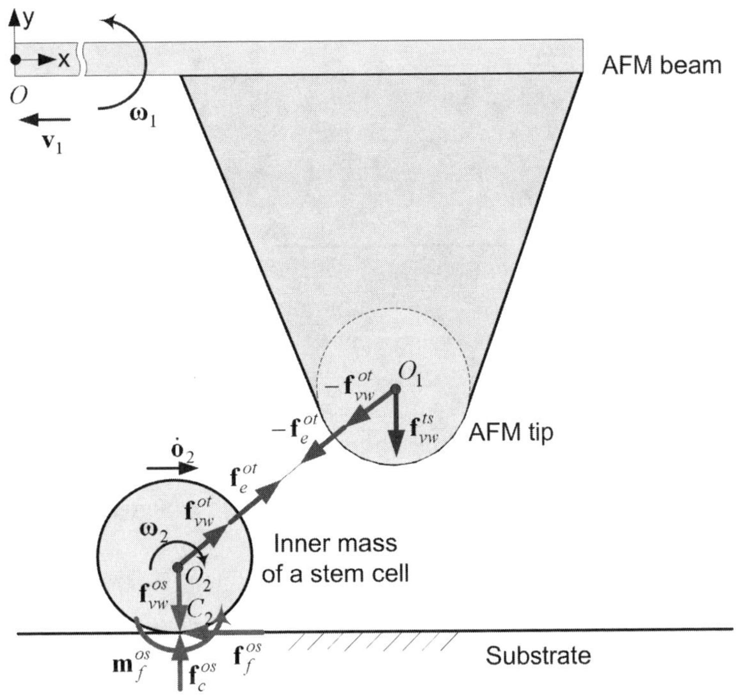

Figure 4.2 Dynamics notation before contact.

micro-/nanoscale object. There are van der Waals force and electrostatic forces between the AFM tip and the micro-/nanoscale object before contact. During contact, the contact force between the AFM tip and the object appears while the electrostatic force vanishes. The interactions between the micro-/nanoscale object and the substrate may include van de Waals force, contact force, sliding friction force, and rolling friction moment. Note that the forces acting on the substrate by the AFM tip and the micro-/nanoscale object are neglected since the substrate is always assumed to be fixed. The AFM tip and the connecting beam are assumed to be one rigid body pivoting at point. This point is assumed to be controlled by the AFM manipulation system and thus, its position is always known. Based upon the above assumptions, dynamics of the AFM tip can be described by the following motion equation:

$$\mathbf{o}_1 \times \left(\mathbf{f}_{vw}^{ts} - \mathbf{f}_{vw}^{ot} - \mathbf{f}_e^{ot} - \mathbf{f}_c^{ot} \right) - \mathbf{K}\boldsymbol{\theta} = \mathbf{I}_1 \dot{\boldsymbol{\omega}}_1 \tag{4.1}$$

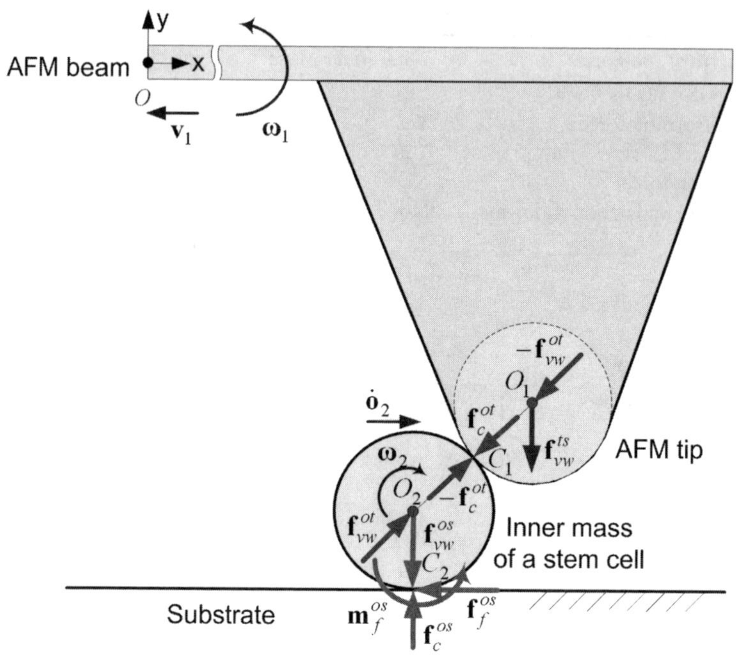

Figure 4.3 Dynamic notation during contact.

The equation of motion of the micro-/nanoscale object is then

$$\begin{aligned} &\mathbf{f}_{vw}^{ot} + \mathbf{f}_{e}^{ot} + \mathbf{f}_{c}^{ot} + \mathbf{f}_{vw}^{os} + \mathbf{f}_{c}^{os} + \mathbf{f}_{f}^{os} = m_2 \ddot{\mathbf{o}}_2 \\ &\mathbf{r}_2 \times \mathbf{f}_f^{os} + \mathbf{m}_f^{os} = \mathbf{I}_2 \dot{\boldsymbol{\omega}}_2 + \boldsymbol{\omega}_2 \times (I_2 \boldsymbol{\omega}_2) \end{aligned} \tag{4.2}$$

Variables and parameters in (4.1) and (4.2) are defined in Table 4.2.

4.3 Modeling of Individual Forces

Interactions among the manipulator tip, the micro-/nanoscale object, and the substrate are subject to several forces, as listed in Table 4.1. Detailed modeling of each of these individual forces is provided in the following sections.

4.3.1 Van der Waals Force

Van der Waals force exists for materials in most environmental conditions. The assumed geometry of the AFM tip apex and the micro-/nanoscale object as modeled here allows applying Hamaker's derivation of van der Waals force between spherical objects [23]. The van der Waals force is defined as [24]:

$$F_{vw} = \frac{32 H r_1^3 r_2^3 (d + r_1 + r_2)}{3d^2 (d + 2r_1)^2 (d + 2r_2)^2 (d + 2(r_1 + r_2))^2} \tag{4.3}$$

Table 4.2 Variables and Parameters in (4.1) and (4.2)

$\mathbf{I}_1$	3x3 inertia tensor of the AFM tip including the beam about its pivoting point
$\boldsymbol{\omega}_1$	3x1 vector of angular velocity of the AFM tip
$\mathbf{v}_1$	3x1 vector of linear velocity of the pivoting point
$\mathbf{o}_1$	3x1 position vector of the geometric center of the spherical shaped AFM tip
$\boldsymbol{\theta}$	3x1 array of three Euler angles representing the orientation of the AFM tip
$\mathbf{K}$	3x3 stiffness matrix representing the rotational stiffness of the AFM at pivoting point
m_2	Mass of the micro-/nanoscale object being manipulated by the AFM tip
$\mathbf{I}_2$	3x3 inertia tensor of the micro-/nanoscale object about its mass center
$\boldsymbol{\omega}_2$	3x1 vector of angular velocity of the micro-/nanoscale object
$\mathbf{o}_2$	3x1 position vector of the mass center of the micro-/nanoscale object
$\mathbf{r}_2$	3x1 vector pointing to the contact point from the mass center of the micro-/nanoscale object
$\mathbf{f}_{vw}^{ts}$	3x1 vector of the van der Waals force exerted on the AFM tip by the substrate
$\mathbf{f}_{vw}^{ot}$	3x1 vector of the van der Waals force exerted on the micro-/nanoscale object by the AFM tip
$\mathbf{f}_{e}^{ot}$	3x1 vector of the electrostatic force exerted on the micro-/nanoscale object by the AFM tip
$\mathbf{f}_{c}^{ot}$	3x1 vector of the contact force exerted on the micro-/nanoscale object by the AFM tip
$\mathbf{f}_{vw}^{os}$	3x1 vector of the van der Waals force exerted on the micro-/nanoscale object by the substrate
$\mathbf{f}_{c}^{os}$	3x1 vector of the contact force exerted on the micro-/nanoscale object by the substrate
$\mathbf{f}_{f}^{os}$	3x1 vector of the sliding friction force exerted on the micro-/nanoscale object by the substrate
$\mathbf{m}_{f}^{os}$	3x1 vector of the rolling friction moment exerted on the micro-/nanoscale object by the substrate

where H is the Hamaker constant for the materials involved, r_1 and r_2 are the radii of the AFM tip apex and the micro-/nanoscale object, respectively, and d is the minimum distance between the AFM tip apex and the micro-/nanoscale object. The van der Waals force between the micro-/nanoscale object and the substrate can be derived from (4.3) by taking the limit as r_1 approaches infinity, which yields

$$F_{vw} = \frac{2Hr_2^3}{3d^2(d+2r_2)^2} \tag{4.4}$$

4.3.2 Electrostatic Force

Electrostatic forces exist among the AFM tip, the micro-/nanoscale object, and the substrate as long as they have electric potential differences. The magnitude and the distribution of the electric charges on the AFM tip and the micro-/nanoscale object are assumed to be identical and the electrostatic effect of the AFM tip is treated as a point charge located at the center of the AFM tip's spherical profile. Since the micro-/nanoscale object continues in contact with the substrate, the electrostatic force between them can be neglected [3]. The electric charges of the micro-/nanoscale object and substrate are mainly concentrated on the micro-/nanoscale object. Therefore, the electrostatic force between the AFM tip and the substrate does not need to be modeled. The electrostatic force between the AFM tip and the micro-/nanoscale object is modeled as [24]:

$$F_e = \frac{4\pi\varepsilon_0 V^2 r_1^2 r_2 (d + r_1 + r_2)}{[(d + r_1)(d + r_1 + 2r_2)]^2} \tag{4.5}$$

where ε_0 is the electrical permittivity of a vacuum, and V is the voltage bias of the AFM's spherical tip.

4.3.3 Contact Force

Contact forces are very important in nanomanipulation. The compliance-based contact forces are modeled in this study. Plastic deformations are not considered in the modeling. The Hertz contact model [25] is used here, which can be expressed in terms of the surface deformation and its time derivative $\dot{\delta}$ as follows:

$$F_c = \begin{cases} k_c |\delta|_{\frac{3}{2}} - c_c\dot{\delta} & if\ \delta \le 0 (\text{in contact}) \\ 0 & \text{otherwise (no contact)} \end{cases} \tag{4.6}$$

where k_c and c_c are the contact stiffness and damping coefficients, respectively. These two parameters reflect the mechanical properties of the contacting objects. Note that k_c may vary nonlinearly with respect to the surface deformation δ. In this study, we assume that k_c is a constant, which is a widely accepted approximation if the deformation is small [25].

4.3.4 Sliding Friction Force

Friction force always exists between two contact surfaces that have relative motion or tend to move relatively. During nanomanipulation, the friction between the micro-/nanoscale object and the substrate may have a significant effect in the resulting motion of the micro-/nanoscale object. Therefore, accurate modeling of friction force is necessary for high-fidelity dynamic simulation of nanomanipulation. Extended from [26], the sliding friction force can be represented as

$$\mathbf{F}_f = -k_t \mathbf{s} - c_t \dot{\mathbf{s}} \tag{4.7}$$

where $\mathbf{s}$ and $\dot{\mathbf{s}}$ are the bristle's deflection and its time derivative, and k_t and c_t are the bristle stiffness and damping coefficients, respectively. The bristle's deflection $\mathbf{s}$ is defined as

$$\mathbf{s}(t) = \begin{cases} \mathbf{s}(t_0) + \int_{t0}^{t} \mathbf{v}_t\, dt & if\ |\mathbf{s}| < s_{\max} \\ s_{\max} \dfrac{\mathbf{v}_t}{|\mathbf{v}_t|} & \text{otherwise} \end{cases} \tag{4.8}$$

where t_0 is the starting time of this particular contact, t is the current time, s_{max} is the maximum bristle deflection, and $\mathbf{v}_t$ is the tangential velocity between the contact objects at the contact point. Moreover, the maximum deflection s_{max} is defined as

$$s_{max} = \frac{\mu_t N}{k_t} \tag{4.9}$$

where μ_t is the translational friction coefficient and N is the magnitude of the normal contact force at the corresponding contact point.

4.3.5 Rolling Friction Moment

When an object rolls on a surface with or without sliding, it may subject to a rolling friction. Similar as the sliding friction force modeled in (4.7), the rolling friction moment can be modeled as

$$\mathbf{M}_f = -k_r\theta - c_r\dot{\theta} \tag{4.10}$$

where θ and $\dot{\theta}$ are the bristle's deflection and its time derivative, and k_r and c_r are the bristle stiffness and damping coefficients, respectively. The bristle's deflection θ is defined as

$$\theta(t) = \begin{cases} \theta(t_0) + \int_{t_0}^{t} \dot{\theta} dt & if\,|\theta| < \theta_{max} \\ \theta_{max} \dfrac{\theta}{|\dot{\theta}|} & \text{otherwise} \end{cases} \tag{4.11}$$

where θ_{max} is the maximum bristle deflection and $\dot{\theta}$ is the angular velocity between the contact bodies. Moreover, the maximum deflection θ_{max} is defined as

$$\theta_{max} = \frac{\mu_r r_2 N}{k_r} \tag{4.12}$$

where μ_r is the rotational friction coefficient [28].

4.4 Numerical Simulation

Experimental validation of the contact-dynamics based model of nanomanipulation is a major challenge due to lack of adequate experimental data. With the currently available sensing technologies, it is very difficult to sense local contact forces and surface deflections at nanoscale. As a minimal effort of verification, we provide a comparison between our simulation output and an experimental result published in [27] at resultant force level. The comparison case is about an AFM tip contacting a flat sample surface. The experimental results of the case were presented in Figure 6 of [27]. Our simulation results for the same case are plotted in Figure 4.4. First, let us explain the approaching phase that is plotted with a solid line. The AFM tip approaches the surface with small force until it almost touches the surface (i.e., at

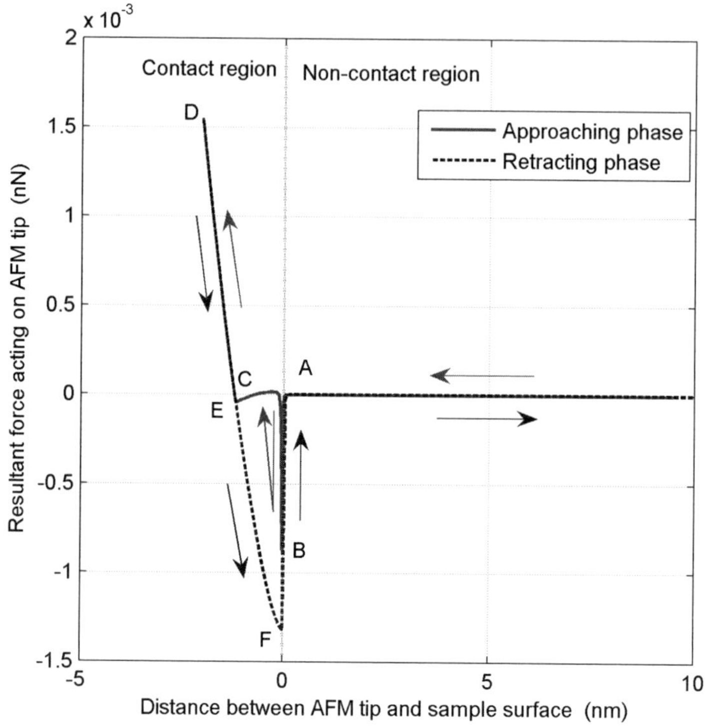

Figure 4.4 Force-distance relation during the approaching and retracting phases.

point A where the distance between the AFM tip and the surface is close to zero). After that point, the tip force significantly increases due to the contribution of van der Waals force. The tip force starts to drop back shortly after contact at point B. This change is due to the arising contact force that starts to balance the van der Waals force. This balancing status is kept for a short period after touching (between points B and C). Then the contact force breaks the balance and starts to increase while the AFM tip continues pushing toward the surface. The AFM tip stops pushing and starts to retract at point D. The retraction phase is plotted with a dashed line. In the first part of the retraction phase (from point D to point E), the tip force drops due to the drop of the dominant contact force. After point E, the tip force becomes negative because the rising van der Waals force dominates the tip force. This situation is kept until point F where the van der Waals force starts to quickly vanish due to distance increase. It can be seen that the simulation results during both the approaching and retracting phases match the published experimental results very well. Although this is only an isolated example and the comparison is still in the resultant force level, it tells us that the proposed modeling approach is capable of reproducing some experimental results.

4.5 Simulation of the Manipulation of Stem Cell Inner Mass

Though still in its early stage, stem-cell-based therapies are advancing rapidly. It is a major area of investigation in cancer research, because the activity of genes that are essential for the process of cell division and differentiation is abnormal in some medical conditions, such as cancer. An important engineering issue for stem-cell-based therapies is stem cell manipulation. The goal is to manipulate stem cells easily and reproducibly for successful differentiation, transplantation, and engraftment.

Furthermore, experimentally fusing adult cells with embryonic cells can produce adult cells with embryonic properties. This experimental manipulation of adult cells and forcing them to take on characteristics of embryonic cells is a powerful tool to study factors necessary for reprogramming the cell nucleus of an adult cell to adopt a less committed state. With this approach, it is possible to create personalized embryonic stem cells in the future by adding reprogramming factors to adult cells. These cells can then be amplified for therapeutic applications through the growth process of cell culture.

One key step for an embryonic stem-cell-based therapy is to harvest inner mass from a stem cell. Advanced medical devices are just starting to be proposed in this field. Automated devices for harvesting human embryonic stem cell inner mass is highly expected. Such a device requires integrating a good dynamics-model-based feedback control system for controlling the stem cell manipulation and inner mass harvesting. In this section, we will discuss how the inner mass manipulation can be simulated using the contact dynamics modeling method discussed in Section 4.3. The model can help propose optimal control strategies to manipulate the inner mass with minimal damage.

The dynamics model was implemented on MATLAB for simulation. Although the presented contact dynamics model is for a general 3-D case, the AFM in this example is assumed to move in a 2-D space only. The apex of the AFM tip is assumed to have a circular shape, as shown in Figures 4.5 and 4.6 and the inner

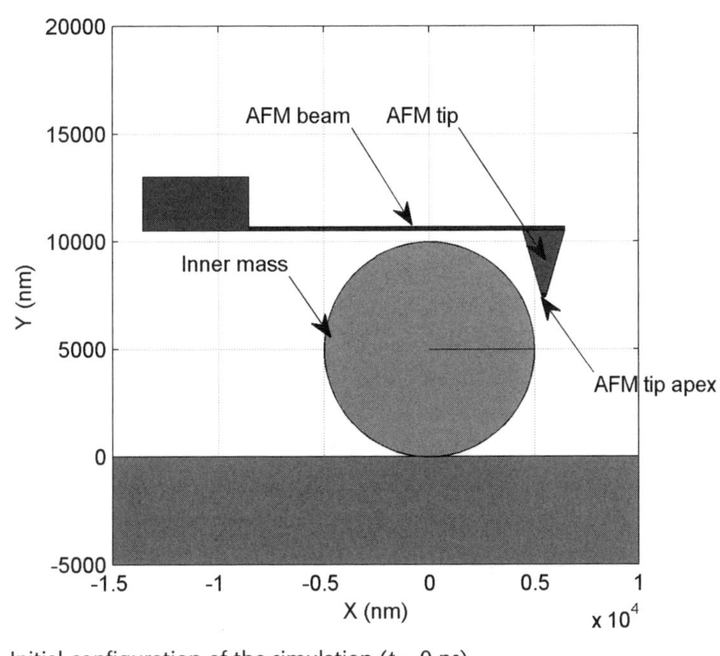

Figure 4.5 Initial configuration of the simulation (t = 0 ns).

mass of a stem cell is also assumed to be circular in the 2-D space. The values of all the basic model parameters used in the simulation example are listed in Table 4.3. The AFM tip is controlled to move towards the inner mass that is initially still. After the contact between the AFM tip and the inner mass occurs, the inner mass will be moved along with the AFM tip due to the pulling of the AFM tip.

Figures 4.5 and 4.6 show the initial and the final configurations of the AFM tip and the inner mass, respectively. By comparing the two figures, one can see that the inner mass indeed moved to the left by about 5,000 nm during the manipulation. The amount of rotation of the inner mass can be clearly seen from its initial and final configurations. This rotation is due to the existence of the rolling friction between the inner mass and its contacting substrate. The simulated motion and force responses between t = 1,000 ns and t = 1,020 ns are plotted in Figures 4.7 to 4.11. The interaction forces change very little before and after the plotted time period, and thus, they are not plotted in order to make the plotted results more visible.

Figure 4.7 shows the position and velocity of the AFM tip. At the beginning of the simulation, the AFM tip moved towards the inner mass along the x direction (from left to right horizontally) with a constant velocity of 1 nm/ns. The velocity of the AFM tip in the y direction (upward vertically) and the angular velocity of the AFM beam about the z-axis were almost zero because the interaction forces (van der Waals and electrostatic forces) acting on the AFM tip were very small due to the far distance between the AFM tip and the inner mass at that time. When the AFM tip and inner mass were about to contact, the interaction forces between them increased drastically. Consequently, the position and velocity of AFM tip also changed drastically, namely, the AFM tip was quickly bent down toward the inner mass until they made contact. This phenomenon can be clearly seen in the animated

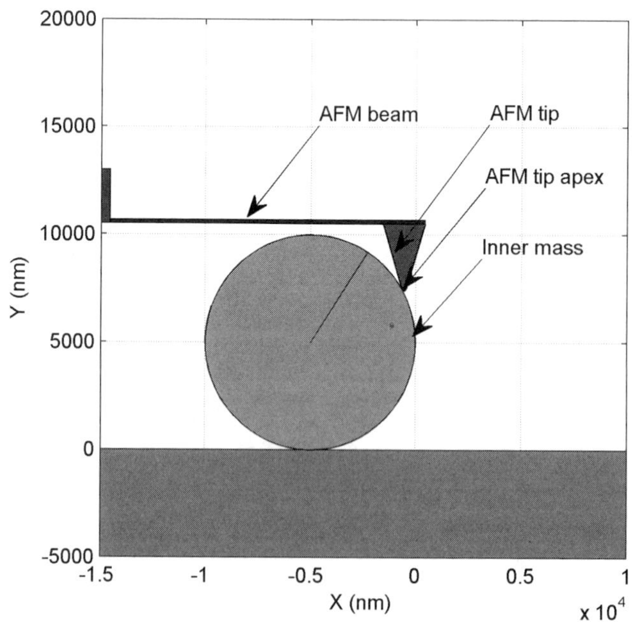

Figure 4.6 Final configuration of the simulation (t = 6,000 ns).

Table 4.3 Model Parameter Values Used in the Simulation

Parameter	*Symbol*	*Value (Unit)*
Hamaker constant	H	2.63×10^{-10} (nJ)
Contact stiffness coefficient	k_c	5.0×10^{-1} (nN/nm)
Contact damping coefficient	c_c	2.0×10^{-1} (nN-ns/nm)
Translational friction coefficient	μ_t	5.0×10^{-2}
Translational bristle stiffness coefficient	k_t	8.0×10^{-1} (nN/nm)
Translational bristle damping coefficient	c_t	5.0×10^{-2} (nN-ns/nm)
Rotational friction coefficient	μ_r	1.0×10^{-3}
Rotational bristle stiffness coefficient	k_r	5.0×10^{-2} (nN/rad)
Rotational bristle damping coefficient	c_r	1.0×10^{-3} (nN-ns/nm)
Electrical permittivity of a vacuum	ε_0	8.85×10^{-12} (nF/nm)
Voltage bias of the AFM tip	V	25 (V)
Radius of the AFM tip	r_1	100 (nm)
Radius of the inner mass	r_2	5000 (nm)
Mass of the inner mass	m_2	1.0×10^{-3} (nKg)
Moment of inertia of AFM cantilever and tip	I_1	8.0×10^{3} (nKg-nm^2)

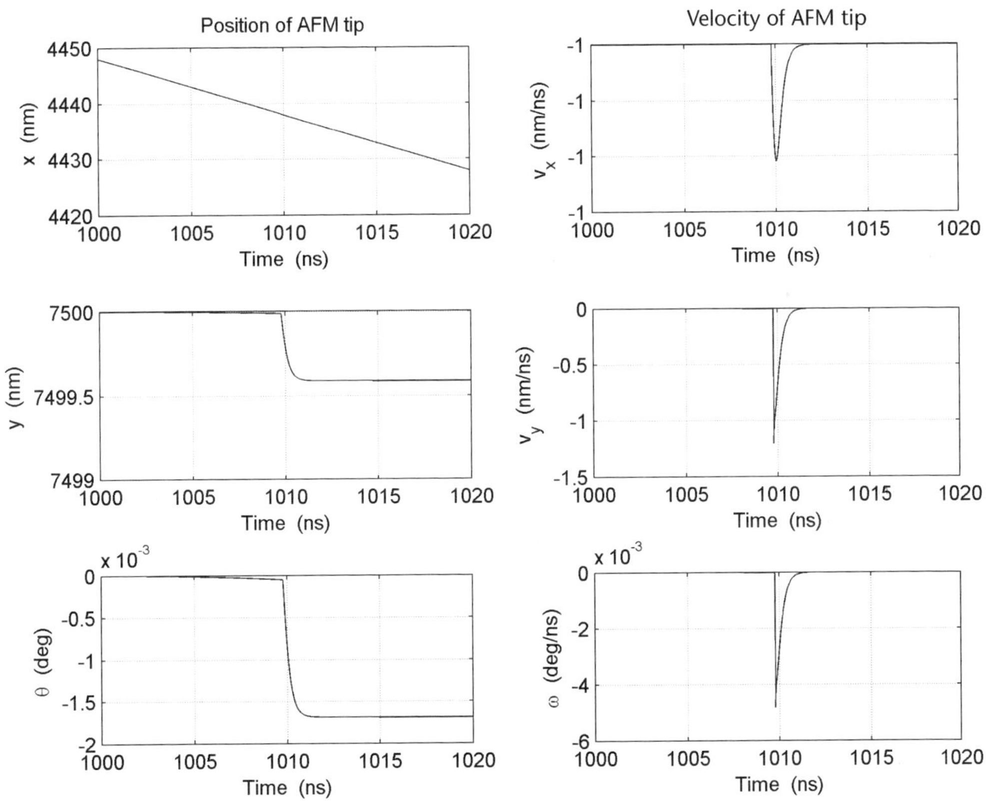

Figure 4.7 Position and velocity of the AFM tip.

simulation result. After the first contact between the AFM tip and the inner mass, the velocity of the AFM tip settled down quickly.

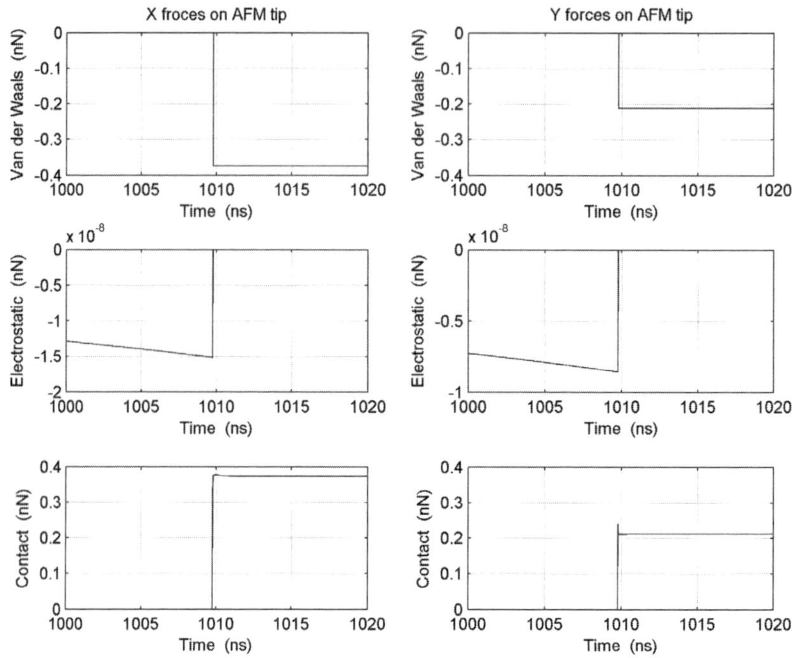

Figure 4.8 Forces on the AFM tip.

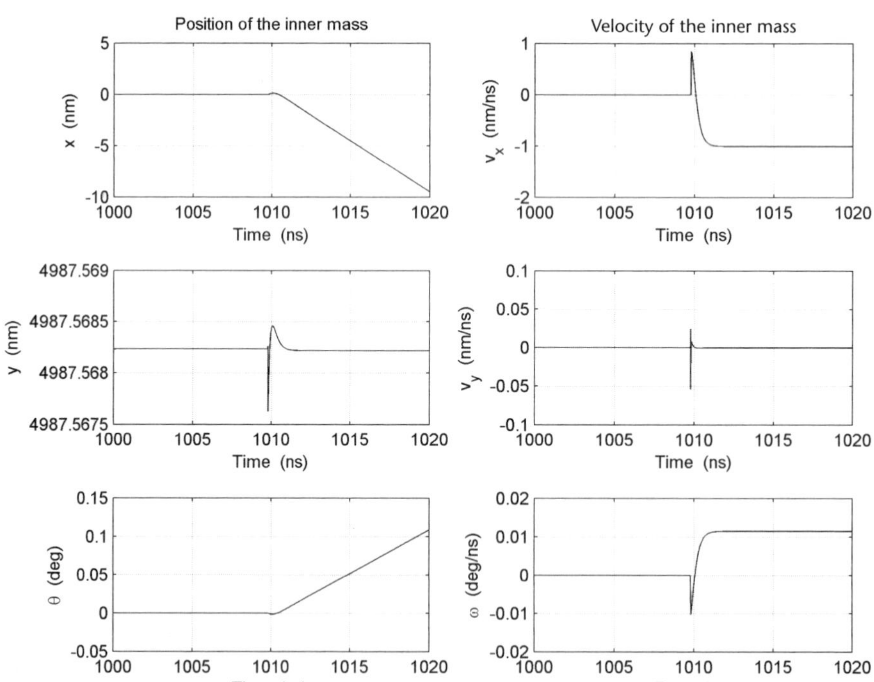

Figure 4.9 Position and velocity of the inner mass.

Figure 4.8 shows the forces acting on the AFM tip. The subfigures in the first row show the x component (left subfigure) and y component (right subfigure) of the

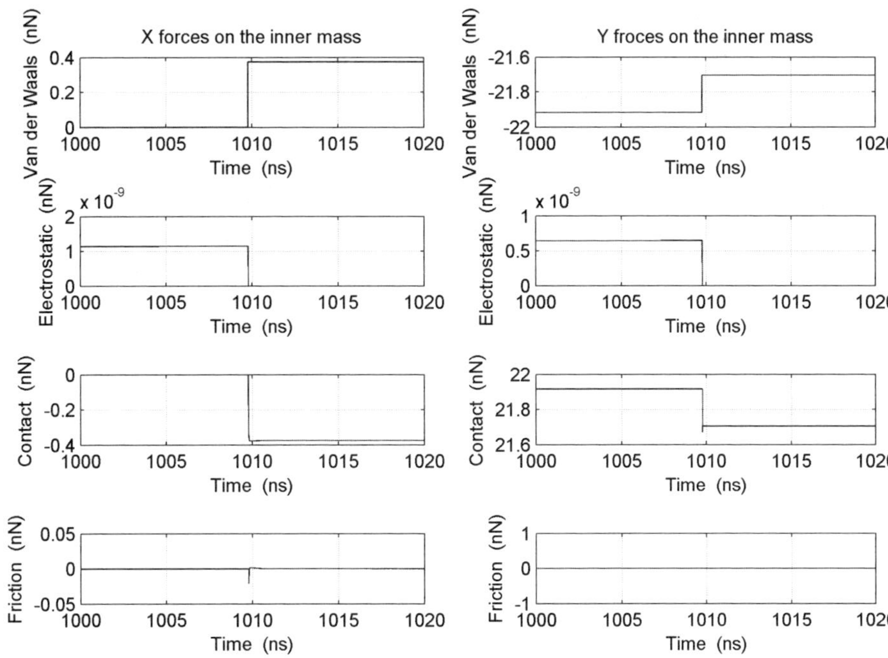

Figure 4.10 Forces on the inner mass.

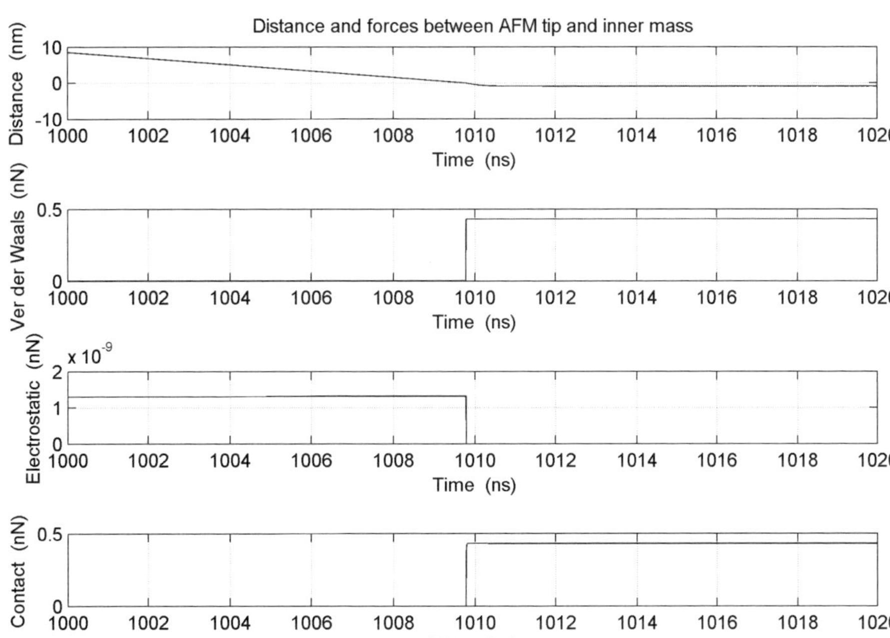

Figure 4.11 Distance and forces between AFM tip and inner mass.

van der Waals force acting on the AFM tip. Similarly, the subfigures in the last two rows show the electrostatic force and contact force, respectively. At the beginning of the simulation, all the forces acting on the tip were very small because the AFM tip was far away from the inner mass. A short time before the first contact happened between the AFM tip and inner mass, the van der Waals and electrostatic forces increased drastically. The contact force increased quickly after the contact occurred. A short time after their first contact, the contact force and the other forces were balanced, and thus they became constant in the plots. The friction force between the AFM tip and the inner mass was not included in this simulation model. It is emphasized that, although the contact and van der Waals forces look likely to have a sudden jump near the first contact time, they were continuously rising in a short time period. This fact can be clearly seen when the plots are zoomed in.

Figure 4.9 shows the position and velocity of the inner mass. At the beginning of the simulation, the inner mass slowly rolled towards the positive x direction due to the van der Waals and electrostatic forces exerted on it by the AFM tip. When it got very close to the AFM tip, the speed of the inner mass increased drastically (in the positive x direction), and thus it was quickly pulled toward the AFM tip. Once it made a contact with the AFM tip, its velocity in x direction quickly reduced to zero and then jumped to the negative value that is the speed of the AFM tip. After the impact, the inner mass was pulled by the AFM manipulator to move along the negative x direction.

Figure 4.10 shows the forces acting on the inner mass. At the beginning of the simulation, the van der Waals and the electrostatic forces were relatively small and the friction force was in the negative x direction. When the distance between the AFM tip and the inner mass was small, the van der Waals and electrostatic forces increased quickly until the first contact between the two objects. Once the contact occurred, the contact force increased drastically and the friction force changed direction due to the contact. A short while later, all the forces acting on the inner mass were balanced. As a result, the inner mass was pulled (while keeping in contact with the AFM tip) and moved with the AFM tip along the negative x direction. Since the inner mass was in pure rolling motion thereafter, the sliding friction force between it and the substrate became zero.

Figure 4.11 shows the relationship of the distance and the forces between the AFM tip and the inner mass. It is clear that the van der Waals and the electrostatic forces were relatively small when the AFM tip and the inner mass were far away from each other. The forces became larger when they were getting closer. Note that the force plots clearly show the nonlinear characteristics of the forces. The impulsive nature of the contact force can be seen at the time when the first contact happened. Again, shortly after the first contact occurred, all the forces rapidly settled down to some constant values.

4.6 Concluding Remarks

In this chapter, a compliance-based contact dynamics modeling approach for AFM-based manipulation of a micro-/nanoscale object for potential medical applications such as the retrieval of the inner mass of a stem cell was presented. In the dynamics model, the van der Waals force, electrostatic force, and friction force have been taken into consideration. By applying the compliance-based contact dynamics approach, the repulsion between the AFM tip and the inner mass is modeled as contact force by considering stiffness, damping, and frictional properties of the interacting objects. A simulation example was presented, where the dynamic response in a 2-D nanomanipulation of a circular inner mass using an AFM-based manipulator was simulated. Simulation results demonstrated that such a model can reasonably predict the nanoscale dynamic behavior as observed from an experiment of AFM based nanomanipulation. Comparison between our simulation results in a special case and some published experimental data was included to show the validity of the discussed modeling approach.

Nanoscale manipulation of biological objects for medical applications raises many challenges, especially in considering dynamics aspects of the manipulation process. Mathematical modeling, control, and simulation are promising approaches to studying the problem. The proposed contact dynamics based modeling approach provides a useful discussion platform. The model may not only be used to understand the dynamics of a stem cell's inner mass manipulation but also cell manipulation in cell sorting, sample collection in tissue engineering, and nanoscale manipulation in regenerative medicine. This study just opens an initial discussion in the topic. We expect more active discussions and development to come in the near future.

References

[1] McNally, H., et al., "Self-Assembly of Micro- and Nano-scale Particles Using Bioinspired Events," *Applied Surface Science*, Vol. 214, No. 1, 2003, pp. 109–119.

[2] Zhang, M., O. Ma, and X. Diao, "Dynamic Modeling and Analysis of Oligo DNA Microarray Spotting," *IEEE Transactions on Automation Science and Engineering*, Vol. 3, No. 2, 2006, pp. 159–168.

[3] Sitti, M., and H. Hashimoto, "Controlled Pushing of Nanoparticles: Modeling and Experiments," *IEEE/ASME Transactions on Mechatronics*, Vol. 5, No. 2, 2000, pp. 199–211.

[4] Li, G., et al., "Assembly of Nanostructure Using AFM Based Nanomanipulation System," *Proc. IEEE Int. Conf. on Robotics and Automation*, New Orleans, LA, 2004, pp. 428–433.

[5] Chan, H. Y., et al., "A Deterministic Process for Fabrication and Assembly of Single Carbon Nanotube Based Devices," *Proc. 5th IEEE Conf. on Nanotechnology*, Nagoya, Japan, 2005, pp. 907–910.

[6] Eigler, D. M., and E. K. Schweitzer, "Positioning Single Atoms with a Scanning Tunneling Microscope," *Nature*, Vol. 343, 1990, pp. 524–526.

[7] Requicha, et al., "Nanorobotic Assembly of Two-Dimensional Structures," *Proc. IEEE Int. Conf. Robotics and Automation*, Leuven, Belgium, 1998, pp. 3368–3374.

[8] Li, Y., and Q. Xu, "A Novel Design and Analysis of a 2-DOF Compliant Parallel Micromanipulator for Nanomanipulation," *IEEE Transactions on Automation Science and Engineering*, Vol. 3, No. 3, 2006, pp. 248–254.

[9] Zhang, J., et al., "Adaptable End Effector for Atomic Force Microscopy Based Nanomanipulation," *IEEE Transactions on Nanotechnology*, Vol. 5, No. 6, 2006, pp. 628–642.

[10] Yang, Q., and S. Jagannathan, "Atomic Force Microscope-based Nanomanipulation with Drift Compensation," *Int. Journal of Nanotechnology*, Vol. 3, No. 4, 2006, pp. 527–544.

[11] Mokaberi, B., and A.A.G. Requicha, "Drift Compensation for Automatic Nanomanipulation with Scanning Probe Microscopes," *IEEE Transactions on Automation Science and Engineering*, Vol. 3, No. 3, 2006, pp. 199–207.

[12] Tafazzoli, A., C. Pawashe, and M. Sitti, "Atomic Force Microscope Based Two-Dimensional Assembly of Micro/Nanoparticles," *Proc. IEEE Int. Symposium on Assembly and Task Planning*, Montreal, QC, Canada, 2005, pp. 230–235.

[13] Pawashe, C., and M. Sitti, "Two-Dimensional Vision-Based Autonomous Microparticle Manipulation Using a Nanoprobe," *J. Micromechatronics*, Vol. 3, No. 3, 2006, pp. 285–306.

[14] Sitti, M., and H. Hashimoto, "Teleoperated Touch Feedback from the Surfaces at the Nanoscale: Modeling and Experiments," *IEEE/ASME Transactions on Mechatronics*, Vol. 8, No. 2, 2003, pp. 287–298.

[15] Li, G., et al., "Modeling of 3-D Interactive Forces in Nanomanipulation," *Proc. 2003 IEEE Int. Conf. on Intelligent Robots and Systems*, Las Vegas, NV, 2003, pp. 2127–2132.

[16] Sitti, M., "Micro- and Nano-Scale Robotics," *Proc. 2004 American Control Conference*, Boston, MA, 2004, pp. 1–8.

[17] Nain, A., and M. Sitti, "3-D Nanoscale Manufacturing by Nanoprobes Based Controlled Pulling of Liquid Polymers," *Proc. IEEE Nanotechnology Conference*, San Francisco, CA, 2003, pp. 60–63.

[18] Li, G., et al., "Development of Augmented Reality System for AFM-based Nanomanipulation," *IEEE/ASME Transactions on Mechatronics*, Vol. 9, No. 2, 2004, pp. 358–365.

[19] Vogl, W., V.K-L. Ma, and M. Sitti, "Augmented Reality User Interface for an Atomic Force Microscope-Based Nanorobotic System," *IEEE Transactions on Nanotechnology*, Vol. 5, No. 4, 2006, pp. 397–406.

[20] Sitti, M., "Survey of Nanomanipulation Systems," *Proc. 1st IEEE Conf. on Nanotechnology*, Aaui, HI, Oct. 28–30, 2001, pp. 75–80.

[21] Sitti, M., "NSF Workshop on Future Directions in Nano-Scale Systems, Dynamics and Control," *NSF SDC Workshop*, 2003.

[22] Weisenhorn, A.L., et al., "Measuring Adhesion, Attraction, and Repulsion Between Surfaces in Liquids with an Atomic-force Microscope," *Phys. Rev. B*, Vol. 45, No. 19, 1992, pp. 11226–11232.

[23] Hamaker H. C., "The London–Van der Waals Attraction Between Spherical Particles," *Physica IV*, Vol. 13, No. 10, 1937, pp. 1058–1072.

[24] Rohrer, B. R., et al., "Nanomanipulation by Dielectrophoresis: Gripping and Releasing Objects Using a Charged Probe, " *IEEE Transactions on Automation Science and Engineering*, 2004.

[25] Johnson, K. L., *Contact Mechanics*, London, UK: Cambridge University Press, 1985.

[26] Haessig, D. A., and B. Friedland, "On the Modeling and Simulation of Friction," *ASME Transactions, Journal of Dynamic Systems, Measurements, and Controls,* Vol. 113, pp. 354–362.

[27] Sitti, M., and H. Hashimoto, "Tele-Nanorobotics Using Atomic Force Microscope as a Robot and Sensor," *J. Advanced Robotics*, Vol. 13, No. 4, 1999, pp. 417–436.

[28] Ma, O., X. Diao, and M. Zhang, "Simulation of Nanomanipulation Using Compliance-Based Contact Dynamics Modeling Technique," *Proc. ASME Int. Design Engineering Technical Conferences and Computers and Information Conference (IDETC)*, Las Vegas, NV, September 4–7, 2007, Paper # DETC2007/VIB-35691.

CHAPTER 5

Modeling and Mathematical Analysis of Swarms of Microscopic Robots for Medical Diagnostics

Tad Hogg

Molecular electronics, motors, and chemical sensors could enable constructing machines able to sense, compute, and act in micron-scale environments. Such microscopic machines could simultaneously monitor entire populations of cells individually in vivo. These microscopic robots require robust control programs suited to the biophysics of their task environments and the robots' limited physical and computational abilities. Distributed control is well-suited to these capabilities by emphasizing locally available information and achieving overall objectives through self-organization of the collection of robots.

This chapter reviews methods of evaluating plausible behaviors for microscopic robots in light of physical constraints due to operation in fluids at low Reynolds number, diffusion-limited sensing, and thermal noise from Brownian motion. One method uses differential equations to describe both robot internal state and interactions with spatially variable fields such as chemical concentration.

This method is illustrated in the context of a prototypical biomedical task, which requires control decisions on millisecond time scales. This task is monitoring for patterns of chemicals in a flowing fluid released at chemically distinctive sites. Information collected from a large number of such devices allows estimating properties of cell-sized chemical sources in a macroscopic volume. Specifically, a simple model of such devices moving with the fluid flow in small blood vessels estimates their ability to detect chemicals released by tissues in response to localized injury or infection. The resulting behaviors illustrate trade-offs among speed, accuracy, and resource use.

We find the devices can readily discriminate a single cell-sized chemical source from the background chemical concentration, providing high-resolution sensing in both time and space. By contrast, such a source would be difficult to distinguish from the background when diluted throughout the blood volume as obtained with a blood sample.

5.1 Microscopic Robots

Microscopic robots, of sizes comparable to bacteria, could move through the tiniest blood vessels. Thus the robots could pass within a few cell diameters of most cells in large organisms via their circulatory systems to perform a variety of biological research and medical tasks. For instance, robots and nanoscale-structured materials inside the body could significantly improve disease diagnosis and treatment [1–4].

Initial tasks for microscopic robots include in vitro research via simultaneous monitoring of chemical signals exchanged among many bacteria in a biofilm. The robots could operate in multicellular organisms for tasks ranging from passively circulating sensors to microsurgery at cellular scales.

Realizing these benefits requires fabricating the robots cheaply, in large numbers, and with sufficient capabilities. Such fabrication is beyond current technology. Nevertheless, ongoing progress in engineering nanoscale devices could eventually enable production of such robots. One approach to creating microscopic programmable machines is engineering biological systems such as bacteria executing simple programs [5], and DNA computers responding to logical combinations of chemicals [6].

However, biological organisms have limited material properties and computational speed. Instead we focus on machines based on plausible extensions of current molecular-scale electronics, sensors, and motors [7–14]. These devices could provide components for stronger and faster microscopic robots than is possible with biological organisms. Thus the focus here is on nonbiological robots containing nanoscale sensors and electronics, along with a power source, within a protective shell. As technology improves, such robots could be supplemented with other capabilities such as communication and locomotion.

Because we cannot yet fabricate microscopic robots with molecular electronics components, estimates of their performance rely on plausible extrapolations from current technology. The focus in this chapter is on biomedical applications requiring only modest hardware capabilities, which will be easier to fabricate than more capable robots. Designing controls for microscopic robots is a key challenge: not only enabling useful performance but also compensating for their limited computation, locomotion, or communication abilities.

Distributed control is well-suited to these capabilities by emphasizing locally available information and achieving overall objectives through self-organization of the collection of robots. Theoretical studies allow developing such controls and estimating their performance prior to fabrication, thereby indicating design trade-offs among hardware capabilities, control methods, and task performance. Such studies of microscopic robots complement analyses of individual nanoscale devices [14, 15], and indicate even modest capabilities enable a range of novel applications.

The operation of microscopic robots differs significantly from larger robots [16], especially for biomedical applications. First, the physical environment is dominated by viscous fluid flow. Second, thermal noise is a significant source of sensor error, and Brownian motion limits the ability to follow precisely specified paths. Third, relevant objects are often recognizable via chemical signatures rather than, say, visual markings or specific shapes. Fourth, the tasks involve large numbers of robots, each with limited abilities. Moreover, a task will generally only require a modest fraction of the robots to respond appropriately, not for all, or even most, robots to do so. Thus controls using random variations are likely to be effective simply due to the large number of robots. This observation contrasts with teams of larger robots with relatively few members: incorrect behavior by even a single robot can significantly decrease team performance. These features suggest reactive distributed control is particularly well-suited for microscopic robots.

Organisms contain many microenvironments, with distinct physical, chemical, and biological properties. Often, precise quantitative values of properties relevant for robot control will not be known a priori. This observation suggests a multistage protocol for using the robots. First, an information-gathering stage with passive robots placed into the organism (e.g., through the circulatory system) to measure relevant properties [17]. The information from these robots, in conjunction with conventional diagnostics at larger scales, could then determine appropriate controls for further actions in subsequent stages of operation.

5.2 Evaluating Collective Robot Performance

Because microscopic robots can not yet be fabricated and quantitative biophysical properties of many microenvironments are not precisely known, performance studies must rely on plausible models of both the machines and their task environments [1, 23, 24]. Microorganisms, which face physical microenvironments similar to those of future microscopic robots, give some guidelines for feasible behaviors.

Cellular automata are is technique to evaluate collective robot behavior. For example, a two-dimensional scenario shows how robots could assemble structures [25] using local rules. Such models can help understand structures formed at various scales through simple local rules and some random motions [26, 27]. However, cellular automata models either ignore or greatly simplify physical behaviors such as fluid flow.

Another analysis technique considers swarms [28], which are well-suited to microscopic robots with their limited physical and computational capabilities and large numbers. Most swarm studies focus on macroscopic robots or behaviors in abstract spaces [29] that do not specifically include physical properties unique to microscopic robots. In spite of the simplified physics, these studies show how local interactions among robots lead to various collective behaviors and provide broad design guidelines.

Simulations including physical properties of microscopic robots and their environments can evaluate performance of robots with various capabilities. Simple

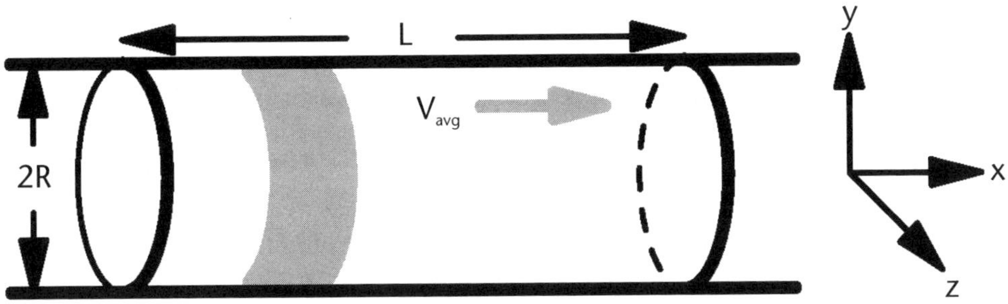

Figure 5.1 Schematic illustration of the task geometry as a vessel, of length L and radius R. Fluid flows in the positive x-direction with average velocity v_{avg}. The gray area is the source region wrapped around the surface of the pipe.

models, such as a two-dimensional simulation of chemotaxis [30], provide insight into robots finding microscopic chemical sources. A more elaborate simulator [31] includes three-dimensional motions in viscous fluids, Brownian motion, and environments with numerous cell-sized objects, though without accounting for how they change the fluid flow. Studies of hydrodynamic interactions [32] among moving devices include more accurate fluid effects.

Another approach to robot behaviors employs a stochastic mathematical framework for distributed computational systems [33, 34]. This method directly evaluates average behaviors of many robots through differential equations determined from the state transitions used in the robot control programs. Direct evaluation of average behavior avoids the numerous repeated runs of a simulation needed to obtain the same result. This approach is best suited for simple control strategies, with minimal dependencies on events in individual robot histories.

Microscopic robots, with limited computational capabilities, will likely use relatively simple reactive controls for which this analytic approach is ideally suited. Moreover, these robots will often act in environments with spatially varying fields, such as chemical concentrations and fluid velocities. Even at micron scales, the molecular nature of these quantities can be approximated as continuous fields with behavior governed by partial differential equations. For application to microscopic robots, this approximation extends to the robots themselves, treating their locations as a continuous concentration field, and their various control states as corresponding to different fields, much as multiple reacting chemicals are described by separate concentration fields. This continuum approximation for average behavior of the robots will not be as accurate as when applied to chemicals or fluids, but nevertheless gives a simple approach to average behaviors for large numbers of robots responding to spatial fields. One example of this approach is following chemical gradients in one dimension without fluid flow [35].

Cellular automata, swarms, physically based simulations, and stochastic analysis are all useful tools for evaluating the behaviors of microscopic robots. One example is evaluating the feasibility of rapid, fine-scale response to chemical events too small for detection with conventional methods, including sensor noise inherent in the discrete molecular nature of low concentrations. This chapter examines this issue in a prototypical task using the stochastic analysis approach. This method allows incorporating more realistic physics than used with cellular automata studies, and is computationally simpler than repeated simulations to obtain average behaviors. This technique is limited in requiring approximations for dependencies introduced by the robot history, but readily incorporates physically realistic models of sensor noise and consequent mistakes in the robot control decisions. The stochastic analysis indicates plausible performance, on average, and thereby suggests scenarios suited for further, more detailed, simulation studies.

5.3 Modeling Behavior of Microscopic Robots

Estimating behavior of microscopic robots requires physical models of task environments and robot capabilities. For initial studies, order of magnitude estimates are sufficient to identify tasks for which the robots could perform well. Minimal capabilities needed for biomedical tasks include chemical sensing, computation, and power [21]. Additional capabilities, enabling more sophisticated applications, include communication and locomotion. This section describes approximate models for robot behaviors.

5.3.1 Fluid Flow and Geometry

Biomedical applications will often involve robots operating in fluids. Viscosity dominates the robot motion, with different physical behaviors than for larger organisms and robots [36–40]. The ratio of inertial to viscous forces for an object of size s moving with velocity v through a fluid with viscosity η and density ρ is the Reynolds number Re = $s\eta\, v/\rho$. Due to power constraints on the devices and to avoid damage to biological tissue, the device speeds through the fluid will typically be fairly slow, below about a centimeter per second [1]. Using typical values for density, viscosity (e.g., of water or blood plasma) and device speeds, motion of a micron-sized robot has Re <<1, so viscous forces dominate.

We can simplify analysis with a suitable choice of geometry. In particular, considering reduced dimensionality for fluid flow significantly reduces the computational cost of evaluating behaviors. One such approach is considering extended channels where fluid flow is essentially two-dimensional. Another approach, and the one taken here, considers axial symmetric geometry. For low Reynolds number flow relevant to microscopic robots, such simplified geometries also provide reasonable approximations for some more complicated cases such as branching vessels [21]. Specifically, we consider robots operating in a long cylindrical pipe. The flow in a pipe of uniform radius R containing only the fluid has a parabolic velocity profile: velocity at distance r from the center of the pipe is

$$v(r) = 2v_{avg}\left(1-(r/R)^2\right) \tag{5.1}$$

where v_{avg} is the average speed of fluid in the pipe.

Fluid drag moves robots in the fluid. Objects, such as robots, moving in the fluid alter the fluid's velocity to some extent. In principle, the differential equation analysis approach can readily include effects objects in the fluid, but the numerical solution becomes significantly more complicated. Instead of the parabolic velocity profile (5.1), fluid flow changes with time as the objects move through the vessel. Simultaneously, the object motion is determined by the drag force from the fluid. Evaluating the flow requires solving the Navier-Stokes equation. For simplicity, and due to the relatively small volume of the vessel occupied by the robots, we ignore these changes and treat the fluid flow as constant in time and given by (5.1).

We suppose the robots have neutral buoyancy and move passively in the fluid. An approximation for robots without active locomotion is to assume they move with the same velocity as fluid would have at the center of the robot if the robot were not there (i.e., with speed given by (5.1) at their centers). This approximation neglects the change in fluid flow due to the robots, and is reasonable for estimating detection performance when the robots are at low enough density to be spaced apart many times their size, as is the case for the example presented here. Numerical evaluation of the fluid forces on the robots for the parameters of Table 5.1 show the robots indeed move close to this speed when the spacing between robots is many times their size. Closer packing leads to more complex motion due to hydrodynamic interactions [32, 41].

5.3.2 Chemical Sensing

Large-scale robots often use sonar or cameras to sense their environment. These sensors locate objects from a distance, and involve sophisticated algorithms with extensive computational requirements. In contrast, microscopic robots for biological applications will mainly use chemical sensors (e.g., the selective binding of molecules to receptors altering the electrical characteristics of nanoscale wires).

Current molecular electronics [14] and nanoscale sensors [42–44] indicate plausible sensor capabilities. At low concentrations, sensor performance is primarily limited by the time for molecules to diffuse to the sensor and statistical fluctuations in the number of molecules encountered are a major source of noise.

To approximate sensor behavior, note the diffusive capture rate γ for a sphere of radius a in a region with concentration C is [45]

$$\gamma = 4\pi Dac \tag{5.2}$$

where D is the diffusion coefficient of the chemical. Even when sensors cover only a relatively small fraction of the device surface, the capture rate is almost this large [45]. Nonspherical devices have similar capture rates so (5.2) is a reasonable approximation for a variety of designs.

We treat detection as due to absorbing spheres with concentration C at the location of the center of the sphere for (5.2), assuming the robot does not significantly alter the average concentration around it. Unlike previous studies of this scenario [17, 21] we include some of the effect of objects, such as blood cells, moving in the fluid via an increase in the effective diffusion constant for both chemicals and robots. Specifically, the hydrodynamic effect of rotating blood cells moving in the fluid greatly increases the diffusion coefficient of smaller objects in the fluid, by an amount $D_{\text{rotate}} = 5\times10^{-10}$ m^2/s [46]. This assumes the cell motion randomizes the locations of the chemical molecules and robots, ignoring systematic hydrodynamic interactions that could bias location of robots (e.g., toward walls).

5.4 Task Scenario

We illustrate the analysis approach with a prototypical task for microscopic robots: identifying a cell-sized source releasing chemicals into a small blood vessel. This scenario illustrates a basic capability for the robots: identifying small chemically distinctive regions, with high sensitivity due to the robots' ability to get close (within a few cell diameters) to a source. Microscopic robots acting independently to detect specific patterns of chemicals are analogous to swarms [28] used in foraging, surveillance, or search tasks. Even without locomotion capabilities, large numbers of such devices could move through tissues by flowing passively in moving fluids (e.g., through blood vessels or with lymph fluid). As the robots move, they can monitor for preprogrammed patterns of chemical concentrations.

Chemicals with high concentrations are readily detected with the simple procedure of analyzing a blood sample. Thus the chemicals of interest for microscopic robot applications will generally have low concentrations. With sufficiently low concentrations and small sources, the devices are likely to only encounter a few molecules while passing near the source, leading to significant statistical fluctuations in number of detections.

We can consider this task from the perspective of stages of operation discussed in the introduction. For a diagnostic first stage, the robots need only store events in their memory for later retrieval, when a much more capable conventional computer can process the information. For an active second stage where robots react to their detections (e.g., to aggregate at a source location or release a drug), the robots would need to determine themselves when they are near a suitable location. In this latter case, the robots would need simple control decision procedure, within the limits of local information available to them and their computational capacity. Such a control program could involve comparing counts to various thresholds, which were selected by analysis of a previous diagnostic stage of operation.

5.4.1 Example Task Environment

Tissue microenvironments vary considerably in their physical and chemical properties. As a specific example illustrating the capabilities of passive motion by microscopic sensors, we consider a task environment consisting of a macroscopic tissue volume V containing a single microscopic source producing a particular chemical (or combination of chemicals) while the rest of the tissue has this chemical at much lower concentrations. This tissue volume contains a large number of blood vessels, and we focus on chemical detection in the small vessels, since they allow exchange of chemicals with surrounding tissue, and account for most of the surface area. A rough model of the small vessels is each has length L and they occur throughout the tissue volume with number density ρ_{vessel}. Localization to volume V could be due to a distinctive chemical environment (e.g., high oxygen concentrations in the lungs), an externally supplied signal (e.g., ultrasound) detectable by sensors passing through vessels within the volume, or a combination of both methods. The devices are active only when they detect they are in the specified region.

Robots moving with fluid in the vessels will, for the most part, be in vessels containing only the background concentration, providing numerous opportunities for incorrectly interpreting background concentration as source signals. These false positives are spurious detections due to statistical fluctuations from the background concentration of the chemical. Although such detections can be rare for individual devices, when applied to tasks involving small sources in a large tissue volume, the number of opportunities for false positive responses can be orders of magnitude larger than the opportunities for true positive detections. Thus even a low false positive rate can lead to many more false positive detections than true positives.

The task scenario examined in this chapter thus includes estimating both true and false positive rates, addressing the question of whether simple controls can achieve a good trade-off of both a high true positive rate and low false positive rate.

For simplicity, we consider a vessel containing only flowing fluid, robots, and a diffusing chemical arising from a source area on the vessel wall. This scenario produces a static concentration of the chemical throughout the vessel, thereby simplifying the analysis. We examine the rate at which robots find the source and the false positive rate as functions of the detection threshold used in a simple control rule computed locally by each robot.

Figure 5.1 shows the task geometry: a segment of the vessel with a source region on the wall emitting a chemical into the fluid. Robots continually enter one end of the vessel with the fluid flow. The robot density in Table 5.1 corresponds to 10^9 robots in the entire 5-liter blood volume of a typical adult, an example of medical applications using a huge number of microscopic robots [1]. These robots use only about 10^{-6} of the vessel volume, far less than the 20% to 40% occupied by blood cells. The total mass of all the robots is about 4 mg.

The scenario for microscopic robots examined here is detecting small areas of infection or injury. The chemicals arise from the initial immunological response at the injured area and enter nearby small blood vessels to recruit white blood cells [47]. We consider a typical protein produced in response to injury, with concentration near the injured tissue of about 30 ng/ml and background concentration in the bloodstream about 300 times smaller. These chemicals, called chemokines, are proteins with molecular weight around 10^4 daltons (equal to about 2×10^{-23}kg). These values lead to the parameters for the chemical given in Table 5.1, with chemical

Table 5.1 Parameters for the Environment, Robots, and the Chemical Signal

Parameter	*Value*
Tissue, vessels, and source	
Vessel radius	$R = 5\mu$m
Vessel length	$L = 1000\mu$m
Number density of vessels in tissue	$\rho_{vessel} = 5 \times 10^{11}/\text{m}^3$
Tissue volume	$V = 10^{-6}\text{m}^3$
Source length	$L_{source} = 30\mu$m
Fluid	
Fluid density	$\rho = 10^3\text{kg/m}^3$
Fluid viscosity	$\eta = 10^{-3}\text{kg/m/s}$
Average fluid velocity	$v_{avg} = 10^{-3}\text{m/s}$
Fluid temperature	T = 310K
Diffusion due to cells in fluid	$D_{rotate} = 5 \times 10^{-10}\text{m}^2/\text{s}$
Robots	
Robot radius	$a = 1\mu$m
Number density of robots	$\rho_{robot} = 2 \times 10^{11}$ robot/m^3
Robot diffusion coefficient	$D_{robot} = 7.6 \times 10^{-14}$ m^2/s
Chemical signal	
Production flux at target	$F_{source} = 5.6 \times 10^{13}$ molecule/s/m^2
Diffusion coefficient	$D = 10^{-10}$ m^2/s
Concentration near source	$C_{source} = 1.8 \times 10^{18}$ molecule /m^3
Background concentration	$C_{background} = 6 \times 10^{15}$ molecule /m^3

The robots are spheres with radius a. The chemical signal concentrations correspond to a typical 10-kilodalton chemokine molecule, with mass concentrations near the source and background (i.e., far from the source) equal to $3{\times}10^{-5}$ kg/m^3 and 10^{-7} kg/m^3, respectively. C_{source} is equivalent to a three nanomolar solution. The $\rho_{vessel}V$ small vessels in the tissue volume occupy about a fraction $\rho_{vessel}\pi R^2 L = 4\%$ of the volume.

concentrations well above the demonstrated sensitivity of nanoscale chemical sensors [43, 44]. This example incorporates features relevant for medical applications: a chemical indicating an area of interest, diffusion into flowing fluid, and a prior background level of the chemical limiting sensor discrimination.

5.4.2 Diffusion of Robots and Chemicals

Diffusion arising from Brownian motion is somewhat noticeable for microscopic robots, and significant for molecules. The diffusion coefficient D, depending on an object's size, characterizes the resulting random motion, with root-mean-square displacement of $\sqrt{6Dt}$ in a time t [45]. Brownian motion also randomly alters robot orientation. The chemical concentration C is governed by the diffusion equation [45]

$$\frac{\partial C}{\partial t} = -\nabla \cdot F \tag{5.3}$$

where $F = -D \operatorname{grad} C + vC$ is the chemical flux (i.e., the rate at which molecules pass through a unit area), and v is the fluid velocity vector. The first term in the flux is diffusion, which acts to reduce concentration gradients, and the second term is motion of the chemical with the fluid.

We suppose the source produces the chemical uniformly with flux F_{source}. To evaluate high-resolution sensing capabilities, we suppose the chemical source is small, with characteristic size L_{source} as small as individual cells. Total target surface area is $2\pi R L_{\text{source}} = 9.4 \times 10^{-10}\text{m}^2$, about the same as the surface area of a single endothelial cell lining a blood vessel. The value for F_{source} in Table 5.1 corresponds to 5×10^4 molecules/s from the source area as a whole. This flux was chosen to make the concentration at the source area equal to that given in Table 5.1. The background concentration is the level of the chemical when there is no injury.

The concentration near the source eventually reaches a steady-state given by the solution of (5.3) with the right-hand side set to zero. The concentration decreases with distance from the source and high concentration occurs downstream of the source due to the fluid flow. We use this concentration profile for evaluating the robots' performance.

To evaluate the approximation for capture rate, we compare it with actual behavior for a sphere moving in the center of a vessel. The numerical solution shows a somewhat higher actual count rate, by about a factor of 2. But in practice not all of the sphere's surface would have sensors, which would multiply by a factor somewhat less than 1 due to not having all of the sphere's surface covered with receptors. These factors thus tend to cancel out, giving a result closer to approximation of fully absorbing sphere using concentration just at center of the sphere. Figure 5.2 illustrates the effect of this approximation. In this case, the fluid pushes the sphere at a speed $1.9v_{\text{avg}}$, only slightly slower than the speed of the fluid at the center of the vessel, $2v_{\text{avg}}$, when the sphere is not present. Thus in this case, the approximation for how the fluid moves the sphere and how it absorbs the chemical are adequate for order of magnitude estimates of performance.

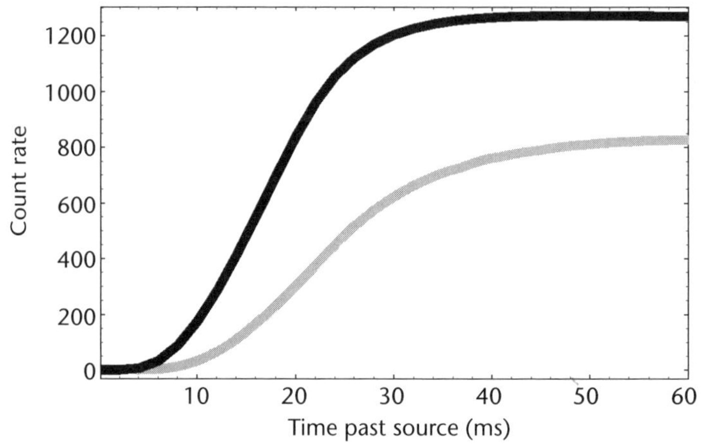

Figure 5.2 Comparison of chemical absorption rate for a sphere pushed by fluid along the center of the vessel (black) with approximation based on concentration at location of center of the sphere when it is not in the fluid (gray). The axes show the time past the upstream edge of the source in milliseconds and the rate chemical molecules are absorbed by the sphere, in counts per second.

5.4.3 Control

The limited capabilities of the robots and the need to react on millisecond time scales require simple controls based on local information. For chemical detection at low concentrations, the main sensor limitation is the time required for molecules to diffuse to the sensor. Thus the detections are a stochastic process.

For the chemical detection task, the key control design issue is a rule whereby a robot determines it has passed a source. Ideally, this rule would be simple to compute (due to limited computational capability of the robot), effectively distinguish passing a source (true positive) from detection due to background concentration (false positive), and allow decision while the robot is still relatively near the source (e.g., for subsequent tasks requiring the robot to move to the source). The key observational distinction between background and the much higher source concentration is the difference in count rates from molecules of the target chemical reaching the robot.

Due to statistical variation in the counts, the robot needs to accumulate counts over enough time to reliably distinguish true from false positives, but a short enough time that decisions are made while still close to the source. The choice of measurement time must balance having enough time to receive adequate counts, thereby reducing errors due to statistical fluctuations, while still responding before the robot has moved far downstream of the source where response would give poor localization of the source or, if robots are to take some action near the source,

Table 5.2 Robot Control Parameters for Chemical Signal Detection

Parameter	*Value*
Decrease rate	$\alpha = 1{,}000/\text{s}$
Detection threshold	$K = 10$

require moving upstream against the fluid flow. Moreover, far downstream of the source the concentration from the target is small so additional measurement time is less useful.

A simple decision criterion is if a sufficient number of detections occur in a short time interval. Quantitatively, such a rule considers a source detected if the robot observes at least $C_{\text{threshold}}$ counts in a measurement time interval T_{measure}. A device passing through a vessel with a source will have about $L_{\text{source}}/v_{\text{avg}} = 30$ ms with high concentration, so a measurement time of roughly this magnitude or smaller allows making a decision while still fairly near the source. A low value for $C_{\text{threshold}}$ will produce many false positives, while a high value means many robots will pass the source without detecting it (i.e., false negatives). False negatives increase the time required for detection while false positives could lead to inappropriate subsequent activities (e.g., releasing a drug to treat the injury or infection at a location where it will not be effective).

The threshold control rule is simple and exploits the key observational difference between source and background concentration. However, this rule depends on the history of the received counts over the past T_{measure} time, which significantly increases the complexity of the analysis if robots reaching a particular location are likely to have followed significantly different paths over the last T_{measure} time interval to reach that location.

When robots operate in fluid without many other objects, the diffusion due to Brownian motion is quite small: giving displacements of order $\sqrt{6D_{\text{robot}}T_{\text{measure}}}$, which is less than 0.1 μm, considerably smaller than the vessel diameter. Thus in this case, robots reaching a location have followed largely similar paths, allowing a reasonable approximation to ignore this variation in analyzing the behavior [21]. However, with the larger effective diffusion $D_{\text{eff}} = D_{\text{robot}} + D_{\text{rotate}}$ due to motion of cells in the fluid, the corresponding displacement is comparable to the vessel diameter. In this case, robots arrive at a given location along significantly different paths during the prior T_{measure} time interval.

Thus the analysis should account for these different histories. This analysis is greatly simplified by using a somewhat different control rule to exploit the observational difference in count rates near and far from the source. Instead of a list of the times, counts were detected during the past T_{measure} time interval (allowing the robot to update the number of counts received in comparison to the threshold $C_{\text{threshold}}$), the robot just maintains an integer count k, initially set to 0. This integer is updated probabilistically. In a short time Δt, if the robot detects a chemical (which occurs with probability $\gamma\Delta t$ based on the concentration at the robot's location, estimated from (5.2)), then it increases k by 1. Otherwise (i.e., most of the time), with probability $\alpha k\Delta t$, it decreases k by 1. When passing through a region of constant concentration, so γ is constant, on average, these changes balance when $k = \gamma/\alpha$. Thus a high chemical count rate γ or a low control decrease rate α tend to give large values of k for the robots. The detection rule using k is whether k reaches a specified threshold value K.

The control parameters $1/\alpha$ and K are analogous to T_{measure} and $C_{\text{threshold}}$ used with the threshold control rule. In both cases, the decision rule relies on detecting many molecules in a fairly short time. The probabilistic control rule has two advantages for the analysis of behavior. First is its use of just a single integer k to characterize the robot, rather than a list of the arrival times of all counts over the previous $T_{measure}$ time interval. Second, the probabilistic decrease in k makes the control a Markov process, since the arrival of molecules at the sensor are independent events.

The effective diffusion for the chemical, $D + D_{rotate}$, gives higher flux to the passing device, via (5.2), than just the intrinsic chemical diffusion D. So robots get enough counts for reasonable statistics in a shorter time, $1/\alpha = 1$ ms, than the $T_{measure} = 10$ ms measurement time used with just the intrinsic diffusion [21].

5.4.4 Analysis of Behavior

Task performance depends on the rate robots detect the source as they move past it, and the rate robots incorrectly conclude they detect the source due to background concentration of the chemical. From the values of Table 5.1, robots enter any given vessel at an average rate

$$\omega_{robot} = \rho r_{obot} \pi \, R2 v_{avg} = 0.016/s \tag{5.4}$$

and the rate robots enter (and leave) any of the small vessels within the tissue volume is

$$\Omega_{robot} = \rho_{vessel} V \, \omega_{robot} = 8 \times 10^3/s \tag{5.5}$$

A robot encounters changing chemical concentration as it moves past the source. The robot motion and change in internal state (in this case the counter k) is described by coupled equations. These equations separate the movement in space from changes in internal state, a general simplification for describing robots interacting with spatially changing fields [35], in this case the fluid velocity and chemical concentration fields.

Let R_k be the concentration (i.e., number of robots per unit volume, with integer value k for their counter), and let R_{detect} be the concentration of robots in the DETECT state (i.e., those for which k reached the threshold K. The effective diffusion rate for the robots is $D_{eff} = D_{robot} + D_{rotate}$. With nonuniform chemical concentration, the robot states vary with location.

For simplicity, we consider steady-state behavior, which is governed by

$$\begin{aligned} &\nabla\left(D_{eff} \nabla R_k - v R_k\right) + \gamma R_{k-1} - (\gamma + \alpha k) R_k + \alpha(k+1) R_{k+1} = 0 \\ &\nabla\left(D_{eff} \nabla R_{detect} - v R_{detect}\right) + \gamma R_{K-1} = 0 \end{aligned} \tag{5.6}$$

for $k = 0,\ldots,K-1$ and the values R_{-1} and R_K are defined to be 0.

Robots enter the vessel with the fluid with $k = 0$ (neglecting the tiny fraction of robots with larger k due to background concentration), pass the source, and leave at the outlet. So with $R = \Sigma R_k + R_{detect}$ the total number of devices, we can take the entering flux of devices to be the inlet flux boundary condition for R_0 and $R_1 = \ldots = R_{K-1} = R_{detect} = 0$ on the inlet. This entering flux is proportional to the fluid velocity across the channel, (5.1). The integral of this flux over the channel inlet is ω_{robot}, given by (5.4).

To complete this set of equations, (5.3) with diffusion $D + D_{\text{rotate}}$ gives the chemical concentration in the fluid, and (5.2) gives the corresponding γ for use in (5.6).

These equations incorporate the physics approximations described above: we use fluid speed in an empty vessel, assume robots move with this fluid speed at the location of the robot's center, and the flux of chemical to the robot equals that of a sphere in a large medium with concentration equal to that at the location of the robot's center.

The rate robots detect the source is the total rate robots switch from $k = K-1$ to $k = K$; that is,

$$\omega_{\text{source}} = \int \gamma R_{K-1} dV \tag{5.7}$$

where the integral is over the interior volume of the vessel containing the source.

Similarly, response of robots to background concentration is evaluated with (5.6) but using a constant $\gamma = 45/\text{s}$ corresponding to the background concentration (i.e., (5.2) evaluated with $C_{\text{background}}$ and effective chemical diffusion $D + D_{\text{rotate}}$ from Table 5.1). In this case, the uniform concentration leads to spatially uniform R_k values. Thus the spatial variation in the equations drops out and (5.6) becomes a set of linear algebra equations. We must consider time-dependent behavior to evaluate the rate at which robots reach the detection threshold $k = K$. That is, unlike robots passing the source where γ is large only near the source, a steady background γ means that every robot will eventually reach $k = K$ after enough time. Since γ is small compared to α, robots typically must wait a long time for enough counts in a short time to have k reach the threshold value K. In particular, the fraction of robots reaching the threshold grows as $1\text{-}e^{-\lambda t}$ for times t large compared to $1/\alpha$ with $\lambda=\alpha(\gamma/\alpha)^K/(K\text{-}1)!$ which is about 10^{-16}s for the parameters of Table 5.2 and γ at the background concentration. The background concentration extends throughout the tissue volume giving many opportunities for false positives.

For a diagnostic task, we pick control parameters of Table 5.2 and a time T for sensors to accumulate counts. The expected number of robots reporting detections from the source is then $\omega_{\text{source}}T$, neglecting the small additional contribution from the background concentration near the source. The actual number of detecting robots is a Poisson process. In particular, the probability at least one robot passing the source during time T detects the source is $1 - \exp(-\omega_{\text{source}}T)$.

For the false positives, as long as T is large compared to $1/\alpha$, each robot encountering just background concentration has probability of about $1\text{-}e^{-\lambda T}$ to reach the detection threshold and thereby produce a false positive. Thus the expected number of false positive detections is approximately

$$\Omega_{robot}\left(1 - e^{-\lambda T}\right) \tag{5.8}$$

with Ω_{robot} given by (5.5).

5.4.5 Detection Performance

For diagnostics, an indication of performance is comparing the likelihood of true and false positives. In particular, identifying choices of control parameters giving both a high chance of detecting a source and a low chance of false positives. In general this involves a trade-off: lower thresholds K or longer count accumulation times (i.e., smaller α) make detections more likely, for both true and false positives. Provided concentration near the source is significantly larger than background, and fluid speeds are small enough to allow robots sufficient time to collect at least a few counts as they pass the source, it is possible to select control parameters giving high chance of true positive detection while false positives remain low.

As an example, evaluating the above expressions for the parameters of Table 5.1 and Table 5.2 gives about 98% of the robots passing the source detect it, for a detection rate of about 0.015/s. By contrast, these control parameters give false positives at a rate of only about 10^{-11}/s, in spite of the much larger number of opportunities for false positives compared to the single vessel with the source. This difference is due to the ability of robots to pass close to the source, where concentration is significantly higher than background. The difference in rates gives a wide range for T with high chance of true positive detection and low false positives. For example, T = 300s gives a 99% chance at least one robot detects the source if it is there (true positive), while the probability any robot makes a detection when there is no source is less than 10^{-8}.

The stochastic analysis approach used here can estimate other aspects of robot performance, such as the average distance to the source if and when a passing robot detects it, based on evaluating $\int x\gamma R_{K-1} dV$ to determine the average value of x at which detection occurs. Using x^2 instead allows computing the deviation in detection locations. In the example discussed here, the detection occurs mostly within about 10 μm of the downstream end of the source.

In summary, simple control allows fast and accurate detection of even a single cell-sized source within a macroscopic tissue volume. The key feature enabling this performance is the robots' ability to pass close to individual cells, where concentration from released chemicals is much higher than in fluid far from the cell.

This approach also allows evaluating performance for various concentrations near the source and background [17], providing estimates of the limits of reliable discrimination of these devices.

For comparison, instead of using microscopic sensors, one could use a blood sample extracted from the body, which is a routine medical procedure. Conventional laboratory analysis outside the body could then attempt to detect the chemical in the sample. However, such a sample will represent the average concentration of the chemical in the full blood volume, leading to a much lower concentration from the source, about 10^{-4} of the background concentration for this scenario [17]. Thus the additional chemical released by the source would be difficult to detect against small variations in background concentration.

5.5 Discussion

Beyond the diagnostic task discussed above, robots able to locate chemically distinctive microenvironments in the body could have capabilities to modify those environments. For instance, the devices could carry specific drugs to deliver only to cells matching a prespecified chemical profile [1, 19] as an extension of a recent in vitro demonstration of this capability using DNA computers [6]. The robots could also use additional sensing modalities and communication to identify correlations among sources, or use locomotion to follow chemical gradients [21]. The behavior of microscopic active swimmers [48] raises additional control issues to exploit the hydrodynamic interactions among swimming objects as they aggregate so the distance between devices becomes only a few times their size [32].

Robots aggregated at chemically identified targets could perform precise microsurgery at the scale of individual cells. Since biological processes often involve activities at molecular, cell, tissue, and organ levels, such microsurgery could complement conventional surgery at larger scales. For instance, a few millimeter-scale manipulators, built from micromachine (MEMS) technology, and a population of microscopic devices could act simultaneously at tissue and cellular size scales. An example involving microsurgery for nerve repair with plausible biophysical parameters indicates the potential for significant improvement in both speed and accuracy compared to the larger-scale machines acting alone [22, 49].

Plausible capabilities for microscopic robots suggest a range of novel applications in biomedical research and medicine. Sensing and acting with micron spatial resolution and millisecond timing allows access to activities of individual cells. The large numbers of robots enable such activities simultaneously on a large population of cells in multicellular organisms. In particular, the small size of these robots allows access to tissue through blood vessels. Thus a device passes within a few tens of microns of essentially every cell in the tissue in a time ranging from tens of seconds to minutes, depending on the number of devices used.

As described above, this access to cells allows rapid, chemical sensing with much higher resolution than possible with in vitro sample analysis. Furthermore, the precision of localization and the robots' programmability gives them a degree of flexibility to alter microenvironments (e.g., by releasing drugs) well beyond what is possible with either large-scale surgery or nonprogrammable chemically targeted drug delivery. The full range of biomedical situations that could benefit from this flexibility (e.g., nerve repair [22]), remains to be seen.

With many devices in the tissue but only a few in the proper context to perform task, false positives are a significant issue. In some situations, these false positives may just amount to a waste of resources (e.g., power). But in other cases, too many false positives could be more serious, such as leading to aggregation blocking blood vessels or incorrect diagnosis.

The sensing task described in this chapter highlights key control principles for microscopic robots. Specifically, by considering the overall task in a series of stages, the person deploying the robots remains in the decision loop, especially for the key decision of whether to proceed with manipulation (e.g., release a drug) based on diagnostic information reported by the devices. Information retrieved during treatment can also indicate how well the procedure is proceeding and provide high-resolution documentation of what was done to help improve future treatments. More generally, this hybrid control illustrates an important approach to using self-organized systems: use local, distributed control to achieve robust self-organized behaviors on small scales in space and time, combined with feedback from a slower, larger central control (e.g., a person) to verify performance and consider global constraints not easily incorporated within local controllers.

The performance estimates for the sensing task show devices with limited capabilities—specifically, without locomotion or communication with other devices—can nevertheless rapidly detect chemical sources as small as a single cell. The devices use their small size and large numbers to allow at least a few to get close to the source, where concentration is much higher than background. This chapter also illustrates use of an analysis technique for average behavior of microscopic robots that readily incorporates spatially variable fields in the environment. Such fields are of major significance for microscopic robots, in contrast to their usual limited importance for larger robots.

As a caveat on the results, the model examined here treats the location of the chemical source as independent of the properties and flow rates in the vessel containing the source. Systematic variation in the density and organization of the vessels will increase the variation in detected values. For instance, the tissue could have correlations between vessel density and the chemical sources (e.g., if those chemicals enhance or inhibit growth of new vessels). Accurate inference requires models of how the chemicals move through the tissue to nearby blood vessels. Chemicals could react after release from the source to change the concentrations with distance from the source. Furthermore, other chemicals, unrelated to the desired detection task, may have similar sensor binding characteristics as the desired chemical. These other chemicals will increase the effective background rate, depending on the selectivity of the sensor.

Nevertheless, the simple model discussed here indicates the devices could have high discrimination for sources as small as single cells. Thus even with some unmodeled sources of variability, good performance could still be achieved by extending the sensing time or using more sophisticated inference methods. Moreover, with some localization during operation, the devices themselves could estimate some of this variation (e.g., changes in density of vessels in different tissue regions), and these estimates could improve the inference instead of relying on average or estimated values for the tissue structure.

Instead of flowing with fluid, the sensors could be implanted at specific locations of interest to collect data in their local environments, and later retrieved. This approach does not take full advantage of the sensor size: it could be difficult to identify interesting locations at cell-size resolutions and implant the devices accurately. Nevertheless, such implants could be useful by providing local signals to indicate regions of interest to other sensors passing nearby in a moving fluid.

Safety is important for medical applications of microscopic robots. Thus, evaluating a control protocol should consider its accuracy allowing for errors, failures of individual devices, or variations in environmental parameters. For the simple distributed sensing discussed in this chapter, statistical aggregation of many devices' measurements provides robustness against these variations, a technique recently illustrated using DNA computing to respond to patterns of chemicals [6]. Furthermore, the devices must be compatible with their biological environment [50] for enough time to complete their task. Appropriately engineered surface coatings and structures should allow sufficient biocompatibility during robot operation [51, 52]. However, even if individual devices are inert, too many in the circulation would be harmful. From Table 5.1, sensors occupy a fraction 10^{-6} of the volume inside the vessels. This value is well below the fraction, about 10^{-3}, of micron-size particles experimentally demonstrated to be safely tolerated in the circulatory system of at least some mammals [51]. Thus the number of robots used in the protocol of this chapter is unlikely to be a safety issue.

Despite the simplifications used to model sensor behavior, the estimates obtained in this chapter with plausible biophysical parameters show high-resolution sensing is possible with passive device motion in the circulatory system, even without communication capabilities. Thus relatively modest hardware capabilities could provide useful in vivo sensing capabilities far more flexible and specific than larger-scale devices. Research studies of tissue microenvironments with such robots will improve knowledge of their biophysical parameters, and hence enable better inferences from the data collected by these devices. The improved understanding will, in turn, indicate distributed controls suitable for more capable devices and appropriate trade-offs between scale and capability for hybrid systems combining coarse centralized control with the flexibility of self-organization within the biological microenvironments.

More generally, the analysis method using differential equations and approximating the robot locations as a continuous concentration field is useful for designing and evaluating controls for large collections of weakly interacting swarm robots. Microscopic robots are a particularly appealing application since they will often interact with spatially-varying fields best modeled with partial differential equations. In this situation, the behavior of the physical fields and the changing internal control states of the robots are both described with the same formulation.

Acknowledgments

I have benefited from discussions with Philip J. Kuekes.

References

[1] Freitas, Jr. , R. A., *Nanomedicine, Volume I: Basic Capabilities*, Georgetown, TX: Landes Bioscience, 1999. Available at www.nanomedicine.com/NMI.htm.

[2] Morris, K., "Macrodoctor, Come Meet the Nanodoctors," The Lancet, Vol. 357, March 10, 2001, p. 778.

[3] NIH, *National Institutes of Health Roadmap: Nanomedicine*, 2003. Available at nihroadmap.nih.gov/nanomedicine/index.asp.

[4] Keszler , Balazs, L., et al., "Molecular Engineering in Nanotechnology: Structure and Composition of Multifunctional Devices for Medical Application," *Proc. of the Ninth Foresight Conference on Molecular Nanotechnology*, 2001.

[5] Andrianantoandro, E., et al., "Synthetic Biology: New Engineering Rules for an Emerging Discipline," *Molecular Systems Biology*, Vol. 2(msb4100073), pp. E1–E14, 2006.

[6] Benenson, Y., et al., "An Autonomous Molecular Computer for Logical Control of Gene Expression," *Nature*, Vol. 429, 2004, pp. 423–429.

[7] Berna, J.,, et al. "Macroscopic Transport by Synthetic Molecular Machines," *Nature Materials*, Vol. 4, 2005, pp. 704–710.

[8] Collier, C. P., et al., "Electronically Configurable Molecular-Based Logic Gates," *Science*, Vol. 285, 1999, 391–394.

[9] Craighead, H. G., "Nanoelectromechanical Systems," *Science*, Vol. 290, 2000, pp. 1532–1535.

[10] Howard, J., "Molecular Motors: Structural Adaptations to Cellular Functions," *Nature*, Vol. 389, 1997, pp. 561–567.

[11] Fritz, J., et al., "Translating Biomolecular Recognition Into Nanomechanics," *Science*, Vol. 288, 2000, pp. 316–318.

[12] Montemagno, C., and G. Bachand, "Constructing Nanomechanical Devices Powered by Biomolecular Motors," *Nanotechnology*, Vol. 10, 1999, pp. 225–231.

[13] Soong, R. K., et al., "Powering an Inorganic Nanodevice with a Biomolecular Motor," *Science*, Vol. 290, 2000, pp. 1555–1558.

[14] Wang, S.-Y., and R. S. Williams, editors, *Nanoelectronics*, Volume 80, (special issue of Applied Physics A), Springer, March 2005.

[15] McCurdy, C. W., et al., "Theory and Modeling in Nanoscience," Workshop report, www.science.doe.gov/bes/reports/files/tmn rpt.pdf, US Dept. of Energy, 2002.

[16] Mataric, M., "Minimizing Complexity in Controlling a Mobile Robot Population," *Proc. of the 1992 IEEE Intl. Conf. on Robotics and Automation*, 1992, pp. 830–835.

[17] Hogg, T. and P. J. Kuekes, "Mobile Microscopic Sensors for High-Resolution in Vivo Diagnostics," *Nanomedicine: Nanotechnology, Biology, and Medicine*, Vol. 2, 2006, pp. 239–247.

[18] Natterer, F., *The Mathematics of Computerized Tomography*. Philadelphia: Society for Industrial and Applied Math (SIAM), 2001.

[19] Freitas Jr, R. A., "Pharmacytes: An Ideal Vehicle for Targeted Drug Delivery," *Journal of Nanoscience and Nanotechnology*, Vol. 6, 2006, 2769–2775.

[20] Casal, A., et al., "Nanorobots as Cellular Assistants in Inflammatory Responses," in *Proc. of the 2003 Stanford Biomedical Computation Symposium* (BCATS2003), p. 62, J. Shapiro, (ed.), Oct. 2003, Available at bcats.stanford.edu.

[21] Hogg, T., "Coordinating Microscopic Robots in Viscous Fluids," *Autonomous Agents and Multi-Agent Systems*, Vol. 14 No. 3, 2007, pp. 271–305.

[22] Hogg, T., and D. W. Sretavan, "Controlling Tiny Multi-Scale Robots for Nerve Repair," *Proc. of the 20th Natl. Conf. on Artificial Intelligence* (AAAI2005), AAAI Press, 2005, pp. 1286–1291.

[23] Drexler, K. E., *Nanosystems: Molecular Machinery, Manufacturing, and Computation*, New York: John Wiley, 1992.

[24] Requicha, A. A. G., "Nanorobots, NEMS and Nanoassembly," *Proceedings of the IEEE*, Vol. 91, 2003, pp. 1922–1933.

[25] Arbuckle, D., and A. A. G. Requicha, "Active Self-Assembly," *Proc. of the IEEE Intl. Conf. on Robotics and Automation*, 2004, pp. 896–901.

[26] Whitesides, G. M., and B. Grzybowski, "Self-Assembly at all Scales," *Science*, Vol. 295, 2002, pp. 2418–2421.

[27] Griffith, S., et al., "Robotics: Self-Replication from Random Parts," *Nature*, Vol. 437, 2005, pp. 636.

[28] Bonabeau, E., et al., Swarm Intelligence: From Natural to Artificial Systems, Oxford: Oxford University Press, 1999.

[29] Gazi, V., and K. M. Passino, "Stability Analysis of Social Foraging Swarms," *IEEE Trans. on Systems, Man and Cybernetics*, Vol. B34, 2004, pp. 539–557.

[30] Dhariwal, A., G. S. Gaurav, and A. A. G. Requicha, "Bacterium-Inspired Robots for Environmental Monitoring," *Proc. of the IEEE Intl. Conf. on Robotics and Automation*, 2004, pp. 1436–1443.

[31] Cavalcanti, A., and R. A. Freitas, Jr., "Autonomous Multi-Robot Sensor-Based Cooperation for Nanomedicine," *Intl. J. of Nonlinear Sciences and Numerical Simulation*, Vol. 3, 2002, pp. 743–746.

[32] Hernandez-Ortiz, J. P, C. G. Stoltz, and M. D. Graham, "Transport and Collective Dynamics in Suspensions of Confined Swimming Particles," *Physical Review Letters*, Vol. 95, No. 20, 2005, epub 204501.

[33] Hogg, T., and B. A. Huberman, "Dynamics of Large Autonomous Computational Systems," in Collectives and the Design of ComplexSystems, K. Tumer and D. Wolpert (eds.), New York: Springer, 2004, pp. 295–315.

[34] Lerman, K., et al., "A Macroscopic Analytical Model of Collaboration in Distributed Robotic Systems," *Artificial Life*, Vol. 7, 2001, pp. 375–393.

[35] Galstyan, A., et al., "Modeling and Mathematical Analysis of Swarms of Microscopic Robots," in *Proc. of the IEEE Swarm Intelligence Symposium* (SIS2005), pp. 201–208, P. Arabshahi and A. Martinoli (eds.), 2005.

[36] Purcell, E. M., "Life at Low Reynolds Number," *American Journal of Physics*, Vol. 45, 1977, pp. 3–11.

[37] Vogel, S., *Life in Moving Fluids, 2nd Edition*, Princeton University Press, 1994.

[38] Fung, Y. C., *Biomechanics: Circulation*, New York: Springer, 1997.

[39] Karniadakis, G. E. M., and A. Beskok, *Micro Flows: Fundamentals and Simulation,* Berlin: Springer, 2002.

[40] Squires, T. M., and S. R. Quake, "Micro Fluidics: Fluid Physics at the Nanoliter Scale," *Reviews of Modern Physics*, Vol. 77, 2005, pp. 977–1026.

[41] Riedel, I. H. ,et al., "A Self-Organized Vortex Array of Hydrodynamically Entrained Sperm Cells," *Science*, Vol. 309, 2005, pp. 300–303.

[42] Li, Z., et al., "Silicon Nanowires for Sequence-Specific DNA Sensing: Device Fabrication and Simulation," *Applied Physics A*, Vol. 80, 2005, pp. 1257–1263.

[43] Patolsky, F., and C. M. Lieber. "Nanowire Nanosensors," *Materials Today*, Vol. 8, April 2005, pp. 20–28.

[44] Sheehan, P. E., and L. J. Whitman, "Detection Limits for Nanoscale Biosensors," *Nano Letters*, Vol. 5, No. 4, 2005, 803–807.

[45] Berg, H. C., *Random Walks in Biology, 2nd Edition*, Princeton University Press, 1993.

[46] Keller, K. H., "Effect of Fluid Shear on Mass Transport in Flowing Blood," *Proc. of Federation of American Societies for Experimental Biology*, September–October 1971, pp. 1591–1599.

[47] Janeway, C. A., et al., *Immunobiology: The Immune System in Health and Disease, 5th Edition*, Garland, 2001.

[48] Dreyfus, R., et al., "Microscopic Artificial Swimmers," *Nature*, Vol. 437, 2005, pp. 862–865.

[49] Sretavan, D., et al., "Microscale Surgery on Axons for Nerve Injury Treatment," *Neurosurgery*, Vol. 57, No. 4, 2005, 635–646.

[50] Nel, A., et al., "Toxic Potential of Materials at the Nanolevel," *Science*, Vol. 311, 2006, pp.622–627.

[51] Freitas Jr, R. A., *Nanomedicine, Volume IIA: Biocompatibility*, Georgetown, TX: Landes Bioscience, 2003. Available at www.nanomedicine.com/NMIIA.htm.

[52] Schrand, A. M., et al., "Are Diamond Nanoparticles Cytotoxic?" *Journal of Physical Chemistry B*, Vol. 111, 2007, pp. 2–7.

CHAPTER 6

Medical and Biometric Identification for Pattern Recognition and Data Fusion with Quantitative Measuring

Zhengmao Ye, Hua Cao, Sitharama Iyengar, and Habib Mohamadian

Automated identification is crucial to numerous challenging issues in the real world, such as medical diagnosis, homeland security and information assurance. This chapter introduces various systematic modeling, fusion, and identification methodologies for automation in medical and biomedical fields. Qualitative and quantitative approaches are applied to evaluate each method being presented. Theoretical formulations, numerical simulations, and practical implementations are presented. Informative conclusions are made accordingly and future research directions are proposed.

6.1 Introduction

In comparison with conventional medical treatment and biometric verification methods, advanced modern medical and biomedical identification have significant advantages over classic forms, which can be used to avoid loopholes and indicate unique patterns for all individuals. Medical and biometric identification provide typical examples of achieving accurate medical decision making and personal identity recognition, so it will be investigated in detail throughout this chapter. Biometric identification is based on fingerprints, retina patterns, facial images, and voice detections, among which digital fingerprint identification is most frequently used and the retina pattern recognition is best known for its uniqueness. Thus, both approaches will be analyzed in order to show how to apply identification techniques to medical and biomedical systems. In this chapter, automated modeling, registration, fusion, and identification are conducted on a basis of both fingerprint and retina images. Both linear modeling and nonlinear modeling are tested to represent the actual patterns individually.

Image registration and image fusion are then implemented for information integration. With respect to system identification, methodologies of principal component analysis (PCA), nonlinear component analysis (NCA), independent component analysis (ICA), 1-D discrete wavelet transform, and 2-D discrete wavelet transform have been presented for pattern recognition. Decision making can be made partially in terms of some qualitative outcomes from processing of the gray level and true color images as well as associated numerical simulations. To reduce the deficiency involved in different approaches, quantitative measures are also proposed and computed in order to enhance the accuracy of decision making. In particular, the information theory has been introduced to these medical and biomedical practical approaches, where quantity measures of discrete entropy, relative entropy, energy, and mutual information will be calculated and compared. Some practical implementation examples have been provided for the concern of the real-world problem solving. All approaches being discussed are promising to be implemented in real time across the broad medical and biomedical fields.

6.2 Overview of Practical Approaches

6.2.1 Principal Component Analysis

The main applications of principal component analysis (PCA) are linear signal processing, pattern recognition, statistics, data compression, and model prediction.

PCA is a classical statistical approach. It acts as a linear optimal transformation that transforms the data into an orthogonal coordinating system and it also maps multidimensional data into a lower dimension space for analysis to minimize the loss of information. PCA is based on the statistical representation of random variables using the unsupervised transformation procedure to create new variables in directions of maximal variations. It has the advantage of keeping the subspace with the largest variances and neglecting insignificant components with noises or unimportant variances. The variance of any component indicates the spread of component values from its mean value. The first principal component (PC_1) describes the greatest amount of variation under the direction with the greatest variation. Each principal component represents a single coordinate in space, perpendicular to each other. A new variable is generated when an observation is projected on a coordinate where the variance of this variable is maximal among all possible alternatives [2, 9–13]. Singular value decomposition (SVD) and covariance matrix approaches are commonly used. For instance, SVD is used for factorization of the complex arbitrary matrices into principal components in an orthogonal basis. By convention, singular vectors will be determined in a nonincreasing order of singular values, with the highest singular value on top of the matrix, thus diagonal matrix S is uniquely determined. For a square symmetric matrix, SVD is equivalent to the diagonalization of eigenvalue and eigenvector problems. The equation of the SVD approach is formulated as:

$$X = U^{*}S^{*}V^{T} \tag{6.1}$$

where X ($m \times n$) is the data matrix, U ($m \times m$) is a unitary matrix, S ($m \times m$) is a nonnegative diagonal matrix, and V^T ($m \times n$) is a unitary matrix and conjugate transpose of V. The dominating eigenvalue is λ_1. The relationship among different PCs is $\lambda_1 > \lambda_2 > \lambda_3 > \ldots > \lambda_n > \ldots\ldots$, and so on.

6.2.2 Nonlinear Component Analysis

The main applications of nonlinear component analysis (NCA) are nonlinear signal processing, pattern recognition, statistics, data compression, and model prediction.

Kernel principal component analysis is an extension of PCA using techniques of kernel methods. Using a kernel, the linear mapping of PCA is replaced in a reproducing kernel Hilbert space with a nonlinear mapping. The basic idea of NCA is to replace the covariance matrix of PCA with that of NCA.

$$A = E[S(m) \otimes S^T(m)] \tag{6.2}$$

$$C = E\{\Phi[S(m)] \otimes \Phi[S^T(m)]\} \tag{6.3}$$

where Φ is the nonlinear mapping to a feature space with a larger dimensionality than the data space. Principal components are then computed in the corresponding feature space. A Hebbian rule-based NCA can be proposed [2, 6, 7, 9–11].

6.2.3 Independent Component Analysis

Independent component analysis (ICA) applications are blind signal separation, loss free coding, feature extraction, data analysis, image processing, and pattern recognition.

ICA is a computation method for revealing hidden factors within datasets of random variables, signals, and images, which separates a multivariate signal into additive subcomponents assuming mutual statistical independence of non-Gaussian source signals. The ICA approach has the advantage of learning probabilistic model efficiently for large dimensions [2, 13–18]. Now the observed random vector is $X = \{x_1, x_2, ..., x_m\}$, and independent component is $S = \{s_1, s_2, ..., s_n\}$. The objective is to transform the observed data X into maximally independent components S measured by the function of independence, using a linear transformation $S = WX$. Estimation is made using a selected training via unsupervised algorithms.

6.2.3.1 Centering

Prior to the implementation of ICA algorithms, two strategies are used for preprocessing. The first necessary preprocessing is centering. The mean vector of image matrix is subtracted from each vector of the matrix, so that all resulting vectors of the matrix are zero-mean vectors. The matrix itself turns out to be the combination of zero mean vectors. This process can be conducted on a basis of either row vectors or column vectors of the image matrix.

6.2.3.2 Whitening

Whitening the observed image matrix is a critical ICA preprocessing strategy. Immediately after centering, the next objective is to make image matrix components uncorrelated whose variances equal to unity. That is, to make the covariance matrix equal to the identity matrix to guarantee that $E\{XX^T\}=I$ (identity matrix). The eigenvalue and eigenvector decompositions of covariance matrices can be applied to the whitening preprocessing.

$$E\{XX^T\} = EDE^T \tag{6.4}$$

where E is an orthogonal matrix of eigenvectors and D is the diagonal matrix consisting of all corresponding eigenvalues. We have:

$$D=diag\ (d_1, d_2, \ldots, d_n) \tag{6.5}$$

An appropriate whitening matrix is used to transform the mixing matrix into the new orthogonal mixing matrix $\tilde{A}$:

$$\tilde{A} = (ED^{-1/2}E^T)A=VA \tag{6. 6}$$

where $D^{-1/2}$ is the diagonal matrix being formulated as:

$$D^{-1/2}=diag\ (\mathrm{d}_1^{-1/2}, \mathrm{d}_2^{-1/2}, \ldots , d_n^{-1/2}) \tag{6.7}$$

The whitening reduces the number of parameters to be estimated and ensures that the new mixing matrix $\tilde{A}$ should be orthogonal so as to guarantee:

$$\tilde{A}\,\tilde{A}^T = I \tag{6.8}$$

6.2.3.3 Independence Optimization

To reconstruct original independent image patterns after dimension reduction, optimization can be conducted in an iterative manner.

$$Y = WX = W\tilde{A}S = W(VA)S = W(ED^{-1/2}E^T)AS \tag{6.9}$$

where W is an orthogonal matrix subject to training.

Oja's rule can be employed. For all independent vectors of matrix W to be trained, the iteration below is applied followed by normalization, subject to the constraint of $\|w\|=1$. The initial matrix W_0 is selected to be a random orthonormal matrix.

$$w^{k+1} = w^k + \rho\left(x^k - y^k w^k\right)y^k \tag{6.10}$$

Even though independent optimization is conducted in a reduced dimension, dominant independent components still indicate the intrinsic patterns for the patterns being processed.

6.2.4 2-D Discrete Wavelet Transform

The main applications for the 2-D discrete wavelet transform are 2-D signal processing, image processing, and denoising.

In a two-dimensional discrete wavelet transform, the scaled and translated basis functions are defined by:

$$\Phi_{j,m,n}(x, y) = 2^{\,j/2}\, \phi(2^j x - m, 2^j y - n) \tag{6.11}$$

$$\psi^i{}_{j,m,n}(x, y) = 2^{\,j/2}\, \psi^i\, (2^j x - m, 2^j y - n),\ i=\{H, V, D\} \tag{6.12}$$

where index i identifies the directional wavelets in terms of values of horizontal (H), vertical (V), and diagonal (D). The discrete wavelet transforms of function $f(x, y)$ ($M{\times}N$) is formulated as:

$$w_{\phi}(j_0 m,n) = \frac{1}{\sqrt{MN}} \sum_{x=0}^{M-1} \sum_{y=0}^{N-1} f(x,y)\phi_{j_0,m,n}(x,y) \tag{6.13}$$

$$w_{\psi}^{i}(j_0 m,n) = \frac{1}{\sqrt{MN}} \sum_{x=0}^{M-1} \sum_{y=0}^{N-1} f(x,y)\psi_{j,m,n}^{i}(x,y) \tag{6.14}$$

where i={H, V, D}, j_0 is the starting scale, the $w_j(j_0, m, n)$ coefficients define the approximation of $f(x, y)$, $w_{\psi}^{i}(j, m, n)$ coefficients represent the horizontal, vertical and diagonal details for scales j>= j_0. Here j_0 =0 and select $N + M = 2^J$ so that j=0, 1, 2, ..., J-1 and m, n = 0, 1, 2, ... , 2^J -1. Then the $f(x, y)$ is obtained via the inverse discrete wavelet transform [1, 19–20].

$$f(x,y) = \frac{1}{\sqrt{MN}} \sum_{m} \sum_{n} w_{\phi}(j_0, m,n)\phi_{j_0,m,n}(x,y) + \frac{1}{\sqrt{MN}} \sum_{i=H,V,D} \sum_{j=j0}^{\infty} \sum_{m} \sum_{n} w_{\psi}^{i}(j,m,n)\psi_{j,m,n}^{i}(x,y) \tag{6.15}$$

The wavelets are defined by both the scaling function ϕ(x, y) (father wavelet) and wavelet functions $\psi(x, y)$ (mother wavelet) in the discrete time domain. The wavelet function represents a bandpass filter whose bandwidth is reduced to half after each scaling. The output of each level always includes approximation, horizontal detail, vertical detail, and diagonal detail.

6.2.5 Image Processing Background [1, 5]

For binary images, each pixel is stored as a single bit with the value of 0 or 1; 0 represents black while 1 represents white. For grayscale images, each pixel is stored as an 8-bit byte with the value ranging from 0 to 255; 0 represents black while 255 represents white. Other various shades of gray exist in the image with the value ranging from 1 to 254. For true color images, each pixel is composed of three colors—red, green, and blue (RGB values)—at certain proportions. Each color is stored as 8 bits. Therefore, each pixel is stored as 24 bits. A color image can be converted into a grayscale image. Weighting parameters of [0.299, 0.587, 0.114] are the default values in general (Figure 6.1).

6.2.6 Image Registration Models

Transformation is a mapping function between coordinates of two images I_1 and I_2 such that the image I_1 and transformed image I_2 have similar structures or features at the same position. Transformation is either local or global. The objective of image registration is to geometrically align two or more images obtained from single or multiple sensors during certain time interval. Registration is generally categorized as intramodality and multimodality registrations. Intramodality registration

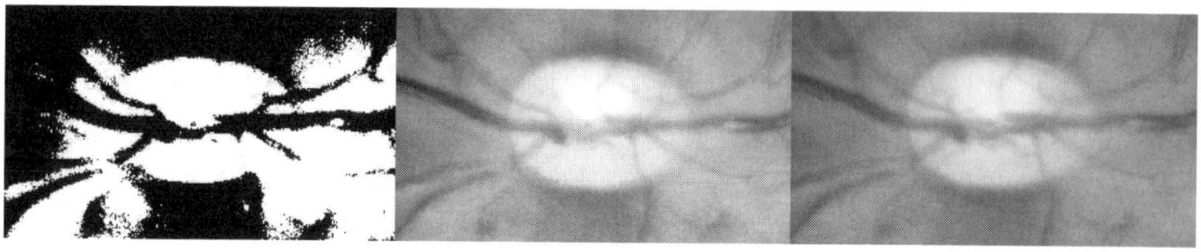

(a) (b) (c)

Figure 6.1 Retina images in (a)black and white (binary, 1 bit), (b)gray level (8 bits), and (c)true color (24 bits).

Table 6.1 Basic Transformation in Image Processing

Transformation	*Description*
Rotation/skew	Points are rotated or turned by an angle θ.
Translation	A linear shift in the position of the vertical and horizontal coordinates of the image in one plane to another set in the same plane.
Scaling	A transformation of the horizontal and vertical coordinate points characterized by scale factors.
Shearing	A transformation in which all points along a line remain fixed while other points are shifted parallel to the line by a distance proportional to their perpendicular distance from the line.

refers to images obtained from the same modality at different time. Multimodality registration covers images obtained from different modalities at certain time. Most popular registration models are the translation model, affine model, rigid transformation model, and quadratic model [29–36].

6.2.7 Area-Based Intramodality Registration: Cross Correlation and Fourier Transform [29-36]

Area-based registration is usually used for intramodality image cases. In the area-based method, a small window of image points is statistically compared with windows of the same size in the reference image. Normalized cross correlation is used to measure alignment. Landmark control points are taken from the center of matched windows. On the other hand, area-based registration and fusion work efficiently for intramodality but not for multimodality image registration. Compared to area-based registration, feature-based is more appropriate to multimodality images. Most area-based registration and fusion method can be conducted in three steps: (1) Perform initial processing of the original image pairs; (2) conduct cross-correlation registration using the Fourier transform; and (3) optimize image fusion and feature extraction.

Maximizing of the cross correlation yields the rotational difference. Suppose $I_{(1)}(x, y)$ is a spatial domain, and $F_{(1)}(u, v)$ is the Fourier transform, we have,

$$F_{(1)}(u,v) = \int\int I_{(1)}(x,y)e^{-2\pi j(ux+vy)}dxdy \tag{6.16}$$

and the inverse Fourier transform:

$$I_{(1)}(x,y) = F_{(1)}^{-1}(u,v) = \int\int F_{(1)}(u,v)e^{2\pi j(ux+vy)}dudv \tag{6.17}$$

The cross-correlation function is defined as:

$$I_{(1)}(x,y) \otimes I_{(2)}(x,y) = \int\int I_1^*(\alpha,\beta)I_{(2)}(x+\alpha, y+\beta)d\alpha d\text{B} \tag{6.18}$$

Cross-correlation (⊗) of two images is equivalent to the multiplication of Fourier transform of one function and complex conjugate (*) of the Fourier transform of another.

6.2.8 Feature-Based Multimodality Registration [26–28]

Feature-based registration extracts and matches the common structures from two images. Feature-based registration and fusion maximize the objective function related to features. Fusion accuracy is estimated by minimizing registration errors.

6.2.8.1 SSD Minimization Algorithm

$$SSD = \sum_{i \in OA}\left(I\left(F\left(F(x;\Delta M);M\right)\right) - I(X)\right)^2 \tag{6.19}$$

where OA is the overlapping area of I_r and I_i, M is the parameter matrix and F is the transformation function.

6.2.8.2 Newton's Iterative Optimization

$$M_{n+1} = M_n + (H_n)^{-1}\nabla_M F_n \tag{6.20}$$

where $\nabla_M F_n$ is the gradient vector of F with respect to M at the nth iteration.

Feature-based registration and fusion approaches are conducted in three steps: (1) Build a multiresolution map for each image; (2) conduct feature control points selection at each featured area in the map; and (3) create a new image based on the fused multiresolution map. In special cases, area-based and feature-based registration and image fusion can be combined.

6.2.9 Image Fusion

Image fusion produces a single image from a set of input images. The fused image should contain more complete information that is more useful for human or machine perception. Registering multimodality data will improve reliability by using mutual information and improve capability by providing complementary information [37–44]. Image fusion can be classified into signal-level fusion, pixel-level fusion, symbol-level fusion, and feature-level fusion.

6.3 Practical Implementation of Medical and Biometric System Identification

6.3.1 Linear and Nonlinear Component Analysis Approach

When actual patterns of fingerprints are captured, the information is represented at first by the plaintexts (3-D). After transformation of plaintext data into ciphertexts (data matrices), PCA and NCA can be conducted for decision making. PCA is a quantitatively rigorous method to generate a new set of variables from a large dimension of correlated data. The data that inherently exist throughout each analysis are represented as principal components. For red-green-blue (RGB) true color systems, each color of red, green, and blue appears as its primary spectral components in the Cartesian coordinate system. In each color subspace, the color component is mapped into a cube in which RGB values are at three corners; black lies at the origin and white lies at the corner opposite to the origin. The grayscale lies along with the diagonal line joining the black to white points. Each color is a vector on or inside the cube from the origin. The amounts of red, green, and blue components needed to form any particular color are referred to as trimulus values. The intensity component is the composite color image from three image planes. Color values from the fingerprint images are normalized so that each color represents a value between zero and one. As a result, principal components of various fingerprint patterns are analyzed in terms of each basis color element and the RGB true color. The results of the dominant six PCs are listed in Table 6.2.

Although the PCA sustains the capability to be effective for the analysis of multivariate fingerprint data, NCA has the potential to be more effective than PCA. Using nonlinear operations, principal components can be efficiently computed in the high-dimensional feature spaces, related to the input space by nonlinear mappings. NCA can employ an effective training algorithm where the data dimensionality of the feature space can be huge. The method being developed is capable of iteratively seeking for a reconstruction in the data space for any point in the original feature space. The idea of NCA is to replace the PCA covariance matrix with nonlinear mapping.

$$C = E\left\{\Phi[S(m)] \otimes \Phi\left[S^T(m)\right]\right\} \tag{6.21}$$

where Φ is a fixed nonlinear mapping to feature space with larger dimensionality than that of data space. Principal components are then computed in that feature space. A simple Hebbian rule-based NCA is proposed here whose single linear neu-

Table 6.2 Variance Percentages of PCs in True Color Systems

Variance Percent	*Red (%) Component*	*Green (%) Component*	*Blue (%) Component*	*RGB (%) Intensity*
PC 1	33.25	33.12	33.86	33.41
PC 2	3.69	3.7	3.65	3.68
PC 3	3.11	3.13	3.07	3.11
PC 4	2.92	2.93	2.89	2.91
PC 5	2.79	2.8	2.76	2.78
PC 6	2.46	2.46	2.43	2.45

tron model is expanded into a feedforward network with a single layer of the linear neutrons (Figure 6.2).

6.3.2 Independent Component Analysis Approach

Since PCA is a linear approach and most engineering problems are nonlinear, sometimes linear PCA methods are inadequate when nonlinearities are involved in the data. Considering limitations of PCA for analyzing complex and nonlinear patterns, rather than NCA, another natural extension to independent component analysis (ICA) has been developed. ICA allows modeling and interpretation of a larger matrix by individual components that are independent to each other. ICA is a statistical tool that can be utilized to perform the same sort of function that PCA possesses using different techniques for data analysis. ICA is an alternative multivariate analysis technique that has the potential to be more effective than PCA. ICA separates the latent variables (independent components) from an array of mixed variables and uses those latent variables to draw conclusions even though they cannot be directly observed. In general, ICA is solved by a three-stage algorithm, consisting of centering, whitening, and independent optimization algorithm, assuming that source data are mutually and statistically independent. It provides a computational technique for creating new data matrices by separating independent sources. It is a blind decomposition of a multidimensional dataset made of unknown components that provides a better decomposition than PCA.

In this application, independence optimization using Oja's rules is proposed. To reconstruct original independent image patterns after dimension reduction, optimization can be conducted in an iterative manner. Oja's rule has been employed and extended to matrix computation. For all independent vectors of matrix W to be trained, the following iteration has been applied, which is followed by normalization, subject to the constraint of $\|w\| = 1$. The initial matrix W_0 is selected as the random orthonormal matrix and W is an orthogonal matrix subject to training. We have

$$Y = WX = W\tilde{A}S = W(VA)S = W(ED^{-1/2}E^T)AS \qquad (6.22)$$

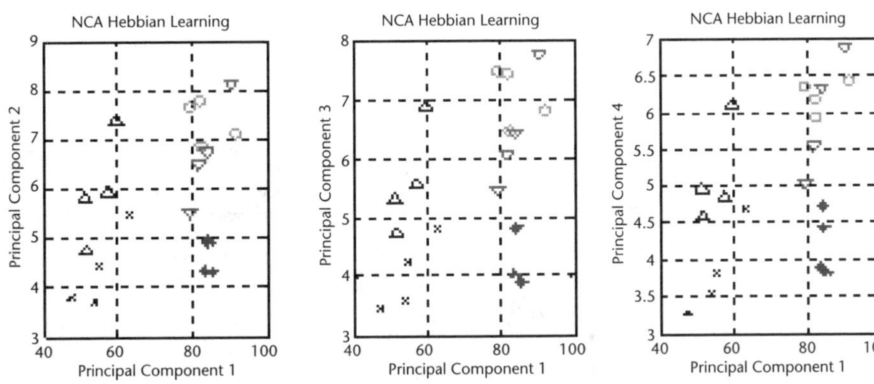

Figure 6.2 Scattering plot of true color component using NCA.

$$w^{k+1} = w^k + \rho\left(x^k - y^k w^k\right)y^k \tag{6.23}$$

Even though independent optimization is conducted in a reduced dimension, dominant independent components still indicate those intrinsic patterns. Similar to the sequence of those in PCA, ICA optimization based on all three components of the trimulus color system is made. The ICA optimization of the RGB intensity component has also been conducted. All independent patterns in four cases are captured and some dominant independent patterns are shown in Figures 6.3 and 6.4, corresponding to several independent patterns from red component, green component, blue component, and RGB intensity component of the fingerprint image, respectively.

6.3.3 2-D Discrete Wavelet Transform Approach

Biometric fingerprint verification has a variety of real-world applications that involve complex pattern recognition and signal and image-processing methodologies. The data exists inherently throughout each analysis for any particular individual sample. A simple and effective way to measure the information in a fingerprint image has been proposed. In this case, the 2-D discrete wavelet transform is presented to investigate some critical aspects of fingerprint identification via formulations at the level 1 and level 2. At the level 1 of the discrete wavelet transform, the original image is divided into four parts: approximation, horizontal detail, vertical detail, and diagonal detail, where the size of each part is reduced by a downsampling factor of two. At level 2 of the discrete wavelet transform, the approximation part at level 1 is further decomposed into another four components (Figure 6.5).

6.3.3.1 Histograms and Probability Distribution Functions

For the 2-D discrete wavelet transform, gray-level images are selected. The occurrence of the gray level is described as a co-occurrence matrix of relative frequencies. The occurrence probability function of the gray level is estimated from its histogram, which is formulated as:

$$p_i(k) = \frac{h_i(k)}{\sum h_i(k)} \tag{6.24}$$

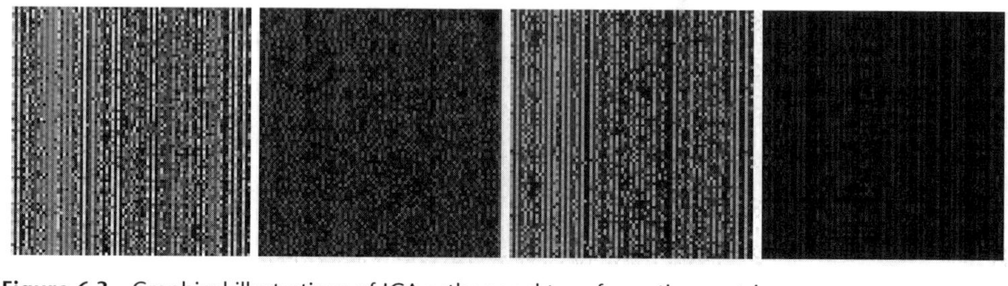

Figure 6.3 Graphical illustrations of ICA orthogonal transformation matrices.

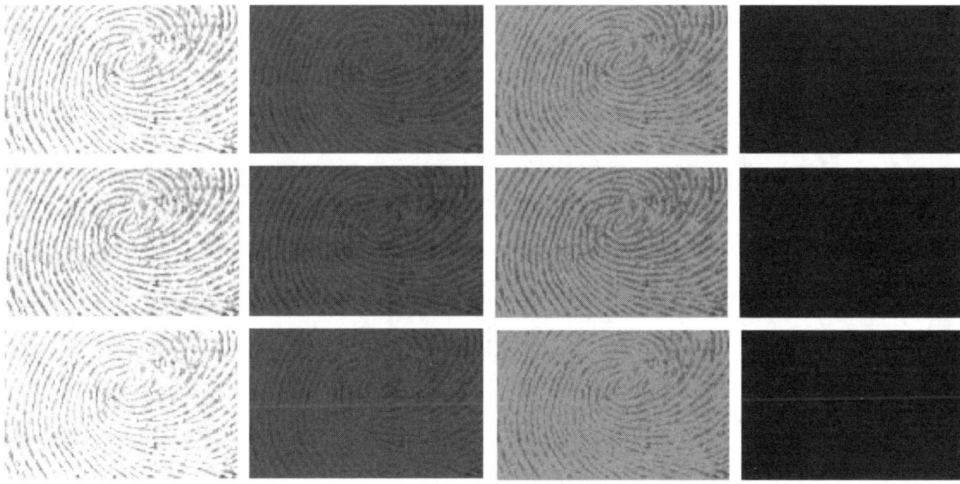

Figure 6.4 Independent components of a fingerprint pattern.

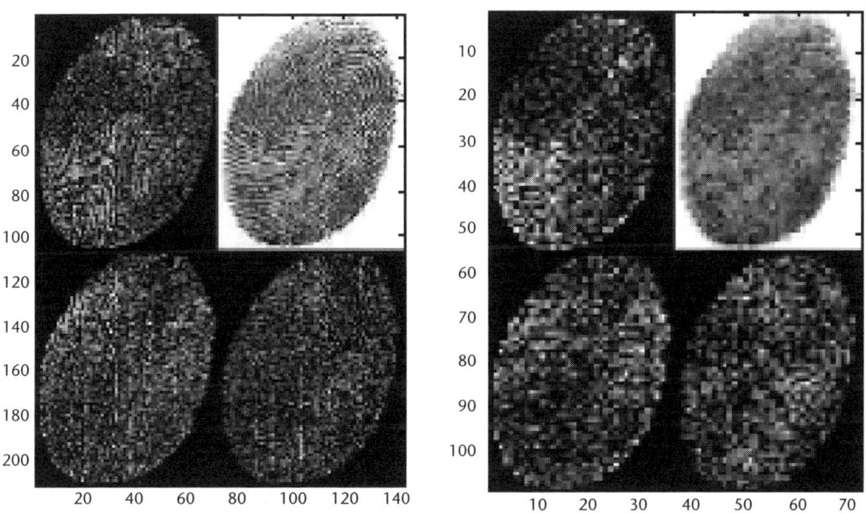

Figure 6.5 2-D discrete wavelet transform at level 1 and level 2.

where i represents the gray level, $p_i(k)$ is the probability distribution function, and $h_i(k)$ is the histogram function. The histograms of approximations and details (at level 1 and level 2) are plotted in Figure 6.6.

6.3.3.2 Wavelet Transform Gray-Level Energy

Using the 2-D discrete wavelet transform, approximations and details at each level have been obtained. The next step is to compute quantity measures of these images. The gray-level energy measure indicates how gray levels are distributed.

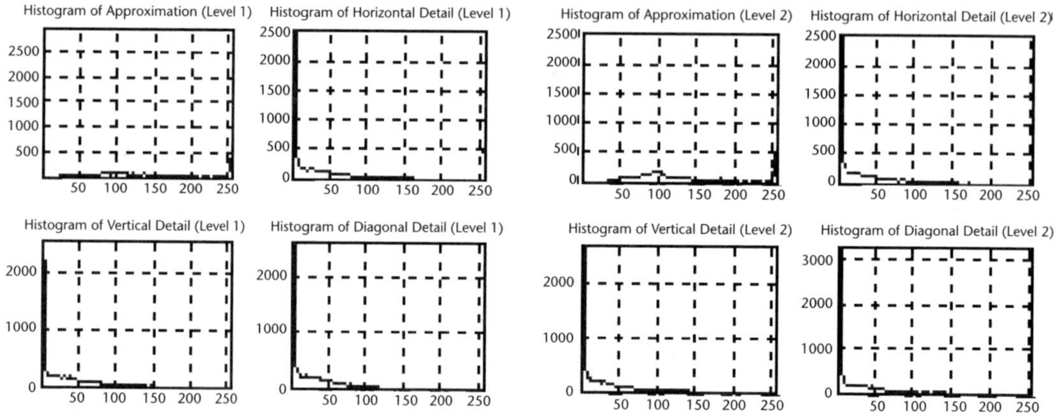

Figure 6.6 Histogram plot of wavelet transform at levels 1 and 2.

$$E(x) = \sum_{i=1}^{k} [p(i)]^2 \tag{6.25}$$

In its formulation, $E(x)$ represents the gray-level energy with 256 bins and $p(i)$ refers to the probability distribution functions of different gray levels, which contain the histogram counts. For any constant value of the gray level, the energy measure reaches its maximum value of 1. The larger energy corresponds to a lower number of gray levels and the smaller one corresponds to a higher number.

6.3.3.3 Wavelet Transform Discrete Entropy

Entropy is the measure of the image information content, which can be interpreted as the average uncertainty of the information source. Discrete entropy is the summation of the products of the probability of the outcome multiplied by the log of the inverse of probability of the outcome, taking into considerations of all possible outcomes $\{1, 2, \ldots, n\}$ as the gray level in the event $\{x_1, x_2, \ldots, x_n\}$, where $p(i)$ is the probability at the gray level of i, which contains all the histogram counts. Discrete entropy of approximations and details at each level of the 2-D discrete wavelet transform is

$$H(x) = \sum_{i=1}^{k} p(i)\log_2 \frac{1}{p(i)} = -\sum_{i=1}^{k} p(i)\log_2 p(i) \tag{6.26}$$

Discrete entropy is a statistical measure of randomness. Maximal entropy occurs when all potential outcomes are equal. When the outcome is a certainty, minimal entropy occurs that is equal to zero. A relatively complex image has higher entropy than a relatively simple image. When the pixels in the image are distributed among more gray levels, the values of the corresponding discrete entropy increase. In addition, the measure of proximity between probability density functions of original and decomposed images is described as the relative entropy. The discrete entropy provides a numerical measure between 0 and $\log_2(256) = 8$ bits, from most

Table 6.3 Gray-Level Energy of 2-D Discrete Wavelet Transform

Level One	*Energy*	*Level Two*	*Energy*
Original	0.0489	Reconstruction	0.0406
Approximation	0.0268	Approximation	0.0227
Detail (H)	0.0254	Detail (H)	0.0257
Detail (V)	0.0266	Detail (V)	0.0275
Detail (D)	0.0286	Detail (D)	0.0335

Table 6.4 Entropy of 2-D Discrete Wavelet Transforme

Level One	*Entropy*	*Level Two*	*Entropy*
Original	6.7320	Reconstruction	6.8837
Approximation	7.1930	Approximation	6.9906
Detail (H)	6.6687	Detail (H)	6.5667
Detail (V)	6.6270	Detail (V)	6.5690
Detail (D)	6.2993	Detail (D)	6.4461

informative case to the totally uninformative case. All quantities of the discrete entropy are within a range between 0 and 8, the latter of which is the maximum entropy possible.

6.3.3.4 Wavelet Transform Relative Entropy

Given the discrete probability distributions of two images with the probability functions of p and q. Relative entropy of p with respect to q is then defined as the summation of all possible system states, which is formulated as

$$d = \sum_{i=1}^{k} p(i) \log_2 \frac{p(i)}{q(i)} \tag{6.27}$$

The effect of the 2-D discrete wavelet transform can be quantified by the measure of the relative entropy. In Table 6.5, the relative entropy of decomposed images with respect to original fingerprint images is given.

Table 6.5 Relative Entropy of 2-D Discrete Wavelet Transform

Level 1 with Respect to Raw Image	*Relative Entropy*	*Level 2 with Respect to aw Image*	*Relative Entropy*
Approximation	0.2038	Approximation	0.2801
Detail (horizontal)	3.9457	Detail (horizontal)	3.9547
Detail (vertical)	3.9739	Detail (vertical)	4.1227
Detail (diagonal)			

6.3.3.5 Wavelet Transform Mutual Information

The concept of mutual information $I(X; Y)$ can used to describe how much information one variable tells about the other, which measures the dependence between the two. Its simple relation is formulated as:

$$I(X; Y) = H(X) + H(Y) - H(X, Y) = H(X) - H(X|Y) \tag{6.28}$$

where $I(X, Y)$, $H(X)$ or $H(Y)$, $H(X,Y)$, and $H(X|Y)$ are the mutual information, entropy of X or Y, joint entropy, and conditional entropy, respectively. The discrete formulation of the mutual information can also be described as

$$I(X;Y) = \sum p_{XY}(x,y)\log_2 \frac{p_{XY}(x,y)}{p_X(x)p_Y(y)} \tag{6.29}$$

where $p_{XY}(x, y)$, $p_X(x)$, and $p_Y(y)$ are the joint probability distribution function and probability distributions of x and y.

Essentially, it can be explained as the information that Y can tell about X is the reduction in uncertainty of X due to the existence of Y. It also shows the relative entropy between the joint distribution and product distribution.

6.3.4 Intramodality Area-Based Registration and Fusion

Original retinal images are taken using modified Topcon TRC-50EX Fundus camera systems over a 400- to 720-nm wavelength range (Figure 6.7). The CCD camera is located in the Polaroid mounting part. A 7-nm bandwidth filter is placed in the path of the illumination. Light source is from an internal flash tungsten halo-

Table 6.6 Mutual Information of 2-D Discrete Wavelet Transform

Raw Image with Respect to Approximation (Level 1)	*Raw Image with Respect to Approximation (Level 2)*
I(X; Y)	I(X; Y)
0.3093	0.1516

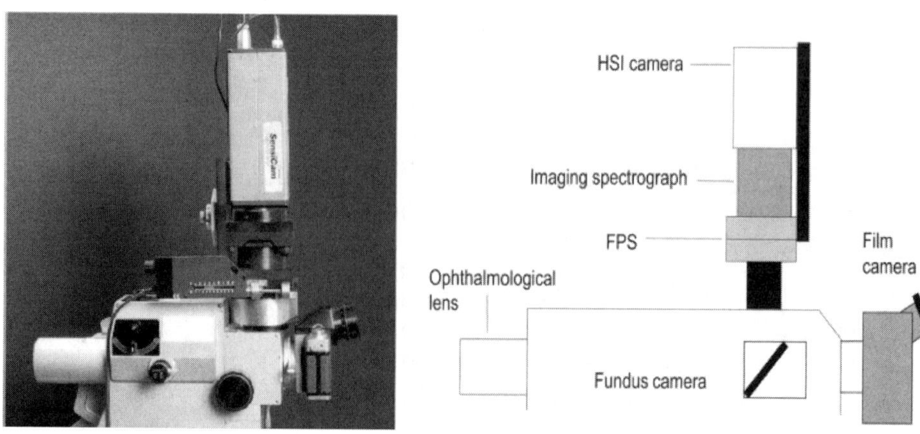

Figure 6.7 Hyperspectral imaging and fundus camera.

gen lamp. The experimental monkeys are both anesthetized with intramuscular ketamine (7–10 mg/kg), xylazine (0.6–1 mg/kg), and intravenous pentobarbital (25–30 mg/kg).

With proper registration of multispectral images, clinicians can determine oxygen saturation in the primate retinal structures on a reasonable time scale, which is used to implement the area-based 2-dimensional cross-correlation image registration. Registration of region of interest (ROI) is the method for alignment of all 2-D intramodality slices in a stack to the first slice using cross correlation. The 2-D affine model is used with only translation, rotation, and scaling being allowed. A small window of points in the input image is statistically compared with windows of the same size in the reference image. The windows surrounding the control points are matched with each other in a hierarchical manner using area cross correlation. It determines registration parameters and transforms the entire image. An operator can select an ROI over a large image so as to lower the computation cost.

The 2-D retinal image registration procedure is applied as follows: Read the reference image and then translate and rotate input images. Next, eliminate translation and allow rotation remains. Rotation of an image in a spatial domain corresponds to a rotation of its Fourier transform by the same angle. Apply the Fourier transform to 2-D cross correlation, whose maximum is the rotational difference. Rotate the input image by the calculated angle. Various moving windows in input images are manipulated in order to determine the x and y translation. The moving window location (u, v) has a normalized correlation with the reference images. The similarity between the reference template (X) and the regions in the input image (Y) are determined using this correlation. Translation is corresponding to a phase shift in the Fourier transform. Apply 2-D cross-correlation transformation on the reference image and rotated one. Find the maximum of cross correlation, which represents the translational difference. Finally the input image is completely registered with the reference image (see Figure 6.8).

6.3.5 Multimodality Feature-Based Image Registration

An appropriate multimodality registration must demonstrate reliability and robustness under a low computational cost. The feature-based registration uses vessel boundaries and strong edges as matching primitives. Retinal vasculature is extracted using a Canny edge detector and the control points are identified using the adaptive exploratory algorithm.

6.3.5.1 Image Binarization Using Discriminant Analysis Method

Image binarization is the first step, which is to provide the input data for a Canny edge detector. The selected Otsu method chooses the optimal threshold to minimize the intraclass variance of the black-and-white pixels, and to maximize the between-class variance in a grayscale image, which is nonparametric and unsupervised. Consider a discrete image with NM pixels, with retinal vessels (objects) in one class and the optic disk (background) in another class. For color images, weights of RGB are corresponding to the three parameters essential for conversion; 0.5236 for red, 0.1232 for green, and 0.3256 for blue are default weight parameters to convert

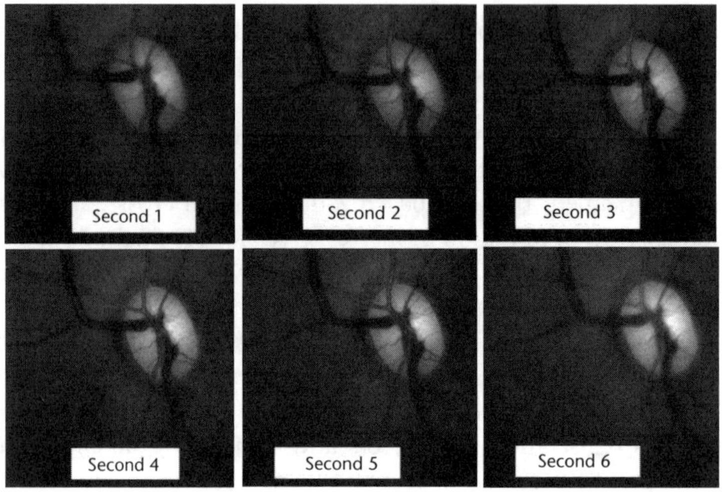

Figure 6.8 Registered input images (input image wavelength = 542, 548, 560, 569, 577, 586).

into gray level at first. Maximal entropy occurs when all potential outcomes are equal. When the color image is converted into grayscale, vessels are darker than background. The darker the vessel, the sharper the contrast. It makes vessel extraction more accurate in the next step.

6.3.5.2 Vasculature Extraction Using Canny Edge Detector

Chain code criteria are the well-known boundary representation technique. A digital curve can be represented by an integer sequence based on the position of the current edge pixel to their eight neighbors at 2-D spatial domain:

$$N_i \in \{1, 2, 3, 4, 5, 6, 7, 8\}$$

where (1-South, 2-North, 3-East, 4-West, 5-Southeast, 6-Northwest, 7-Southwest, 8- Northeast). Canny edge detector is used to extract vasculature.

Two adaptive threshold values (high threshold and low threshold) are used in Canny edge detector approach. The high threshold is used to find the start point of strong edges. Any point that meets the high threshold has to be selected as the edge point. These start points are growing into different directions until there is no edge strength falling below the low threshold.

6.3.5.3 Control Points Selection Using Adaptive Exploratory Algorithm

Control points selection is the essential step of image registration. The pixel processing approach uses matched filtering, segmentations, thinning, bifurcation identification by processing every pixel and imposing numerous operations at each pixel.

Table 6.7 Image Global Entropy for Monkey 1 and Monkey 2

Global Entropy	*Red*	*Green*	*Blue*
Color image (#1 / #2)	4.0098 / 6.9310	5.4289 / 6.9670	6.0919/ 6.5670

Overall it scales poorly with the large image size and can hardly meet short computation requirement. The exploratory algorithm is suitable for real-time live ophthalmic image processing in a large image size. The vessel tracking and tracing is efficiently performed in this adaptive exploratory algorithm without traveling at every pixel by locating an initial point and exploiting the local neighbors to trace vasculature (see Figure 6.9).

6.3.5.4 Control Point Pair Matching

The 2-D affine transformation model requires three pairs of control points $\{(x_i, x_i), (U_i, V_i)\}$ $(i = 1, 2, 3)$. The first element inside each pair is from the reference image and the second one is from the input image. First, images with less number of control points are taken as the grouping base. Suppose image I_1 has n control points, and image I_2 has m control points, and $m < n$, then m will be the group number of control points. Second, all control points in I_1 are combined with each control point in I_2, and $m \times n$ control point pairs are totally obtained. Now calculate the distance $|d|$ between each control point pair within each group using the Euclidean distance. Third, the pair with minimum $|d|$ is picked up inside each group. The assumption is to use the distance as the measurement of the control point pair is that no huge rotation, shearing, or translation occurs on two images, thus, the same features on each image are close to each other. If there are two or more control points in one image matching the same control point in the other image, the true match will always has a smaller $|d|$ than the false match. Finally, three smallest distance $|d|$ control point pairs are chosen as the final (x, y) and (u, v) pairs for the transformation model.

Hence, an automatic approach for multimodality retinal image registration is developed using the feature-based method, which is very efficient and robust to han-

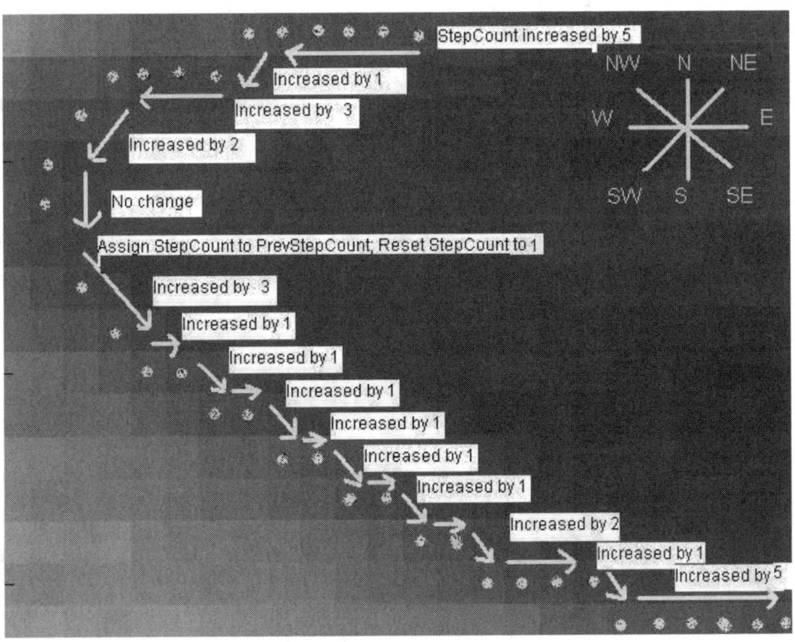

Figure 6.9 Example for step count calculation logic.

dle multisensor retinal image registration. The slowly varying rotation or translation is assumed. For a huge rotation or translation of the input image, an area-based registration method should be applied to align the image beforehand. The reason is that the cross correlation of area-based registration is designed for alignment of single-sensor images with rotation, scaling, or translation. Feature-based registration, however, is more suitable for multisensor registration that only allows small rotation, scaling, or translation. A combination of area-based and feature-based techniques within the framework of an artificial intelligence system may be necessary for the poorly deformed multisensor images. The feature-based multimodality registration algorithm presented serves as fundamental step for the hybrid area-based and feature-based systems.

6.3.6 Optimal Fusion Based on Common Pixel Count Maximization

The optimization algorithm is used to find optimal similarity measure by refining transformation parameters in an ordered way. In the iteration, control points of the reference image are fixed and that of the input image are subject to adjustment. Increase control point's x-coordinate by a step size s initially. If the common pixel count of ROI is increased due to movement, keep increasing by step size s until the number of counts stops increasing. Repeat this sequence for the y-coordinate as well. Second decrease control point's x-coordinate and y-coordinate by a step size s until the number of counts stops increasing. Via changing transformation parameters, the number of counts changes smoothly in a slowing varying mode.

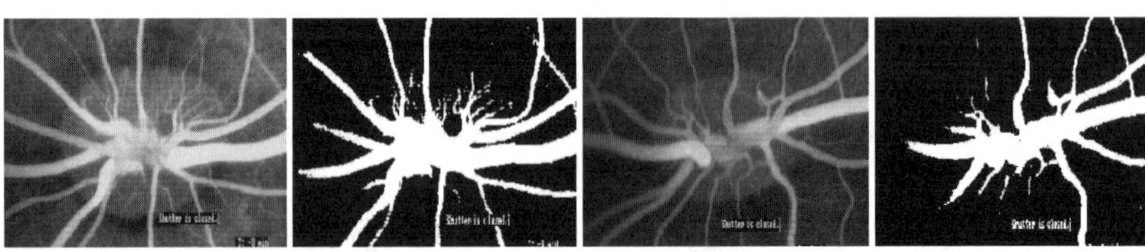

Figure 6.10 Gray level and b/w images using Otsu's threshold (monkey 1 and monkey 2).

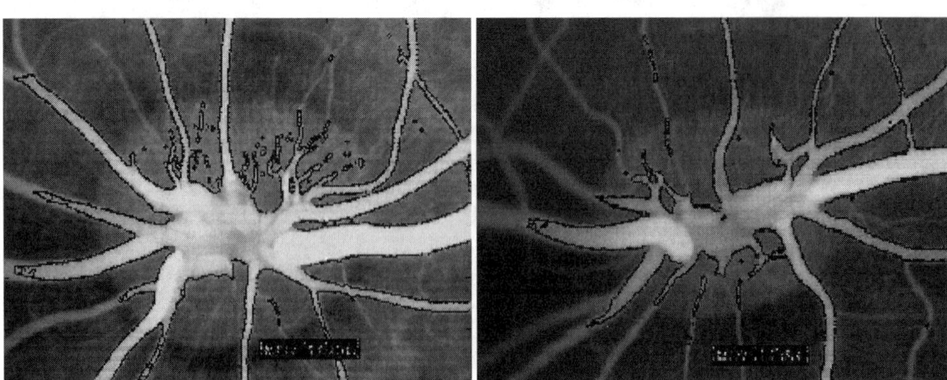

Figure 6.11 Gray-level image control point selection (monkey 1 and monkey 2).

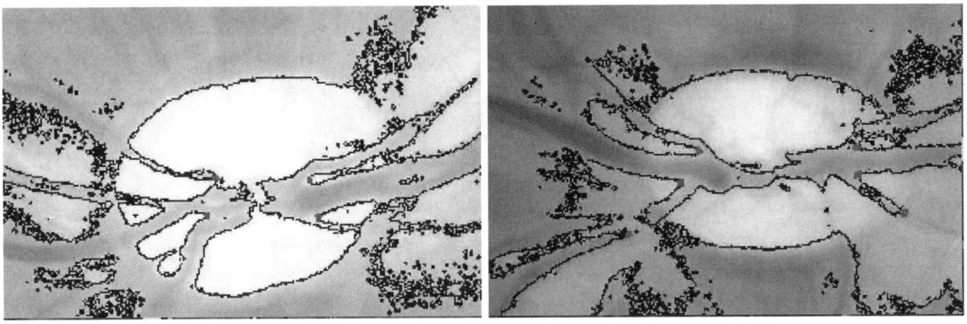

Figure 6.12 True-color image control point selection (monkey 1 and monkey 2).

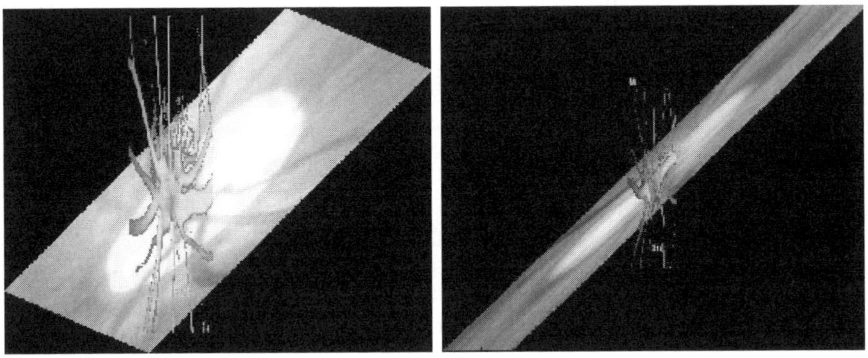

Figure 6.13 Fusion failure due to mismatched control point pairs (1 and 2).

The control point pairs from the registration step are selected. A maximum allowable loop number L is set to avoid redundant computation for mismatched control points, which leads to fusion failure. Increase the x-coordinate and y-coordinate by a step size, respectively, until the number of the common pixel counts of ROI stops increasing. Then decrease the x-coordinate and y-coordinate by a step size as well, respectively, until the number of the common pixel counts of ROI stops increasing. Repeat this sequence across the first, second, and third control points of input images. If the maximum loop number reaches L, restart iteration to avoid fusion failure. Reconstruct the fusion image based on the optimal control point coordinates through the transformation model. Once coordinates of three pair control points are available, the Gaussian matrix can be used to obtain parameters $P \in \{a_1, a_2, a_3, a_4, b_1, b_2\}$.

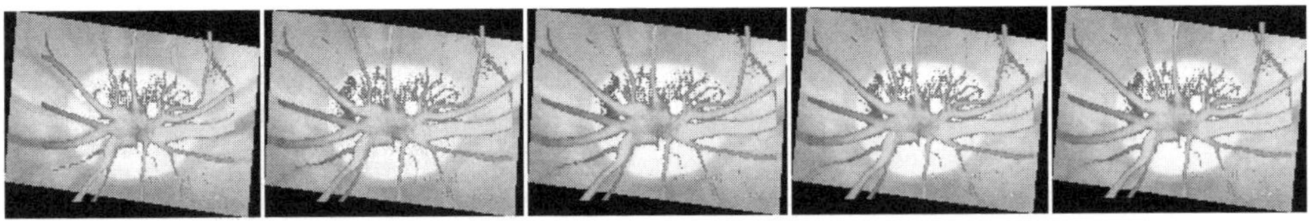

Figure 6.14 Fusion image (monkey 1). Number of counts = 5144, 7396, 7484, 7681, 7732.

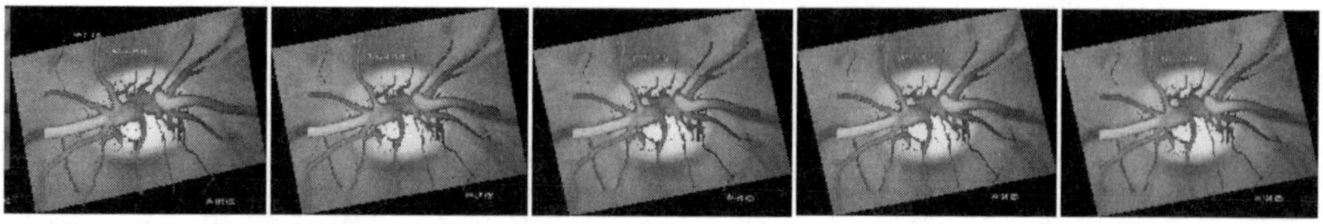

Figure 6.15 Fusion image (monkey 2). Number of counts =30732, 30888, 30914, 31134, 32254.

Via this approach, optimal parameters are determined for cases of both monkeys:

- Monkey 1: {1.0488, -0.1067, -2.0575, 0.1211, 0.9429, -37.7558};
- Monkey 2: {0.9950, 0.1925, -79.1302, -0.1669, 0.8F526, 98.7476}.

References

[1] Gonzalez, R., and R. Woods, *Digital Image Processing, 2nd Edition*, Prentice-Hall, 2002.

[2] Duda, R., P. Hart, and D. Stork, *Pattern Classification, 2nd Edition*, John Wiley & Sons, 2000.

[3] Cover, T., and J. Thomas, *Elements of Information Theory, 2nd Edition*, New York: Wiley, 2005.

[4] MacKay, D., *Information Theory, Inference and Learning Algorithms*, University of Cambridge Press, 2005.

[5] Sonka, M., *Image Processing, Analysis and Machine Vision, 2nd Edition*, Thomson and Cole Press, 1998.

[6] Hykin, S., *Neural Networks–A Comprehensive Foundation, 2nd Edition*, Prentice Hall, 1999.

[7] Hassoun, M., *Fundamentals of Artificial Neural Networks*, MIT Press, 1995.

[8] Vajda, I., *Theory of Statistical Inference and Information*, Dordrecht, Netherlands, Kluwer, 1989.

[9] Keinosuke, F., *Introduction to Statistical Pattern Recognition*, Elsevier, 1990.

[10] Ye, Y., Z. Ye, and J. Luo, "On Linear and Nonlinear Processing of Underwater, Ground, Aerial and Satellite Image," *Proceedings of the 2005 IEEE International Conference on Systems, Man and Cybernetics*, Hawaii, Oct. 10–12, 2005, pp. 3364–3368.

[11] Ye, Z., "Artificial Intelligence Approach for Biomedical Sample Characterization Using Raman Spectroscopy," *IEEE Transactions on Automation Science and Engineering*, Vol. 2, No.1, January, 2005, pp. 67–73.

[12] Ye, Z., and G. Auner, "Principal Component Analysis for Biomedical Sample Identification," *Proceedings of the 2004 IEEE International Conference on Systems, Man & Cybernetics*, Hague, Netherlands, October 10–13, 2004, pp. 1348–1353.

[13] Ye, Z., Y. Ye, and H. Mohamadian, "Biometric Identification via PCA and ICA Based Pattern Recognition," *Proceedings of the 2007 IEEE International Conference on Control and Automation*, Guangzhou, China, May 30– June 1, 2007, pp. 1600–1604.

[14] Hyvärinen, A., J. Karhunen, and E. Oja, *Independent Component Analysis,* John Wiley & Sons, 2001.

[15] Plumbly, M., "Algorithm for Nonnegative Independent Component Analysis," *IEEE Transactions on Neural Networks*, Vol. 14, No. 3, May, 2003, pp. 534–542.

[16] Jenssen, R., and T. Eltoft, "Independent Component Analysis for Texture Segmentation," *Pattern Recognition*, Vol. 36, 2003, pp. 2301–2315.

[17] Plumbly, M., and E. Oja, "Nonnegative PCA Algorithm for Independent Component Analysis," *IEEE Transactions on Neural Networks*, Vol. 15, No. 1, January, 2004, pp. 66–76.

[18] Vrabie, V., J. Mars, and J. Lacoume, "Modified Singular Value Decomposition by Means of Independent Component Analysis," *Signal Processing*, Vol. 84, No. 3, March, 2004, pp. 645–652.

[19] Ye, Z., H. Mohamadian, and Y. Ye, "Information Measures for Biometric Identification via 2D Discrete Wavelet Transform," *Proceedings of the 2007 IEEE International Conference on Automation Science and Engineering*, Scottsdale, AZ, September 22–25, 2007, pp. 835–840.

[20] Stark, J., F. Murtagh, and R. Gastaud, "A New Entropy Measure Based on the Wavelet Transform and Noise Modeling," *IEEE Transactions on Circuits and Systems*, Vol. 45, No. 6, August, 1998, pp. 1118–1124.

[21] Refregier, P., and F. Goudail, "Kullback Relative Entropy and Characterization of Partially Polarizes Optical Waves," *Journal of the Optical Society of America A*, Vol. 23, No. 3, March 2006, pp. 671–678.

[22] Wang, J., Y. Du, and C. Chang; "Relative Entropy-Based Methods for Image Thresholding," *Proceedings of 2002 IEEE International Symposium on Circuits and Systems*, Phoenix, USA, May 26-29, 2002, pp. 265–268.

[23] Fotinos, A., G. Economou, and E. Zigouris, "The Use of Entropy for Color Edge Detection," *Proceedings of the 6th IEEE International Conference on Electronics, Circuits and Systems*, Vol.1, No.1, 1999, pp. 437–40

[24] Maes, F., et al., "Multimodality Image Registration by Maximization of Mutual Information," *IEEE Transactions on Medical Imaging,* Vol. 16, 1997, pp. 187–198.

[25] Stark, J., "Adaptive Image Contrast Enhancement Using Generalizations of Histogram Equalization," *IEEE Transactions on Image Processing*, Vol. 9, No.5, May, 2000, pp. 889–896.

[26] Netsch, T., et al., "Towards Real-Time Multi-Modality 3D Medical Image Registration," *Proceedings of the Eighth International Conference on Computer Vision*, Vancouver, 2001, pp. 718–725.

[27] Chen, H., et al., "A Pyramid Approach for Multi-Modality Image Registration based on Mutual Information," *Proceedings of the 3rd International Conference on Information Fusion*, Paris, France, 2000, pp. 9–15.

[28] Matsopoulos, G., et al, "Multimodal Registration of Retinal Images using Self Organizing Maps," *IEEE Transactions on Medical Imaging,* Vol. 23, No. 12, 2004, pp. 1557–1563.

[29] Chanwimaluang, T., G. Fan, and S. Fransen, "Hybrid Retinal Image Registration," IEEE *Transactions on Information Technology in Biomedicine*, Vol. 10, No. 1, 2006, pp. 129–142.

[30] Ritter, N., et al., "Registration of Stereo and Temporal Images of the Retina," *IEEE Transactions on Medical Imaging*, Vol. 18, No. 5, 1999, pp. 404–418.

[31] Stewart, C., C. Tsai, and B. Roysam, "The Dual-Bootstrap Iterative Closest Point Algorithm with Application to Retinal Image Registration," *IEEE Transactions on Medical Imaging*, Vol. 22, No. 11, 2003, pp. 1379–1394.

[32] Laliberte, F., L. Gagnon, and Y. Sheng; "Registration and Fusion of Retinal Images: A Comparative Study," *Proceedings of the* 2002 *International Conference on Pattern Recognition*, Quebec, Canada, August 11–15, 2002, pp. 715–718.

[33] Laliberte, F. and L. Gagnon, "Registration and Fusion of Retinal Images—An Evaluation Study," *IEEE Transactions on Medical Imaging*, Vol. 22, No. 5, 2003, pp. 661–673.

[34] Noblet, V., et al., "3D Deformable Image Registration: A Topology Preservation Scheme Based on Hierarchical Deformation Models and Interval Analysis Optimization," *IEEE Transactions on Medical Imaging*, Vol. 14, No. 5, May, 2005, pp. 553–566.

[35] Antoine, J. B., et al, "An Overview of Medical Image Registration Methods," *Proceedings of the 1996 Symposium of the Belgian Hospital Physicists Association*, Vol. 12, 1996, pp. 1–22.

[36] Klein, S., M. Staring, and J. Pluim, "Evaluation of Optimisation Methods for Nonrigid Medical Image Registration using Mutual Information and B-Splines," *IEEE Transactions on Image Processing* (in press).

[37] Otsu, N., "A Threshold Selection Method from Gray-Level Histograms," *IEEE Transactions on Systems, Man & Cybernetics*; Vol. 9, 1979, pp. 62–66.

[38] Canny, J. F., "A Computational Approach to Edge Detection," *IEEE Transactions on Pattern Analysis and Machine Intelligence*, Vol. 8, Nov. 1986, pp. 679–698.

[39] Sahoo, P. K., A. A Farag, and Y. P. Yeap, "Threshold Selection Based on Histogram Modeling," *IEEE Transactions on Systems, Man and Cybernetics*, Vol. 1, Oct 1992, pp. 51–356.

[40] Can, A., et al., "Rapid Automated Tracing and Feature Extraction from Retinal Fundus Images Using Direct Exploratory Algorithms," *IEEE Transactions on Information Technology in Biomedicine*, Vol. 3, June, 1999, pp. 125–138.

[41] Chou, K. C., A. S.Willsky, and A. Benveniste, "Multiscale Recursive Estimation, Data Fusion, and Regularization," *IEEE Transactions on Automatic Control*, Vol. 39, 1994, pp. 464–478.

[42] Ma, B., S. Lakshmanan, and A. O. Hero, "Simultaneous Detection of Lane and Pavement Boundaries Using Model-Based Multisensor Fusion," *IEEE Transactions on Intelligent Transportation Systems*, Vol. 1, September 2000, pp. 135–147.

[43] Becker, D., et al., "Image Processing Algorithms for Retinal Montage Synthesis, Mapping, and Real-Time Location Determination," *IEEE Transactions on Biomedical Engineering*, Vol. 45, January, 1998, pp. 105–118.

[44] Chaudhuri, S., S. Chatterjee, and N. Katz, "Detection of Blood Vessels in Retinal Images Using Two-Dimensional Matched Filters," *IEEE Transactions on Medical Imaging*, Vol. 8, September, 1989, pp. 263–269.

CHAPTER 7

Lab-on-a-Chip Automation of Laboratory Diagnostics: Lipoprotein Subclass Separation Automation

David Deng and Mingjun Zhang

This chapter introduces automation techniques for lipoprotein subclass separation. Discussions on fundamentals, and advantages and disadvantages of different lipoprotein subclass separation automation systems are presented along with an example of electrophoresis-assisted lab-on-a-chip automation for lipoprotein subclass separation.

7.1 Introduction

Automation plays an increasingly important role in molecular diagnostics for clinical applications by saving cost, improving assay precision, and reducing operational errors. Promising areas of clinical diagnostics automation include polymerase chain reaction (PCR), quantitative PCR, enzyme-linked immunosorbant assay (ELISA), microfluidics, microarray hybridization, molecular diagnostics (MDx), microchip liquid chromatography, and mass spectrometer (Chip-LC/MS). This chapter reviews existing techniques for lipoprotein subclass separation, and presents an electrophoresis assisted lab-on-a-chip technology for lipoprotein subclass separation automation.

Lab-on-a-chip enables rapid, complex biological sample analysis at a small scale. A lab-on-a-chip system usually can complete quantitative and multianalytes biochemical assays in a compact, easy-to-use platform. For example, microfluidic chips developed for inorganic device processing combined with synthetic chemistry and biochemistry integrate electrical, optical, and physical measurements with fluid handling to create a new class of functional chip [1]. The development may lead to ultrasensitive sensors and medical diagnostic systems for rapid molecular diagnosis.

One critical issue to automate the lab-on-a-chip system is multiscale system integration. Automation for life science instrumentation systems may require new standards and procedures.

In this chapter, we will first review current techniques for lipoprotein subclass classification, and present an automated system for lipoprotein subclass assays to demonstrate how the lab-on-a-chip technology may be integrated with liquid handling systems and software automation for high throughput molecular diagnostics. Experimental results are presented to show the effectiveness of the platform.

7.2 Lipoprotein Subclass Assay

7.2.1 Classifications and Functions of Serum Lipoproteins

Cholesterol is an essential component of the cell membranes and myelin sheaths. It also is a precursor to bile acids and steroid hormones. Most of the cholesterol in the body is not obtained directly from the diet, but rather is synthesized by chemical reaction controlled by proteins. Contrast to the usual thinking, the "good" and "bad" blood cholesterol levels do not refer to different types of cholesterol molecules. In fact, the lipids transport vehicles including lipoprotein particles required for blood stream cholesterol, triglycerides, and other lipids, determine the "good" or "bad" nature of cholesterol.

The five major classes of lipoprotein particles ranging in size from 10 to 1,000 nm in an increasing order are chylomicron (CM), very low density lipoprotein (VLDL), intermediate density lipoprotein (IDL), low density lipoprotein (LDL), and high density lipoprotein (HDL). CM is the largest and the least dense of the lipoproteins. It is produced for transporting dietary triglycerides and cholesterol from intestinal epithelia to liver for further processing and transforming to other lipids. VLDL is smaller and less dense than CM. VLDL assembled in liver contains several types of apolipoproteins (apo) including apo-B100 and apo-C. IDL contains the similar apolipoproteins as VLDL, but smaller and denser than VLDL. IDL is derived from triglyceride depletion of VLDL. The fate of IDLs in the body is either taken up by the liver for reprocessing or become LDL in the blood flow.

The LDLs are smaller and denser than IDLs, VLDLs, and CMs. One LDL particle contains one molecule of apolipoprotein apo-B100. LDL is the main transporter of cholesterol and cholesteryl esters. It makes up more than half of the total lipoprotein in plasma. The clearance of LDL from serum is absorbed by the liver and other tissues via receptor mediated endocytosis. Another lipoprotein containing apo-B100 is lipoprotein (a) [Lp(a)]. Lp(a) is functionally different from LDL. It contains a protein called apolipoprotein(a) [apo-(a), pronounced as "apo little a"], which covalently bound to apo-B. Apo-(a) has been found to have a high sequence homology with plasminogen. Its function is believed to be related to triglyceride metabolism and possibly thrombotic as well as atherogenic pathways.

HDLs are the smallest and densest of all lipoproteins. HDL contains several types of apolipoproteins including apo-AI, II & IV, apo-CI, II & III, apo-D, and apo-E. HDL is produced as a protein-rich particle in the liver and intestine. The HDL protein particle accumulates cholesteryl esters by the esterification of cholesterol by lecithin-cholesterol acyl-transferase (LCAT). LCAT is activated by apo-AI on HDL. HDL can return to the liver where cholesterol is removed by reverse cholesterol transport, thus serving as a scavenger to free cholesterol. The liver can then excrete excess cholesterol in the form of bile acids.

Normally, the good or bad cholesterol terminology refers only to HDL and LDL, respectively. HDL and LDL particles are heterogeneous, and differ in their contents of proteins and lipids, and thus different in clinical significance. In general, the higher the ratio of protein to lipid content the higher the density; similarly, the higher the density of a lipoprotein particle the smaller its size and molecular weight.

As shown in the left of Figure 7.1, HDL and LDL are particles consisting of phospholipid bilayer membrane and a lipid core containing triglycerides, cholesterol, and cholesteryl esters. The main difference between HDL and LDL are lipid contents and apolipoproteins (apo) embedded in the phospholipid membrane. In HDL, apo-A is the main lipoprotein, while LDL contains apo-B100. According to electrophoresis and ultracentrifugation, HDL and LDL each can be further separated into several subclasses. These subclasses have different particle size, density, lipid contents, and apolipoproteins, and thus may be of different atherogneicity. For example, HDL2B, the largest HDL subclass, is believed to be athereo-protective. However, small and dense LDL particles (sd-LDL, Type B) containing high triglycerides larger LDL, are more atherogenic. The atherogeneicity of sd-LDL may be partially attributed to the longer half life in circulating blood flow, easy penetrating to vascular wall due to the smaller size and high affinity to vascular LDL receptors, and susceptibility to chemical modification such as oxidization.

Low HDL (<35 mg/dL), may be associated with increased risk of coronary atherosclerosis, conversely high levels of HDL (>55 mg/dL) appear to have a protective affect. High LDL levels have been shown to correlate with coronary atherosclerosis as discussed below.

7.2.2 Cardiovascular Diseases Diagnosis

An important factor in much of the incidence of heart diseases is high blood cholesterol levels. If there is too much cholesterol in the bloodstream, it can deposit on the walls of arteries and restrict the flow of blood to the heart. Recent studies suggest that abnormal lipid metabolism contributes to the pathogenesis of coronary artery disease (CAD) [2]. For example, elevated serum levels of LDL, low levels of HDL, and high fasting or postprandial levels of TGs, are predictive for CAD. Despite the importance of these lipoproteins in the development of CAD, the extent of cardiovascular events and CAD risk may vary among individuals with similar levels of lipids and other risk factors. For example, patients who had normal total HDL level or total LDL level may still risk heart attack. In fact, more than half of patients with

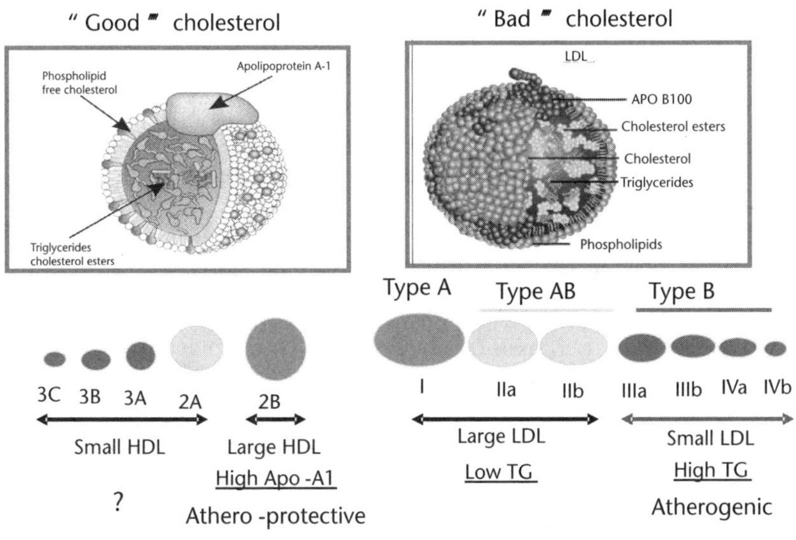

Figure 7.1 Schematic representation of HDL and LDL particles and their heterogeneicity.

CAD events, such as myocardial infarction, unstable angina pectoris, have a normal total cholesterol level.

It is important to understand the variability of influence of lipid metabolism in cardiovascular diseases. Some variability may be due to the heterogeneity within each of the three major lipoprotein classes: HDL, LDL, and VLDL. Each of these lipoproteins is composed of various subclasses that differ in particle diameter, density, lipid composition, and possibly atherogeneicity as discussed above [3]. For example, HDL is believed to be good cholesterol and protects the heart from heart attacks. It mediates reversed cholesterol transportation and inhibits inflammation as well as oxidative stress in the vascular wall. However, HDL contains heterogeneous subclasses, namely larger buoyant HDL2 and smaller HDL3. Accumulative evidence suggests that the larger HDL2 provides more protective effects than the smaller HDL3. Similarly, LDL also contains several subclasses by ultracentrifugation and gradient gel electrophoresis. Plasma LDL is identified to have at least two LDL phenotypes: phenotype A characterized by a predominance of large buoyant LDL (particle size > 25.5 nm), and phenotype B characterized by small dense particles (particle size < 25.5 nm). LDL phenotype B is significantly associated with the increased risk of atherogenic coronary artery disease [4]. Thus further detemination of HDL subclasses and LDL subclasses will improve the assessment of cardiovascular risk and provide information for doctors to design the regime of early patient treatment. Unfortunately, current methods for separation of HDL subclasses and LDL subclasses are either time-consuming and/or expensive. A cost-effective automated system for HDL subclasses and LDL subclass analyses is highly expected for clinical applications.

7.2.3 Current Technologies of Lipoprotein Subclass Assay

Technologies currently used in research and clinical labs to analyze HDL subclasses and LDL subclasses include analytical and density gradient ultracentrifugation (ADGUC), gradient gel electrophoresis (GGE), tube gel electrophoresis (TGE), proton nuclear magnetic resonance spectroscopy (NMR), high performance liquid chromatography (HPLC), and vertical auto profile (VAP) [5].

Analytical ultracentrifugation separates lipoprotein particles based on the density of the particles, and is considered to be the benchmark for determining lipoprotein subfractions [10]. However, technical challenge and labor-intensive procedure of the ADGUC has limited its use mainly in research laboratory.

7.2.3.1 Gradient Gel Electrophoresis

Gradient gel electrophoresis (GGE) on nondenaturing polyacrylamide gels is one of the most widely used methods in clinical labs for plasma and serum LDL and HDL subclass analysis. Literature from various laboratories supports the clinical utility of this procedure [9]. The GGE method is robust, accommodating samples with different levels of cholesterol and triglyceride, and requires less than 100 μL of sample per test.

In general, LDL subclass analysis by GGE is reported as both peak and average particle diameter, in addition to percentage of each fraction (large, intermediate, small, and very small). HDL particles are reported as five subfractions: 2a, 2b, 3a, 3b, and 3c. The concentration of lipid components in each subfraction can be estimated from total LDL or total HDL lipid concentrations. It is believed that HDL2b, the largest HDL subclass, is the most clinically relevant to patients' outcome, while the clinical significance of other HDL subfractions by GGE is unknown.

GGE data analysis is usually based on a direct scan of the electrophoretic pattern from each sample, and determines peak LDL particle diameter and a semiquantitative estimate of the actual distribution of LDL particle sizes.

7.2.3.2 Nuclear Magnetic Resonance

The basic principle of the nuclear magnetic resonance (NMR) technique for lipoprotein subfraction analysis is that each molecule subjected to magnetic field will give a unique signal. A library of signals unique to various human proteins can be created by subjecting plasma with specific proteins of known composition to magnetic resonance imaging (MRI), such as plasma containing known lipoproteins. Quantification is achieved in a three-step process: First, to measure the plasma NMR spectrum, then a computer deconvolutes the spectral data and calculates the subclass concentration.

Lipoprotein particle profiling by NMR has the advantage of providing information about the full spectrum of lipoprotein particles in a single analysis sample. In addition, it is fast and well standardized. However, unlike lipoprotein profiling by GGE, NMR lipoprotein analysis is based on an algorithm generated from a limited number of untreated normal samples. This can be problematic, because it is known that the fatty acid composition of lipoproteins significantly affects the lipoprotein NMR signal [11]. This effect may be further accentuated in specimens with elevated triglyceride. Also, NMR analysis does not account for the contribution of apolipoproteins to lipoprotein particle size. Thus, lipoprotein particles containing the apolipoprotein (a) [Lp(a)] are not distinguished from normal LDL particles.

7.2.3.3 Vertical Auto Profile (VAP)

Another approach to determine lipoprotein subclasses is the vertical auto profile (VAP) method. This method is based on the use of a single-spin, vertical density-gradient ultracentrifugation, and an associated analyzer that measures the cholesterol of the resulting serum fractions. Using a computer program designed to deconvolute the resulting cholesterol profile into lipoprotein, subclasses are developed. The method has the advantage of giving information about VLDL, IDL, Lp(a), LDL, and HDL subclasses from a single analysis.

However, for LDL, this procedure has only a moderate correlation with other methods [12]. As is the case for NMR, the method is available only in one laboratory and therefore has not been independently validated.

7.2.3.4 High Performance Liquid Chromatography

A novel method for assessing lipoprotein subclasses is size-exclusion high performance liquid chromatography (HPLC). This method has been used to determine the changes in LDL and HDL particle sizes of diabetic patients during bezafibrate and pravastatin treatment. It can also be used to separate classes of lipoprotein remnants isolated by immunoseparation techniques into subclasses. The basis for this technique is the use of special columns (Lipopropak from Tosoh, or Superose from Amersham Biosciences), and online methods for measuring column effluent cholesterol and triglyceride.

Currently, the widely used methods in clinical laboratories HDL and LDL subclasses are based on either electrophoresis, such as GGE and TGE, or NMR. Ultracentrifugation is not widely accepted in clinical labs due to the technical challenge and laborious nature of ultracentrifugation. Using GGE is also time-consuming, and not automatable, which may take about one week to run a single round of assay. This lengthy and laborious process of GGE has limited its applications in clinical diagnostic laboratories. Furthermore, the above methods all lack accuracy and consistency. It has been reported that complete agreement in the direct comparison among the four methods (GGE, TGE, NMR, and VAP) on the same set of samples is about 8% (3 out of 40) [5]. Therefore, reproducible, sensitive, and automatable method for the rapid separation of lipoprotein subclasses is needed for clinical laboratories.

7.3 Lab-on-a-Chip Bio-Analyzer for HDL and LDL Subclass Separation

Lab-on-a-chip is a MEMS technology to integrate multiple analysis on a single chip with significantly reduced size (from centimeter to millimeter) using techniques like electrophoresis, chromatography, MS detection, and with a higher degree of automation integrating multiple steps of a complex protocol into a miniaturized system. This brings to the technology the advantage of using extremely small sample volumes (even in picoliter) for assays, and makes virtually everything that is done in a laboratory able to be done on a chip.

Microfluidics is often associated with lab-on-a-chip technology, primarily due to the microfluidic status of the sample involved in the assay. To speed up sample movement in microchannels, external physical methods, such as electrophoresis, may be applied. Electrophoresis can be applied to electrically move the charged proteins under the influence of the electric field, due to the Coulomb force. For protein separation, most proteins are charged when the buffer pH for the interested protein(s) is different from the isoelectric point of corresponding protein molecules.

The integrated Agilent BioAnalyzer system consists of a microfluidic chip, the BioAnalyzer for controlling chip running, and computer software for data processing automation (Figure 7.1). The system integrates an ultrasensitive and flexible CCD camera for fluorescence and chemiluminescence imaging. By integrating lab-on-a-chip with BioAnalyzer, we can have electrophoresis assisted lab-on-a-chip technology for sensitive molecular detection. A wide variety of applications of the system has been reported to analyze DNA, RNA, proteins, and stained cell count [6].

As shown in Figure 7.2, the microfluidic chip used in the Agilent BioAnalyzer system contains 16 wells. The chip accommodates 12 sample wells, 3 gel wells, and a well for a standard (ladder). Each well is connected to a separation channels via

Figure 7.2 Agilent BioAnalyzer system.

microchannels carved on a piece of glass. The microchannels connecting all chip wells are conversed to a point in a short distance to the separation channel in the middle that separates the chip wells on two sides. The detection point is located at the end of the separation channel. When the chip is placed on the BioAnalyzer, 16 pin-electrodes in the electrode cartridge (standard equipment) in the BioAnalyzer are arranged such that they fit in the wells on the chip.

The Agilent BioAnlayzer microchip system includes the BioAnalyzer, microchip, and computer software system for data processing automation.

We use the lab-on-a-chip system shown in Figure 7.3 for lipoprotein separations by electrophoresis. All microchannels of the glass chip are filled with a sieving polymer gel matrix. Samples containing interested analytes (i.e., HDL and/or LDL particles) are diluted in sample buffers and stained with specific fluorescent dyes.

When in operation, the diluted and fluorescence prestained samples are loaded onto sample wells on the chip. The chip is then loaded onto the BioAnalyzer, which controls the voltage of electrical nodes, and measures the abundance of fluorescent-labeled samples. Each well is connected through electrodes in the BioAnalyzer and the buffer/gel matrix filled in the microchannels. The sample analytes (negative charged in optimized buffer conditions) are electro-driven from sample wells through the microchannels to the nearby injection points. By switching the current direction, the sample is electrokinetically injected into the separation channel and moved toward detection point, where the fluorescence signals in the samples are detected. In this way, the sample analytes are electrophoretically separated according to the mass/charge ratio of the analytes. The smaller the analyte particle size and the mass/charge ratio, the faster the analyte runs toward the detector. By continuous monitoring the fluorescent-labeled analyte particles, an electropherogram of analyte particle movements is generated for subclass (subfraction) analysis and quantitation of total analyte concentration. The absolute concentration can be calculated by referring to outputs of the calibrator run on the same chip. The fluores-

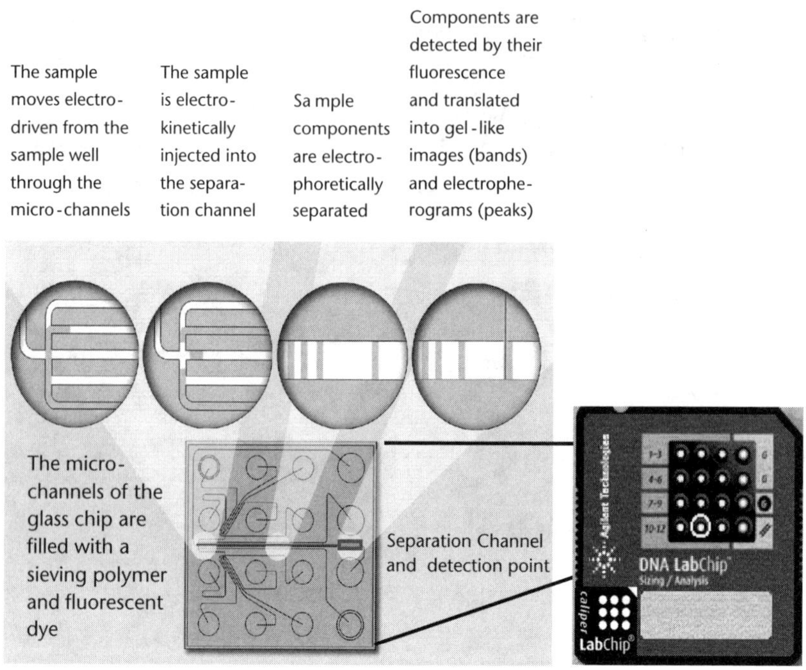

Figure 7.3 The electrophoresis assisted lab-on-a-chip. The microchip (DNA chip as an example) is on the right side. The schematic graph of the chip is on the left, with circles representing each well, and lines indicating microchannels carved on the grass by PDMS Soft Lithography. The separation channel and detection point are highlighted in red.

cence is also translated into gel-like images, which most molecular biologists are familiar with.

7.4 Automated Lipoprotein Subclass Separation

Robustness, high throughput, and low cost are fundamental requirements for an automated system. In addition, an automated process should be simple, concise, and easily manageable. The above electrophoresis assisted lipoprotein subclass separation process can be divided into three steps for automation. As shown in Figure 7.3, the microchannels are first filled with sieving gel matrix by pressure (priming). Then the samples that have been diluted and stained in sample buffer are added onto wells of the chip. The last step is to load the chip into the BioAnalyzer and run the chip to obtain detection results. The detailed experiment procedure is as follows:

- Serum samples are directly mixed with sample buffer containing fluorescent dyes that specifically stain lipids and lipoproteins. The proper detergents were included in the buffer to suppress the analyte-wall interactions. This takes about 10 to 15 minutes.
- The microchip is then primed with separation polymer gel matrix. This process will fill all channels with gel matrix for the separation.

- Samples are loaded onto the 12 wells and the ladder well for calibration. The ladder well in the HDL subfraction assay is used as control well and loaded with control serum (calibrator), which is a known HDL concentration and used as the reference for all serum samples within a chip (intrachip) or among chips (interchips) comparison.
- The chip is run on a BioAnalyzer by applying a preset electrocurrent that determines the migration speed and pattern of the analytes on channels for the sample drawing, injection, and separation.
- Finally, the HDL subclass result can be digitally displayed as distinct peaks and deconvoluted subpeaks in the electropherogram as well as gel-like images.

To automate the above process, we have integrated a robotic liquid handling system and data processing software with the lab-on-a-chip technique. As shown in Figure 7.4, the integrated system includes a computer controlled sample loading station, a BioAnalyzer, and a computer for data processing.

The automated workflow can be described as follows: first, the serum sample tubes, sample dilution buffer tube, and sieving gel matrix tube are placed in the corresponding positions of the sample rack. The Tecan robotic arm then adds serum samples and a sample buffer into a 96-well plate for dilution, fluorescent staining, and mixing. The microchip is primed by the gel matrix. After dilution, the samples are loaded onto corresponding wells as shown in Figure 7.5. The loaded microchip is then moved to the BioAnalyzer by the right robotic arm. The microchip-run procedure will be automatically initiated and electrical voltage is applied to the electrical node at each well. The raw fluorescent data detected by the BioAnalyzer fluorescence reader is sent to a computer for analysis. After the BioAnalyzer finishes the above procedure, the robot moves the microchip out of the BioAnalyzer and replaces it with an electrode cleaning chip containing cleaning solution as shown in Figure 7.6. Water can be used as cleaning solution for the chip to avoid cross contamination. The microchip scanning data from the BioAnalyzer is finally sent to a data processing computer for separating subclass from the lipoprotein. To improve the throughput, the sample rack and the BioAnalzer can start immediately for the next chip without waiting for completion of the current run.

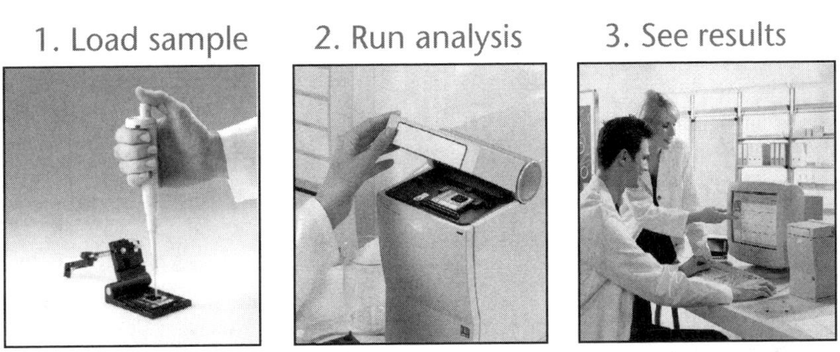

Figure 7.4 Three steps of the lipoprotein subclass separation. First, the microchannels are filled with sieving matrix, then, samples are stained with fluorescent dyes and added to sample wells to be electronically driven from the sample wells through microchannels, and finally, the samples are electrokinetically injected into separation channels and separated by their mass and /or charges.

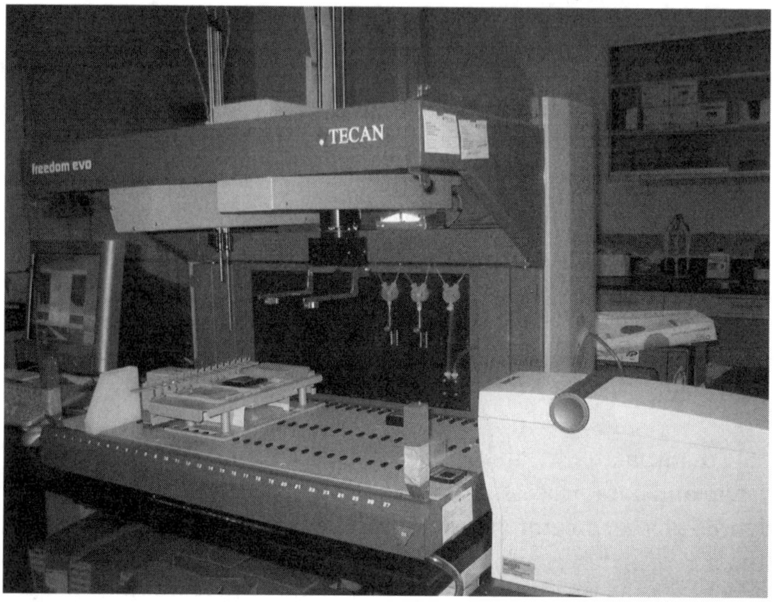

Figure 7.5 The automated electrophoresis assisted lab-on-a-chip system. The system is automated by a Tecan Freedom EVO system. The system consists of a control computer, a sample rack, and is connected with Agilent's BioAnalyzer. The workstation includes two 4-degree-of-freedom robot arms, each of which is an end-effector that can pick sample bottles or the mcirofluidic chip and move around the sample rack.

Figure 7.6 Sample rack to load and move sample around. The far left-hand side shows a 96-well plate of the original samples, the microchip, the sample picking-up tip, and tip washing stations. The right-hand side is for microchip cleaning.

7.4.1 Robotic Liquid Handling for Sample Preparation

As shown in Figures 7.5, 7.6, and 7.7, the robotic liquid handling system include two 4-degree-of-freedom robot arms and a flexible sample rack. The left-hand side of the sample station is used to host the sample plate, microchip channel cleaning station, and microchip. The right-hand side of the rack is occupied by microchip washing station. All the moving parts are along the top hood. The left robotic arm handler has a long tip to take and distribute samples among various channel wells of the chip. The right arm is to open the BioAnalyzer and move the microfluidic chip in and out of the analyzer. The station is controlled by a motion controller operated by a personal computer.

7.4.2 Software Automation

Vast information needs to be processed and synchronized during the sample preparation, scanning, acquisition, and data interpretation. It is important to automate these processes. The following efforts have been made to automate the software components.

Software for microchip lipoprotein subclass separation includes several panels for instrument control, sample information, data display, and data output (Figure 7.8). The software user interface (UI) include several panels. From left to right, columns are for the function selection, display of selected the function, and functional panels (Figure 7.9).

7.4.2.1 Instrument Control

Instrument control UI (Figure 7.9). The upper panel indicates the firmware version, assay class selection, instrument running script used, and so forth. The lower panel

Figure 7.7 Chip loaded to the BioAnalyzer.

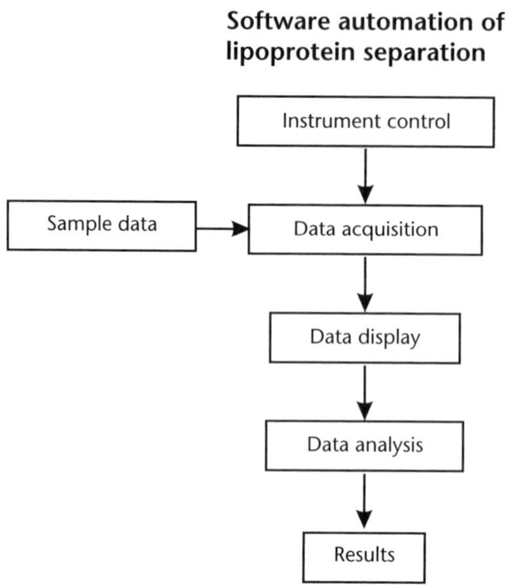

Figure 7.8 Flow chart of software automation of lipoprotein separation.

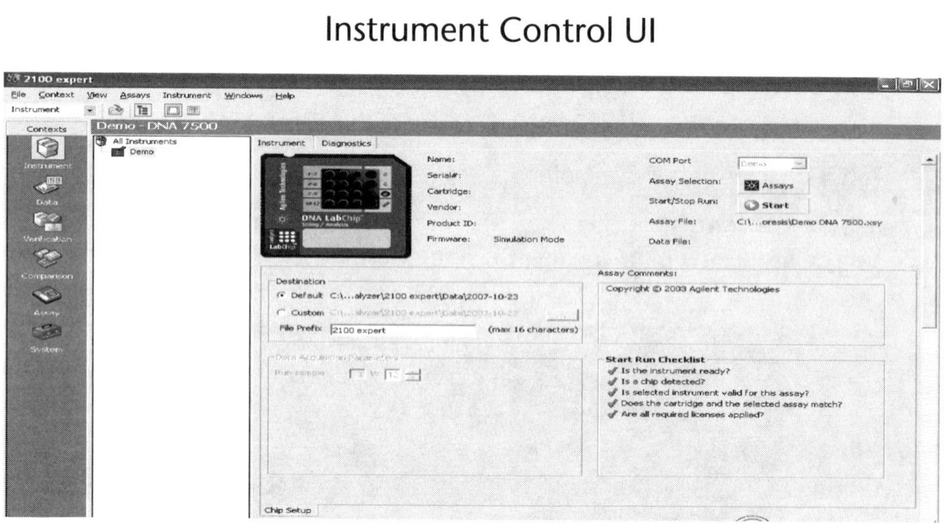

Figure 7.9 The instrument control UI.

indicates the location of assay file and the instrument status. The software will do a self-check to see if the instrument is connected, chip detected, and whether the selected cartridge and assays are matched, and so forth.

7.4.2.2 Sample Information UI

Figure 7.10 shows the user interface (UI) for the microfluidic chip and sample information. The left-hand side provides a browser for different sample annotation in the

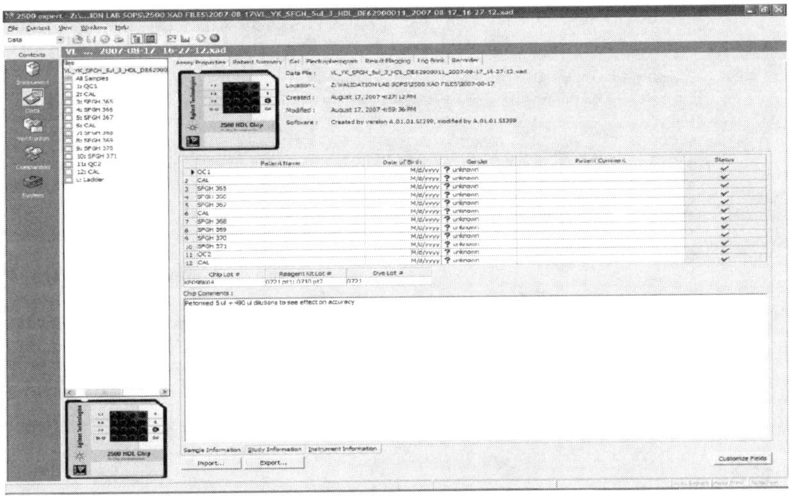

Figure 7.10 Sample information UI.

chip. The table is used for sample information. The check on green is an indication if the sample has been run successfully. The two blanks below the sample information table are for entering chip and reagent lot numbers. The user can also enter assay notes on the lower blank panel.

7.4.2.3 Data and Result Display UI

The assay data and results can be displayed in different ways. The display panels can display samples as an individual sample electropherogram together with a gel-like image, and as a summary of the assay results, such as total HDL concentration, subfractionated HDL as absolute amount and relative percentage of total HDL (Figure 7.11(a)), all sample electropherograms for overview (Figure 7.11(b)), a gel-like image for all samples (Figure 7.11(c)), and overlaid sample electropherograms for sample comparisons within chip (Figure 7.1(d)). Chip-to-chip sample comparison can be done by selection of chip comparison function on the far left panels

Figure 7.11(a) is the UI for one sample of electropherogram and deconvoluted subpeaks of HDL lipoprotein subclass separation. The left-hand tree directory offers the option to browse the folders for each sample run. The left lower panel shows the gel-like imagine of all samples in the same chip with flagging of the chip run status of each sample. For example, if a sample run is out of the preset linear range, it will be flagged as red. The right upper panel is the electropherogram of HDL fluorescent tracing and a gel-like image of the sample. The HDL electropherogram is deconvoluted by the Gaussian fit model to separate total HDL into five subpeaks (subclasses), which will be discussed in more detail later.

Figure 7.11(b) is the gel-like image of the chip run showing that all samples in a chip are aligned with a lower and upper dye alignment marker. HDL peaks are between the lower and upper markers. Figure 7.11(b) shows the overlaid electropherogram of the same sample run on 12 sample wells in the same chip for

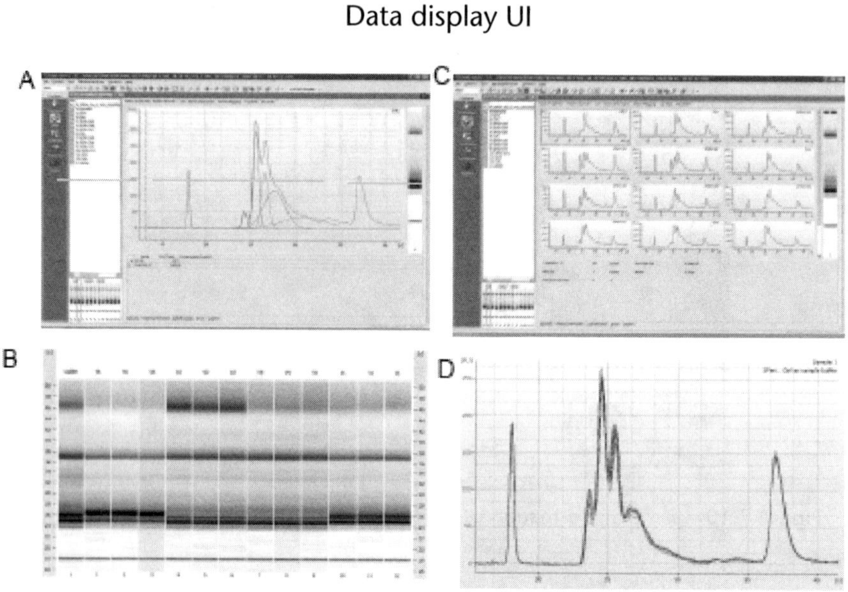

Figure 7.11 Lipoprotein subclass analysis.

the comparison of the reproducibility after the samples were aligned with lower and upper dye markers.

Figure 7.11(c) is the UI for the overview of electropherogram for all samples run on the chip. It can quickly identify sample(s) with abnormal tracing (run). The flags on some of the lanes of gel-like image in the lower left panel suggest some errors on those samples, either no proper chip run, or sample concentration out of the preset linear range. Those samples will need to be re-run.

7.4.2.4 Microfluidic Chip Physical Parameters

In the microfluidic diffusion channel, the fluid flow is laminar with a Reynolds number as shown in the following calculation:

$$\mathrm{Re} = \frac{\rho UH}{v} = 0.0265$$

where U is the fluid velocity, H is the channel height, v is the fluid kinematic viscosity, and ρ is the fluid density.

The transport of target molecules can be modeled as convective diffusion of pieces governed by both molecular diffusion and the convection of target molecules in the direction of the fluid flow. The Peclet number, which is a measure of the relative rate of convective to diffusive transport, can be calculated as

$$Pe = \frac{UH}{D} = \frac{0.000015m/s*0.00003m*10000}{6.6*10^{-6}cm^2/s} = 0.682$$

where D is the diffusion coefficient of serum in aqueous buffer. The value is greater than 1; thus, the transport is dominated by convection.

In the channel, the time τ it takes for a biological molecule to travel over a distance x=14 mm by diffusion is given by Fick's second law as follow [7]:

$$\tau = \frac{x^2}{2D} = \frac{1.96cm^2}{2*6.6*10^{-6}cm^2 / s} = 1.5*10^5 s$$

where D is the diffusion coefficient. Clearly, it is too long if by diffusion only. We need electrophoresis assists the molecule transfer. In the case of electrophoresis transport, the time is given as [8].

$$\tau = \frac{x}{\mu E} = 25(second)$$

where μ is the electrophoretic mobility and E is the electric field strength.

7.5 Experimental Results

The microchip electrophoresis can determine total HDL and HDL subclasses. Total HDL-C is measured by the calculation of total area of HDL peaks. Total HDL concentration of each sample is quantified by comparing the sample total HDL area to the calibrator sample area run in the same chip with known HDL concentration. The known HDL concentration of calibrator is used as a reference point for the chip. HDL will be further subfractionated by deconvolution of HDL peaks into subpeaks, and the concentration of each HDL subfraction can be calculated by the percentage of each subfraction area under curve.

7.5.1 Lab-on-a-Chip Assay of HDL Subclasses

Our microchip electrophoresis assay of HDL subclasses can clearly separate at least three distinct subpeaks of HDL, which shows the heterogeneity of HDL population. Figure 7.12 shows the typical tracing of sample electropherogram, with three different peak areas. The y-axis in the electropherogram is the intensity of fluorescence (arbitrary unit), and the x-axis the migration in seconds with the fast-running molecules on the left side. The first and third sharp peaks around 18 and 38 seconds, respectively, are dye marker peaks (lower and upper markers), used for alignment of all sample lanes in the chip to correct any potential drifting. The marker dyes are small molecules and usually run very consistently, which can be used as sample-to-sample and chip-to-chip normalization. The HDL area is in the middle part as indicated. In the example shown, the HDL has at least four distinguishable subpeaks. The HDL subclass with lower mass/charge ratio runs faster.

To identify the subpeaks of HDL corresponding to classical HDL subclasses, the purified HDL2 and HDL3 subfractions from ultracentrifugation were spiked in serum samples or run alone on microchips. The patterns of HDL2 and HDL3 in the electropherogram are clearly different in term of peak shape and migration time. The HDL2 subfraction ran slower and had lower peak slope, while HDL3 subfraction, on the opposite, ran faster and a steeper peak slope. There is some overlap between the HDL2 and HDL3 peaks, which may suggest some impurity of the

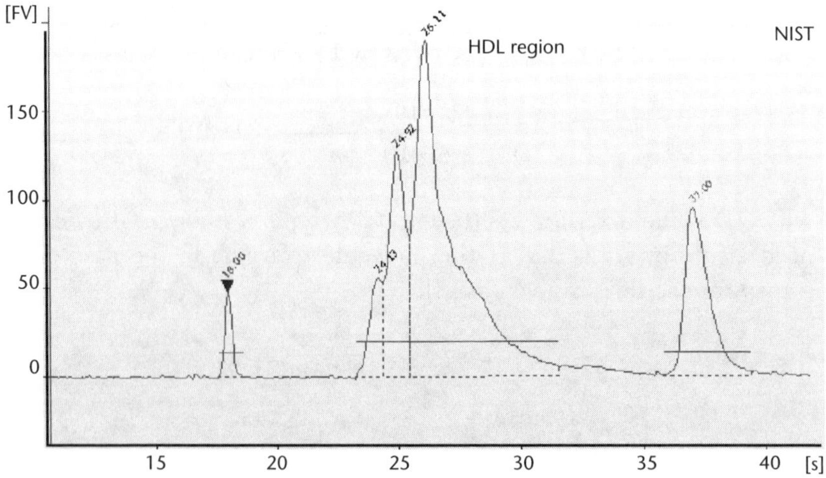

Figure 7.12 Electropherogram of HDL subfractions. At least three subpeaks are separated for typical serum samples.

ultracentrifuged subfractions. When purified HDL2 subfraction was spiked into human serum sample, the slow migrating peak in HDL electropherogram was increased without a significant increase in the fast migrating peaks (Figure 7.13). These results clearly suggest that the microchip assay can separate HDL subclasses. Further, good correlation ($r = 0.9$) was observed by direct comparison of the HDL2 level determined by microchip assay in 30 clinical serum samples to the result by ultracentrifugation.

By applying the Gaussian fit model, microchip HDL can be further deconvoluted into four subpeaks (Figure 7.14). We compared the deconvoluted HDL subfractions peak number 3 (P3) to the HDL2b fraction by ultracentrifugation, and there is excellent correlation between microchip HDL peak 3 and HDL 2b (Figure

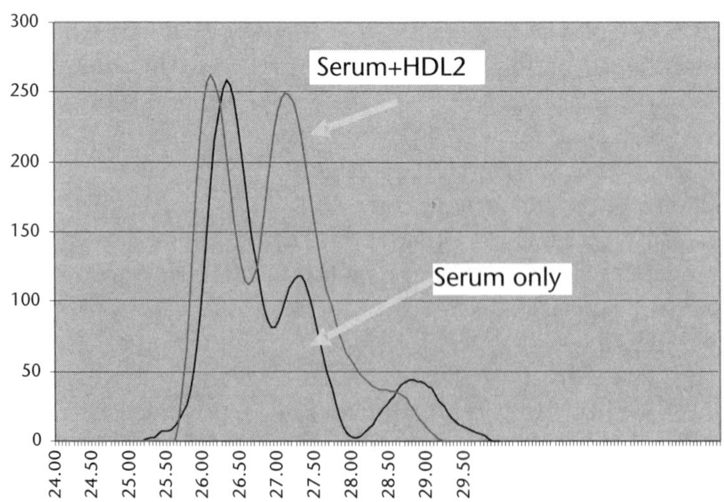

Figure 7.13 HDL subfractionation patterns after spiking in purified HDL2 subfraction in serum.

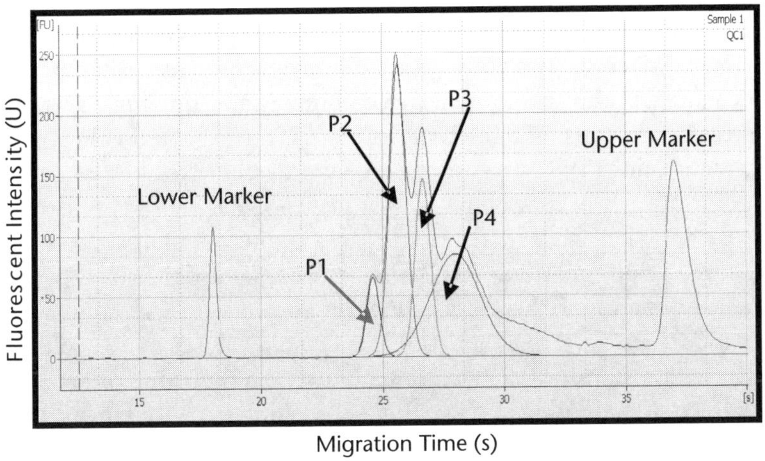

Figure 7.14 HDL subpeaks by deconvolution.

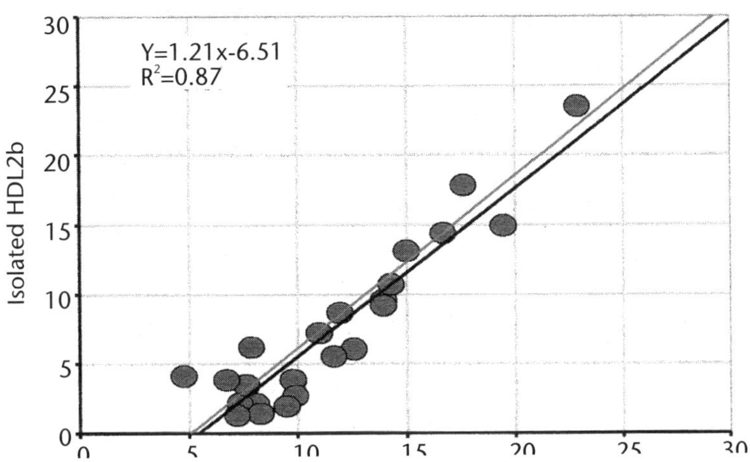

Figure 7.15 Good agreement between HDL peak 3 from the electrophoresis assisted microchip and HDL2b from ultracentrifugation.

7.15). Furthermore, there is significant difference in HDL peak 3 in a retrospective study of selected 500 patients with similar total HDL level between the patient group who were with myocardial infarction (MI) and the patient group who without MI (control). By adding HDL subclasses obtained from microchip electrophoresis into the assessment parameters of cardiac risk, the false positive predicative rate was reduced by more than 25% (unpublished data). These results clearly suggest that subfractioning HDL by the lab-on-a-chip assay will help physicians to increase accurate assessment of cardiac risk.

7.5.2 Quantification of Total HDL Concentration

Total HDL-C is measured by the calculation of total area of HDL peaks. Total HDL concentration of each sample is quantified by comparing the sample total HDL area to calibrator sample area run in the same chip with known HDL concentration. The known HDL concentration of the calibrator is used as a reference point for the chip. To ensure the chip run quality, two control sera, one high and one low HDL concentrations, are also included in the chip. If the reported value of any control serum sample is out of the preset range, the whole chip will be flagged and the values of patient samples in that chip will not be reported.

By applying these quality controls, the microchip assay can accurately estimate the HDL concentration. The known serum samples from 10 to 70 mg/dL determined by the microchip assay, linearly correlate to the HDL values by enzymatic methods, with CV< 5%, r =~ 0.99 and a slope closer to 1.0. We also compared our assay with a standard enzymatic technique on the value of total HDL. Serum samples from 80 individuals, with HDL-C ranging from 26 to 110 mg/dL, were run on the Clinic Analyzer as well as on a Bayer Clinic Analyzer. The correlation of total HDL value is R^2=0.87. The BioAnalyzer thus provides rapid, accurate information on HDL-C and HDL subclass distribution. Its flexibility and ease of use make it ideal for clinical testing.

Figure 7.16 shows linear correlation of known HDL serum samples with the experimental measurement. It clearly indicates that the system works very well.

7.6 Discussion and Conclusion

This chapter has presented an automated system for profiling lipoprotein subfractions by electrophoresis assisted microfluidic lab-on-a-chip technology and the BioAnalyzer. Experimental results show that the system can determine serum total HDL-C, HDL subclasses (HDL2 and HDL3), and LDL subtypes (A Type, B Type, or AB Type). This system can be used for high throughput clinical applications.

The total HDL concentration is calculated based on the area under the curve. The total HDL peaks are normalized with the calibrator (reference serum) run on

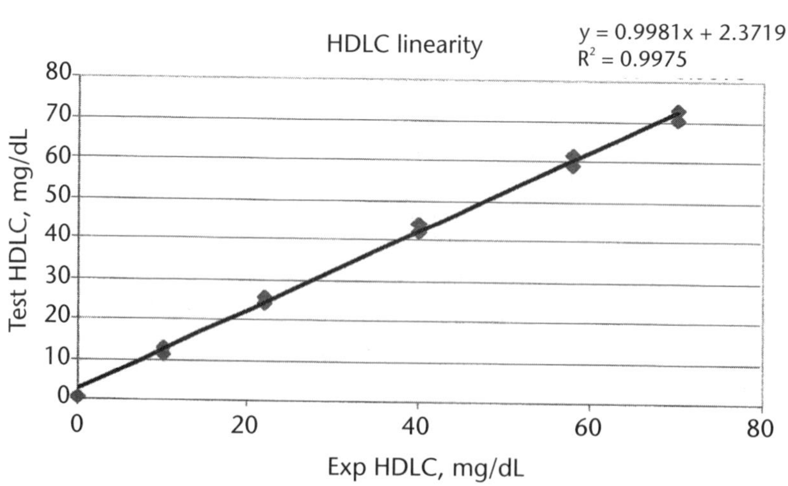

Figure 7.16 Linear correlation between known HDL serum samples of 10 to 70 mg/dL and chip run results.

the ladder well in the same chip. The HDL subclasses (subfractions) are computed by the area of each deconvoluted subpeak and reported as absolute concentration (mg/dL) and as a percentage of total. As a benchmark, the total HDL concentration was compared to values obtained by standard enzymatic method. HDL subclasses were verified with purified HDL subfractions from isopycnic density ultracentrifugation (UC) and gradient gel electrophoresis (GGE), a third widely used method.

Molecular diagnosis automation has been emerging as an exciting area for life science automation. This chapter has demonstrated how microfluidic lab-on-a-chip technology can be integrated with automation techniques to achieve robust and efficient medical automation. As a widely open research area, there are many challenging issues in the field. This raises demanded research topics in the field. We believe the limit of automation in this field has not been reached, and there is still plenty to work on.

References

[1] Craighead, H., "Future Lab-on-a-Chip Technologies for Interrogating Individual Molecules," *Nature,* Vol. 442, No. 7101, pp. 387–393, 2006.

[2] Bates, E. R., "Raising High-Density Lipoprotein Cholesterol and Lowering Low-Density Lipoprotein Cholesterol as Adjunctive Therapy to Coronary Artery Revascularization," *Am J Cardiol,* Vol. 86, No. 12A, 2000, pp. 28L–34L.

[3] Mykkänen, L., et al., "LDL Size and Risk of Coronary Heart Disease in Elderly Men and Women," *Arteriosclerosis, Thrombosis, and Vascular Biology,* Vol. 19, pp. 2742–2748.

[4] Austin, M. A., M. C. King, and K. M. Vranizan et al., "Atherogenic Lipoprotein Phenotype. A Proposed Genetic Marker for Coronary Heart Disease Risk," *Circulation,* Vol. 82, 1990, pp. 495–506.

[5] Ensign, W., N. Hill, and C. B. Heward, "Disparate LDL Phenotypic Classification among 4 Different Methods Assessing LDL Particle Characteristics," *Clinical Chemistry,* Vol. 52, 2006, pp. 1722–1727.

[6] Jain, K. K., "Nanodiagnostics: Application of Nanotechnology in Molecular Diagnostics," *Expert Review of Molecular Diagnostics,* Vol. 3, 2003, pp. 153–161.

[7] Madou M. J., *Fundamentals of Microfabrication, Second Edition,* Boca Raton, FL: CRC Press, 2002.

[8] Eisenberg, D., and D. Crothers, *Physical Chemistry with Applications to the Life Sciences,* Menlo Park, CA: The Benjamin/Cummings Publishing Company, Inc., 1979.

[9] Krauss, R. M., and P. J. Blanche, "Detection and Quantitation of LDL Subfractions," *Curr Opin Lipidol,* Vol. 3, 1992, pp. 377–383.

[10] Langlois, M. R., and V. H. Blaton, "Historical Milestones in Measurement of HDL-Cholesterol: Impact on Clinical and Laboratory Practice," *Clin Chim Acta,* Vol. 369, No. 2, 2006, pp. 168–178.

[11] Bell, J. D., et al., "Effects of n-3 Fatty Acids on the NMR Profile of Plasma Lipoproteins," *J Lipid. Res,* Vol. 37, 1996, 1664–1674.

[12] Chu, J. W., et al., Multiple Lipoprotein Abnormalities Associated with Insulin Resistance in Healthy Volunteers Are Identified by the Vertical Auto Profile-II Methodology," *Clin Chem,* Vol. 49, 2003, pp. 1014–1017.

CHAPTER 8

Clinical Laboratory Automation

Robin A. Felder

This chapter will cover the basics of clinical laboratory automation and describe the various automation technologies and their attributes. This chapter is intended to be a primer for those uninitiated in the basics of clinical laboratory automation. For more in-depth information, the reader is directed to various references.

8.1 Definition of Laboratory Automation

Clinical laboratory automation is defined as "the use of instruments and specimen processing equipment to perform clinical assays with only minimal involvement of the technologist" [1]. Although much has been written on the subject of the automation of the analytical process [1], this chapter will focus on automation of the steps prior to and after the analytical steps such as robotic automation of specimen labeling, transportation, and accessioning (checking the specimen into the laboratory), as well as storage and retrieval.

8.2 History of Clinical Laboratory Automation

Clinical laboratory automation was first conceived of and deployed at the Kochi Medical School (Kochi, Japan) in the early 1980s by the late automation pioneer Masahide Sasaki [2]. Driven by the need to expand laboratory services without substantially increasing the cost of labor, Dr. Sasaki, Mr. Ogura, and their staff of engineers and students assembled conveyor systems, robotic manipulators, and a software system to oversee the automation process. Their first (but ever-evolving system) was first installed in 1982 [3, 4]. The work of Sasaki inspired pioneers such as Dr. Obrien at the Smith Kline Beecham Laboratories, Stuart Wills at Coulter Company (formerly Miami, Florida), Laura Fitzgerald at Johnson & Johnson Clinical Diagnostics, Rodney Markin (University of Nebraska, Omaha, Nebraska), Charles Hawker (ARUP Laboratories, Salt Lake City, Utah), Alan Lloyd (Sonic Healthcare, Sydney, Australia), Ralph Dadoun (St. Mary's Hospital, Quebec, Canada), Dennis Ferkany (BCI), Larry McGuire (Medical Robotics, Lexington, Kentucky), William Neely (Detroit Medical Center, Detroit, Michigan), and many others to lead their companies and laboratories into developing their own versions of laboratory automation, which served as the basis for future developments in this area. These first generation systems, which consisted of various automated workstations connected by a conveyor belt specimen transport system, were enhanced by the emergence of modular automation pioneered by Roche Diagnostics (system manufactured in Japan by Hitachi High Technologies Corporation, Tokyo, Japan), and the A&T Corporation (Kanagawa, Japan). Subsequently, over 800 laboratories [5, 6] have invested over one million dollars in laboratory automation in order to contain labor costs, provide faster turnaround, and to establish a basis for high-quality sample handing for their laboratories. The complexity of automating a clinical laboratory cannot be underestimated, and thus the stepwise process for specifying, purchasing, and installing a laboratory can take several years. Successful automated laboratories have resulted from the combined effort of visionary leadership, and with a well-coordinated staff of laboratory professionals, all laboratory automation leads to improved efficiency. Tatsumi et al. have shown that the success of laboratory automation depend on the type of system purchased and how it is implemented [7]. It is generally recognized that automation of the preanalytical phase of specimen processing will yield the greatest labor savings, and hence the greatest return on investment in automation technology, since this is the most labor-intensive area of the laboratory (Figure 8.1). One can divide the steps in specimen processing into the individual unit operations that lend themselves to automation (Table 8.1).

There is little published literature describing the performance of laboratory automated systems [8–12]. Thus, it is necessary to visit laboratories that are similar in size and scope to what the reader wishes to accomplish in order to successfully implement laboratory automation.

8.2.1 Specimen Labeling

The creation of bar codes and their placement on specimen containers is one of the most effective tools for maximizing the use of robotic automation in the laboratory. Bar codes that are one-dimensional are relatively easy to read by bar code readers placed at various decision points in the automation process. However, tubes generally must be rotated in order to expose the bar code to the reader. The more rotational axis one has to employ in automation results in additional costs and potential

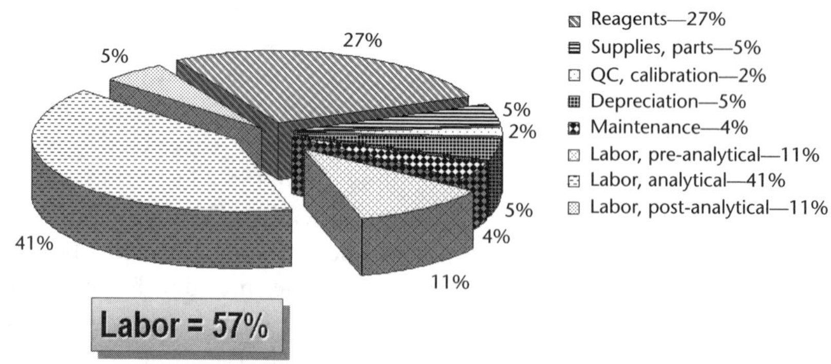

Figure 8.1 The costs of a lab test are broken down into reagent costs, supply and parts, quality control and calibration material, depreciation on equipment, maintenance of analytical equipment, and three phases of labor (pre-analytical, analytical, and post-analytical). Labor constitutes 57% of the cost associated with a laboratory test (Data courtesy of Ortho Clinical Diagnostics).

Table 8.1 List of Individual Laboratory Units Operations (labor requiring steps) in a Clinical Laboratory.

1. Specimen collection
   (a) phlebotomy [14]
   (b) labeling [15]
   (c) transportation [16]
   (d) accessioning
   (a) computer data entry
   (b) Sample integrity checking
2. Preanalytical processing
   (a) Sorting
   (b) centrifuging
   (c) aliquotting
   (d) distributing
3. Analysis
   (a) aspiration
   (b) analysis
   (c) validating
   (d) reporting
4. Postanalysis
   (a) processing
   (b) relabeling
   (c) storage
   (d) disposal

failure points. Two-dimensional codes that are much smaller are beginning to be used since they may be printed on all visible aspects of the tube in order to obviate the need for tube rotation. Next generation systems will employ the use of radio frequency identification chips (RFID) that may be read wirelessly at a distance [17].

8.3 Definitions

A clinical laboratory automation system consists of automated workstations, workcells, and total laboratory automation. Despite the variety of systems on the market, standards have been developed and published by the Clinical Laboratory Standards Institute (CLSI, www.clsi.org)[18].

8.3.1 Workstation

The minimalist automation device is a single-function component that performs one task efficiently. For example, a tube decapper can be considered a single-function device, which will pull the caps off a medical specimen. However, there are varying degrees of versatility found in decappers from being able to only decap one size of a tube with a single kind of closure, to decappers that can pull stoppers, unscrew threaded caps, and punch through rubber stoppers and/or aluminum foil based closures from containers of many sizes. Thus, one must consider carefully what kind of functionality one wishes to obtain. Another example of a laboratory automation workstation includes a phlebotomy tray production station (Figure 8.2), automated

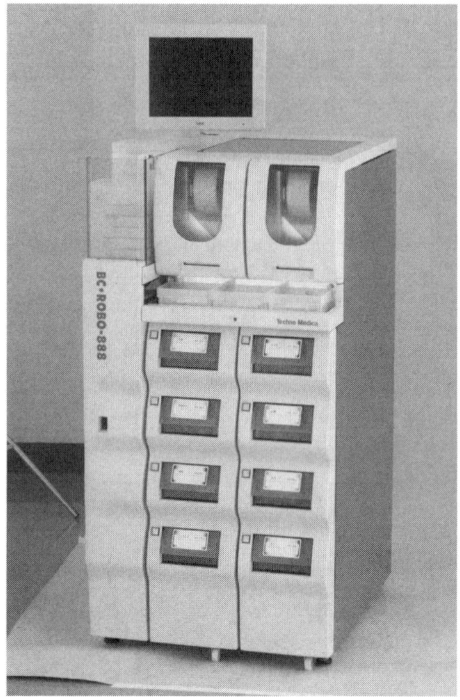

Figure 8.2 The BC Robo 888™ device.

centrifugation stations, sorting stations, decapping stations, bar code labeling stations, sample inspection stations, sample thawing stations [20], cold storage and retrieval stations [19], and preanalytical processing stations [13]. Phlebotomy tray production can be an excellent efficiency tool in phlebotomy stations with fairly high throughput. The BC Robo 888™ device (Figure 8.2), was developed in Japan to rapidly place the appropriate number, type, and labeled blood collection tubes into a tray to be used at the point-of-use for phlebotomy. In a setting where large numbers of patients are expected to show up for blood collection, this system can provide increased throughput as well as reduced errors.

8.3.2 Workcells

A workcell is defined as the integration of several automated workstations to provide a higher level of functionality and varied productivity as compared to a workstation. For example, Ortho Clinical Diagnostics (www.orthoclinical.com) pioneered the Chemistry workcell in the 1990s, which employed the LabInterlink (Omaha, Nebraska) automation technology (under the visionary guidance of Rodney Markin, University of Nebraska). The efficiency afforded by the dry chemistry platform produced by Kodak (Rochester, New York) (and later sold to Ortho Clinical Diagnostics) allowed specimens to be entered into the workcell, decapped, centrifuged, and ultimately transported to an analyzer for analysis. Similar functionality was offered originally by Coulter Corporation that provided centrifugation, transportation, and analysis in the Ektachem analyzers. This approach was ultimately sold to Beckman to form Beckman Coulter's workcell.

8.3.3 Total Laboratory Automation

Workcell technology may be linked together using standard conveyor systems to provide even greater hands-free specimen transportation, queuing, and analysis. Thus the definition of total laboratory automation (TLA) is a system that automates virtually all of the routine tasks associated with sample preparation and analysis using subcomponents that function as a coordinated system. The common components of a TLA include an area where specimens may be stored while waiting for entry into the system (input buffer), a conveyor, an automated centrifuge, a decapper, an aliquotter, a relabeler, analytical stations, an output buffer (may be the same device as the input buffer), and a long-term storage refrigerator/freezer.

TLA was seen by engineers as an interim step in the evolution of the laboratory into a truly integrated automation solution. Unfortunately, market pressures have prevented true integration of analytical systems and the preanalytical automation to support it. Modular automation is defined as interchangeable automation technologies that allow functionality to be created using a plug and play and interchangeable platform. Due to the lack of an agreement by vendors to allow interoperability between automated systems, the modular concept has only evolved within a single vendor's offering to the market platform designed and built by Hitachi, Tokyo, Japan, and sold by Roche Diagnostics (Indianapolis, Indianapolis) as the MODULAR™ workcell, and the A&T Clinilog system (www.aandt.co.jp) designed, in part, by Masahide Sasaki (Figure 8.3). The MODULAR™ and A&T modular systems provide integrated preanalytical and analytical workstations in a flexible format. Modular type technologies allow flexible attachment of numerous preanalytical devices and chemistry and immunoassay instruments as the needs of the laboratory change. One benefit of modular automation is the optimization of sample transportation via rapid conveyance to and from the various modular integrated workstation components.

Figure 8.3 The A&T Modular system was designed as individual laboratory devices (e.g. sample sorter, centrifuge, decapper) and analyzers (e.g. chemistry and immunoassay analyzers) that are connected together to form a modular automated laboratory. (Picture courtesy of A&T Corporation, Japan).

8.4 Pediatric Samples

Automation systems have focused on the routine specimens that are obtained from adults. Pediatric patients present particular challenges to automation, since it is either difficult or medically inadvisable to draw 5 to 10 mL of whole blood (12–16 mm in diameter; 75–100 mm in height). Small volumes are placed in "pediatric tubes" (10–13 mm in diameter) for which engineers have not designed grippers, bar code readers, specimen transport pucks, or centrifuge buckets. Typically, pediatric specimens are centrifuged manually, transferred to small containers that are placed in the top of an adult specimen container. However, there are several analytical systems that can accommodate pediatric containers. For example, the enGen™ automation platform offered by Ortho Clinical Diagnostics (www.orthoclinical.com) will transport small tubes and allow direct sampling from the Ortho clinical analyzers. Small volume point-in-space sampling is accomplished by using slender pipettes that descend faster into the tube during the aspiration process since a narrower tube will empty more quickly than a wide bore tube (Ortho Clinical Diagnostics, personal communication).

8.5 Process Control

Traditional laboratory information systems are versatile databases that manage the gathering of analytical data and providing of standardized reports. However, LIS software is not generally designed to manage the complexities of running automation hardware. Process control software is essential to manage the complex data and processes that an automation system generates. For example, an LIS will organize patient demographics, analytical results, and provide some level of result interpretation. Where LIS software falls short (result interpretation, result validation and verification, and other intelligent data housekeeping chores), middleware has been produced to provide this ability to interpret data. However, the next layer in software management of the automated laboratory is defined as "process control." Process control is the use of software to manage the complexities of sample distribution, analysis, storage, and retrieval, as well as reporting on the full status of the analytical systems in the laboratory. Some of the functionality that must be managed by a process controller include sample routing and sorting as well as reporting the status of each specimen back to the process controller. The many man hours associated with deciding the fate of each specimen can be eliminated through the use of a process controller (routing around the centrifuge, centrifugation temperature, requirement for decapping, repeat testing, add-on tests, and reflex testing). In addition, when a process controller is provided a full bidirectional interface with each analytical instrument, it can provide the conduit for reporting instruments alerts such as low reagents or a breakdown.

Several process controllers are available for the clinical laboratory. Data Innovations have produced a process controller for the Abbott Accelerator™ LAS. Ortho Clinical Diagnostics, Roche, Siemens, and Beckman Coulter have either written or purchased process controllers that manage their respective LAS systems.

8.6 Automated Clinical Laboratory Efficiency and Quality Programs

Two of the world's leading authorities on quality paved the road for improving quality in organizations that produce products (www.juran.com, www.deming.com). Juran published his quality method in 1986, which then led to Deming's famous work on quality in 1987. Six Sigma is essentially building on these two foundations. Lean is an old productivity and quality improvement that has undergone continuous evolution (www.lean.org). Lean clinical laboratories are striving to shorten the time between obtaining the specimen from the patient to reporting a result by eliminating sources of waste, by leveraging technology to maximize the work effort of a company's number one resource, the technologist. In order to constantly achieve higher levels of quality and efficiency, laboratories have to adopt new technologies (like robotics and process management), eliminate waste, and continuously improve.

Juran's work was based on total quality management (TQM) that provided a broad perspective for quality management. TQM requires the organization to understand customer needs and requirements and to focus on the process. TQM was a relatively static quality management process. Therefore, continuous quality improvement (CQI) processes were implemented. CQI requires the user to establish a problem-solving method and emphasizes teamwork and group problem solving. Management must also participate in the quality program. The program called improving organizational performance focuses on management quality control by emphasizing teamwork. Operational performance is an outcomes measuring method. Technology can be used by management as well as the laboratory to improve overall performance.

Six Sigma has its most profitable benefit when combined with an efficiency program such as Lean. Sigma is a statistical term that measures the consistency of a process. Once we have measured the consistency of a process, we can use these measures to predict the error rate (e.g., measuring how often bar codes are placed incorrectly on a tube). Once we have defined a target, then we can measure the standard deviation (or sigma) around the target. A process that achieves a 2-sigma performance has 308,537 defects for every million products. Six Sigma achieves only 3.4 defects per million opportunities or only one error in 300,000 events. Another way of visualizing this difference is to think of 6 sigmas worth of defective DPC immunoassay analyzers out of the 300,000 you might have just sold, which is one bad analyzer sitting in your office. Now, think of 6-sigma defective ultrasound machines. This is more than 20,000, enough to stretch from Paris to Naples. Moving a process from a 3 sigma to a 6 sigma results in a 20,000 times improvement in a process. Sigma values for other industries include 6.3 for domestic airline fatalities but 4.1 for airline baggage handling, 4.2 for the U.S. television industry and 5.9 for the Japanese television industry, and 2.1 for Internal Revenue Service phone-in tax service. Six Sigma is difficult to achieve in clinical laboratories unless there is extensive use of automation. The ARUP Laboratories (Salt Lake City, Utah), have achieved Six Sigma qualities for a number of lost specimens per reportable test. Therefore, organizations that have embraced quality measures, even with the complexity of clinical laboratory analysis, can achieve Six Sigma qualities. While quality is arguably the most important measure of the performance of a laboratory, timeliness and efficiency are also important to support a just-in-time medical system. Thus, the Lean process has been developed to address these specific performance measures.

8.7 Automated Centrifugation

Automation centrifugation allows for continuous flow in an area where bottlenecks occasionally occur. Robotic systems have been developed to automatically remove specimens from a conveyor, load and run the centrifuge, and unload specimens that are then placed back on the track. Most commercial centrifuges will process about 300 specimens per hour; therefore many laboratories purchase two systems. Centrifugation is occasionally performed manually in fully automated laboratories because laboratories that adopt Lean processes can generally perform manual centrifugation more rapidly and efficiently than today's automation systems.

8.8 Automated Decapping and Recapping

Most customers purchase the decapping module because decapping can lead to aerosol generation and repetitive stress injury. Specimen recapping and sealing is an essential tool because it eliminates a boring laboratory job that leads to repetitive stress injury. Many automated decappers can remove the closure from tubes sealed with rubber stoppers, screw caps, and other types of closures.

8.9 Automated Storage and Retrieval

The storage and retrieval module is an absolute requirement for efficient specimen buffering in an automation system, since not all specimens can be assayed in a linear fashion. Buffering modules should not be confused with long-term specimen storage devices, such as automated refrigerators and freezers. Buffers will hold specimens for a defined period of time until presumably no longer needed and then pass them onto the final step in the process. Automated buffers will shunt specimens that need repeat, reflex, or add-on tests back into the conveyance system in order to be reaspirated by the analytical system.

Postanalytical processing is almost as important as preanalytical processing for improving laboratory efficiency because up to 20% of laboratory labor can be spent on postanalytical tasks. Recappers ensure long-term specimen integrity in the refrigerator or freezer. Foil-based recappers allow recapping material to be stored in large quantities on a roll, making it unnecessary to replenish caps regularly. Furthermore, foil-based systems allow piercing of the foil lid for postanalytical reanalysis and then replacement of yet another foil cap on top of the previous recap. Once the specimen has reached its analytical endpoint, it can be placed in an automated long-term storage system at either refrigerated or frozen (–20°C) storage. Systems have been developed for long-term refrigerated storage (e.g., Inpeco SpA, Milan, Italy). In the case of delicate specimens, the BioPhile was produced for automated long-term storage at – 80°C [19]. The BioPhile was intended to be an integral component of a biorepository in which specimens could be stored and retrieved from an – 80°C environment for medical or research purposes [19].

8.10 Automated Aliquotting

Aliquotting is present on some systems but not on others because the goal of a Lean laboratory is to eliminate the need for aliquotting. Rapid sampling and release of specimens to the next task can allow sharing of a single tube throughout the laboratory. However, this does not eliminate the need to draw additional specimens or aliquot a single specimen for sending tests outside the laboratory. Thus, each laboratory must decide if it is more efficient and medically justifiable to draw a second tube for sendouts, or to aliquot within the laboratory. Aliquotters have been developed that aspirate the required aliquot directly through an uncapped tube, or employ the use of a decapper prior to aliquotting.

8.11 Mobile Robotics

Mobile robotics can replace the human-based delivery system, and provide more flexibility than can be afforded by a point-to-point pneumatic tube system. For example, the Helpmate Robot (www.cardinal.com/pyxis), Tug (Aethon Robotics, Pittsburgh, Pennsylvania), and RoboCart (CCRI, Lake Arrowhead, California), are mobile robot systems that have been configured to provide transportation support to the clinical laboratory. Although the Helpmate is the only robot of this class that is designed for routine unattended elevator use, all these systems can provide more consistent point-to-point delivery service than can be provided by even the most diligent human-based transportation system. In a simulation study of the effect of using multiple robot systems in a large medical facility, it was demonstrated that three mobile robots could reduce the cost of delivery by over $250K per year while improving delivery turnaround time by 30% [16]. By equipping mobile robots with automated drop-off and pickup capabilities (US Patent #6,941,762), one can improve the efficiency of mobile robots by over 20% (unpublished studies).

8.12 Point-of-Care Automation

There are two kinds of point-of-care (POC) technologies: either broad menu devices that are attempting to perform up to 80% of central laboratory testing, or single analyte easy-to-use devices. There is no question that point-of-care technology promises to reduce the amount of routine central laboratory chemistry testing. For example, I-Stat (www.i-stat.com) focused on bringing a wide variety of clinical tests onto a relatively simple device. One of the principal technologies that have contributed to the successful automation of point-of-care technology has been the interlinking of POC devices and the laboratory [21].

8.13 System Integration

Increased systems integration will be a key feature of fourth generation automation systems. Automation technical support will be needed to help users integrate their laboratory information system with their lab automation system. Laboratories should look for software systems with elegantly designed user interfaces and common database architectures (for example, SQL). Incorporating many of the LIS functions into the LAS process controller has the benefit of providing a single interface for patient demographics, rules engine, reporting, Internet connectivity, help screens, quality control data, statistical reports, and autovalidation. In commercial production factories, which have managed complex automated processes for decades, it is known that the use of a process control architecture is the most critical component of an automated system. Laboratories should be cautioned that not all LISs will support an efficient interface with the process controller. Therefore, some laboratories strongly consider either purchasing an LIS that supports process control or purchasing a process controller that already includes many of the traditional LIS functions.

The basic dimensions of an automation component are important from a flexibility standpoint. Long track systems with little flexibility will limit the selection of analyzers that might be interfaced if lateral floor space is required near the track (most systems accommodate parallel or perpendicular attachment of analyzers). The importance of vendor adherence to the CLSI standards (Auto 1-5) cannot be overemphasized because these standards have been developed through a rigorous process by automation users [18]. Having an automation system that can manipulate many tube sizes has long been debated. A Lean approach would emphasize the standardization on as few containers as possible since this approach results in the greatest institutional savings and greatest gains in process efficiency. Once the tubes are placed on the automation system, conveyance throughput is important because it can become a rate limiting step over which the laboratory has no further control (a conveyor system usually cannot be speeded up once it is purchased and installed). However, process gains can be realized by the use of intelligent specimen rerouting since this will obviate the need for technologist intervention in the process of deciding where specimens have to be moved in subsequent processes (automatic rerouting for reflex/repeat/dilutions).

A useful feature of some automation systems is the placement of all utilities in the space underneath the track. When utilities are integrated, the finished product is efficient and visually appealing. Analytical systems that eliminate the need for water supplies and waste drains can reduce laboratory renovation costs considerably (required utilities). Automation systems usually require only weekly preventive maintenance. However, the replenishment of tips for aliquotters and caps for recappers may require scheduled daily maintenance (required maintenance). Systems with self-washing probes will save material and labor costs.

In laboratories with a significant amount of sample rerouting, single-tube carriers (or pucks) offer the greatest flexibility in routing specimens and optimizing system throughput. However, most laboratories operate with little variance in test requests, thus no significant degradation in system performance is seen with multiple-tube carriers. Laboratories are advised to use a process simulation tool to determine if the choice of multiple specimen carriers will influence turnaround time significantly. Similarly, a scalable system will be able to accommodate a growth in laboratory business.

8.14 Summary and Conclusion

Laboratory automation is a discipline that combines the knowledge of laboratory science with process engineering. Developing an optimal solution requires careful attention to customer needs, available budget, laboratory infrastructure, and performance requirements. Flexibility must be maintained for future growth and possible need for laboratory consolidation. Fortunately, the relatively simple nature of automation systems allows them to provide laboratories with many years of trouble-free service and a significant improvement in laboratory efficiency and patient safety.

References

[1] Boyd, J. C., and C. Hawker, in *Textbook of Clinical Chemistry and Molecular Diagnostics*, C. A. Burtis, E. R. Ashwood, and D. E. Bruns. (eds.), St. Louis, MO: Elsevier Saunders, 2006.

[2] Sasaki, M., "Total Laboratory Automation in Japan: Past, Present, and the Future," *Clin. Chim. Acta,* Vol. 278, 1998, pp. 217–227.

[3] Sasaki, M., "Building Bridges Toward the 21st Century," *Rinsho Byori,* Vol. 114, 2000, pp. 44–50.

[4] Sasaki, M., "The Robotic System of the Clinical Laboratories," *Rinsho Byori,* Vol. 35, 1987, pp. 1072–1078.

[5] "Laboratory Automation Systems & Workcells," *CAP Today,* March 2007, pp 2–24.

[6] Felder, R. A., "Laboratory Automation Systems Demystified," *CAP Today,* March, 2007, online archive at http://www.cap.org.

[7] Tatsumi, N., K. Okuda, and I. Tsuda, "A New Direction in Automated Laboratory Testing in Japan: Five Years of Experience with Total Laboratory Automation System Management," *Clin. Chim. Acta,* Vol. 290, 1999, pp. 93–108.

[8] Dadoun, R., "Impact on Human Resources: Core Laboratory Versus Laboratory Information System Versus Modular Robotics," *Clin. Lab. Man. Rev.,* Vol. 12, 1998, pp. 248–255.

[9] Seaborg, R. C., B. E. Statland, and R. O. Stallone, "Planning and Implementing Total Laboratory Automation at North Shore–Long Island Jewish Health System Laboratories," *Med. Lab. Obs.,* Vol. 31, 1999, pp. 46–54.

[10] Beckman Coulter Resource Center, http://www.beckmancoulter.com/resourcecenter/literature/Diaglit.

[11] Hawker, C. D., "Laboratory Automation: Total and Subtotal," Clin. Lab. Med., Vol. 27, No. 4, 2007, pp. 749–770.

[12] Hawker, C. D., et al., "Automated Transport and Sorting System in a Large Reference Laboratory: Part 1. Evaluation of Needs and Alternatives and Development of a Plan," *Clin. Chem.,* Vol. 48, No.10, 2002, pp. 1751–1760.

[13] Holman J. W., et al., "Evaluation of an Automated Preanalytical Robotic Workstation at Two Academic Health Centers," *Clin. Chem.,* Vol. 48, No. 3, 2002, pp. 540–548.

[14] Zivanovic, A., and B. L. Davies, "A Robotic System for Blood Sampling," *IEEE Trans. Inf. Technol. Biomed.,* Vol. 4, 2000, pp. 8–14.

[15] Felder, R. A., "Push for Patient Safety is Nudge for Automation," *CAP Today,* Vol. 17, No. 5, 2003, pp. 33–36, 38, 40.

[16] Rossetti, M. D., R. A. Felder, and A. Kumar, "Simulation of Robotic Courier Deliveries in Hospital Distribution Services," *Health Care Manag. Sci.,* Vol. 3, No. 3, 2000, pp. 201–213.

[17] Dinh, A. K., "RFID Systems in Healthcare. Emerging Uses and Potential Issues," *J.* AHIMA, Vol. 79, No. 1, 2008, pp. 62–63.

[18] Hawker, C. D., and M.R. Schlank, "Development of Standards for Laboratory Automation," *Clin. Chem.,* Vol. 46, No. 5, 2000, pp. 746–750.

[19] Felder, R. A., "Automated Storage and Retrieval at -80°c: Managing and Documenting Specimen Security, Integrity, and Accessibility," *Am. Lab.,* Vol. 35, No. 1, 2003, pp. 22–25.

[20] Hawker, C. D., et al., "Development and Validation of an Automated Thawing and Mixing Workcell," *Clin. Chem.,* Vol. 53, No. 12, 2007, pp. 2209–2211.

[21] Felder, R. A., "Robotics and Automated Workstations for Rapid Response Testing," *Am. J. Clin. Pathol.,* Vol. 104, No. 4 (Suppl. 1), 1995, pp. S26–S32.

CHAPTER 9

Pharmacy Automation Technologies

Elizabeth A. Flynn

9.1 Background

The use of automation to distribute medications appears to be an ideal use of technology replacing humans in the performance of repetitive, tiring, memory-intensive record-keeping tasks. However, the holy grail of accurately closing the loop between prescriber ordering and patient receiving the intended medication has yet to be achieved, based on both case reports and observational studies of medication administration errors [1, 2]. The goal of this chapter is to describe case studies of the capacities and limitations of pharmacy automation with respect to efficiency and accuracy of medication distribution.

An automated pharmacy system is defined as a mechanical system that performs operations involving the storage, packaging, dispensing, or distribution of medications while enabling control of the operation and electronic documentation of transactions [3].

Pharmacy automation has primarily been incorporated into hospital, community, and mail-order settings. There are a small, but growing, number of established vendors in each setting. Key criteria for comparing systems in any setting should include consideration of space and facility requirements, accuracy, software compatibility, reliability and downtime (maximum time and manual procedures), training method, impact on labor time, incorporation of bar code technology, packaging required, and ease of integration into the work flow of the medication distribution system [4].

9.2 Technology and Automation: Hospitals

Automated hospital pharmacies may utilize decentralized semiautomated dispensing cabinets, centralized automation, or a combination of both ("hybrid" system). Decentralized systems can involve replacement of medication carts that once delivered a 24-hour supply of medications to each nursing unit by "cartless" unit-based dispensing cabinets in 38% of hospitals [5]. Ninety percent of hospitals with more than 300 beds use semiautomated dispensing cabinets [5.] Unit-based cabinets include the Pyxis Medstation (Cardinal Healthcare), AcuDose-Rx (McKesson), SureMed (Omnicell), MedSelect (AmeriSource Bergen), and MedDispense. A showcase for automated devices is available (http://www.rxshowcase.com/pharmacy_automation_companies.htm). The general procedure for unit-based cabinets is for the nurse to log in to the device and select a function to perform, such as retrieve medications for a patient. After selecting a patient from the list of patients on the nursing unit, the nurse can request all meds scheduled for administration during the next medication pass. Devices should be interfaced with the pharmacy computer system so that only medications that are ordered and approved for a patient are available to the nurse. The nurse then retrieves each dispensed dose and verifies current inventory levels. Devices vary in the number of different drugs and quantities available to the nurse—narcotics typically have their own dedicated storage area, while noncontrolled substances may be stored in areas where as many as 20 different drugs are available for retrieval. It is recommended that the nurse leave each medication in its original package until arriving at the patient's bedside for administration. This ensures accurate drug identification up to the point of administration, and if the patient refuses the medication, it can be returned easily. Bar code verification systems require an intact package for scanning prior to medication administration.

Unit-based cabinets linked to pharmacy-controlled patient medication profiles offer advantages such as improving response time (medications are available on the unit as soon as orders are approved), doses are retrieved only if needed, doses aren't charged until removed, dose credits are minimized, and medication replenishments are more simplified than the traditional cart fill process. Disadvantages include the expense of the systems, nurse delays as they wait to retrieve medications at peak administration times, increased and duplicate inventory throughout the hospital, and similar disadvantages to the floorstock system (potential for error due to access to many medications if not properly controlled) [6].

Centralized systems are typically used to fill a 24-hour supply of patient-specific medications, delivered to nursing units in a cassette of drawers that are exchanged for the set of drawers from the previous 24-hour period. Medications not administered during the previous day are credited to the patient (depending on the billing system), and returned to stock. An important feature of some automated systems is the ability to return items to stock without requiring pharmacy staff to perform this time-consuming function manually. Key characteristics to evaluate on centralized systems include the scope of unit dose packaging accommodated (oral solids, oral liquids, injections, etc.), incorporation of bar code verification, need for special packaging (e.g., overwraps or covers over oddly shaped containers, such as syringes), and scope of products included (to minimize manually picked medications).

Forty percent of hospitals with more than 400 beds have centralized robotic medication processing [5]. There is no pharmacist double-check of doses dispensed from robots in 7% of hospitals [5], while many hospitals check at least 5% of all robot-processed orders, since local regulations require this step. Centralized drug dispensing machines include robot-based systems (McKesson ROBOT-Rx) [7], and high-capacity storage and retrieval electronic shelving systems (Talyst AutoCarousel) [8], all of which are bar code based. Advantages of centralized systems include automation of most if not all of the manual cart fill process, inventory consolidation in one location, and bar code control of medications to improve accuracy and inventory control. Disadvantages include cost of the system and time required for the pharmacy staff to package medications in a format that can be used by the automated system [6] .

9.3 Effect of Automation on Efficiency in Hospitals

There is evidence from one study of the effect of unit-based medication dispensing devices on nurse and pharmacist time. Guerrero and colleagues studied the Baxter Sure-Med system and found that there was no significant change in medication-related time for nurses on a medical unit (20.7% to 18.4% after device was installed) or in a surgical intensive care unit (10.8% to 11.0% after device was installed). Pharmacist clinical time increased after implementation of the system from 36.5% of their shift to 49.1%, while technical tasks consumed less time, changing from 22.1% to 17.4% of the day [9]. The increase in clinical time represents 1 additional hour of time spent on clinical activities, while the change in technical tasks decreased by 22 minutes per day, thus making the installation of automation equivocal with respect to saving labor.

Shirley studied the effect of an automated dispensing system (Pyxis Medstation Rx 1000) on percent of doses that were administered when scheduled. The unit dose cassette exchange system resulted in 59% of 76 doses given when due. After the devices were installed on 11 nursing units, 77% of 87 doses were administered when due (chi square = 5.18, $p = .0235$) [10]. Thus, there was a significant increase in correct administration times following the implementation of automation.

9.4 Effect of Automation on Medication Errors in Hospitals

Manual filling of patient-specific medication doses by technicians have resulted in inpatient dispensing error rates ranging from 0.04% to 2.9% [11–18]. A recent study by Cina and colleagues found that 0.75% of 140,755 manually-filled doses contained dispensing errors that were not detected by pharmacists during their inspection prior to release of the medication to the nurse on the unit [19]. A physician panel judged that 23.5% of the dispensing errors were potential adverse drug events (ADEs), and that 28% could have been serious ADEs, and 0.8% were life-threatening [19]. Poon and colleagues studied the effects of bar code verification of accurate medication retrieval before and after installation of a bar code system in a pharmacy. A carousel system was also compared to manual filling procedures. Target dispensing error rates decreased from a range of 0.25% to 0.71% (depending on the process studied) to between 0.018 to 0.2% ($p < 0.001$). The lowest target dispensing error rates were achieved by using the carousel system in conjunction with bar code checking (scanning one dose per pick), and manual retrieval where every dose picked was scanned using the bar code system. The authors recommended scanning every dose for maximum accuracy (as opposed to scanning one representative dose retrieved from each storage location) [20]. Oswald and Caldwell studied the effect of an automated pharmacy carousel system (APCS) on dispensing error rates in a 613-bed acute and tertiary care university hospital. The manual system was bar code based—a technician used a mobile bar code scanner to scan a label for an item to be filled, and then scanned the bar code on the medication being retrieved. The manual filling of first dose and missing dose requests had a 0.5% dispensing error rate on 422 orders, while the APCS had a 1.2% dispensing error rate on 173 orders. When remeasured several months later, the dispensing error rate was 0.5% on 387 orders. Wrong strength, wrong medication, wrong dose, and wrong dosage form errors were detected. Dispensing error rates for refilling automated dispensing cabinets were 0.4% before and 0.3% approximately 8 months after implementation of the APCS [21]. Thus, in these studies, it appears that error rates are initially higher following the use of automation, probably due to the difficulty in learning how to use the new technology. However, the error rates seen following the continued use of automation approximate those found in manual dispensing.

9.5 Medication Administration Errors in Hospitals

In the early 1980s, Barker and colleagues [22] did a crossover study to measure the effect of an automated bedside medication dispensing machine on medication administration errors. The system sounded an alert to notify nurses when medications were due to be administered, and allowed access to only those medications that were due to be given at that time. The bedside device system had a significantly lower observation-based medication administration error rate (10.6%) compared to the previous unit dose system (15.9%). All types of errors were detected while using the automated system (i.e., omissions, unordered drug errors, extra doses, wrong doses, wrong route, and wrong time errors), but at lower rates for most categories than for the previous system.

Cooper and colleagues [23] compared observation-based medication administration error rates for three different systems designed to package pharmaceuticals to facilitate their delivery to patients: a blister card, an ATC-212 single-unit dose packager, and the ATC Plus multidose packager in a 300-bed multisite nursing home. The ATC Plus system packaged all oral solid doses due to be given at the same time in the same package, which enabled the nurse to select one package instead of multiple packages. The machines packaged the medications in a central location, and a pharmacist inspected each package for accuracy before it was released to the nursing unit. The blister card system had an error rate of 8.0% on 265 orders, the single-unit dose package system had a 2.5% error rate on 287 orders, and no errors were detected with the multidose ATC Plus system on 265 orders. Error rates from these systems usually result from incorrect manual loading of the dispensing device prior to packaging the drug or two units placed in the same package.

A medication administration error rate of 6.9% was measured by observation for two nursing units using the Pyxis MedStation system [24, 25]. The MedStation system did not have a link to the patient's list of approved medications in the hospital's computer system, which allowed nurses to obtain any medication stored in the device for any patient. The error rate for all doses retrieved from the MedStation was 17% (21 errors per 123 opportunities), while the error rate for doses retrieved from the traditional patient medication drawer on a unit dose cart was 5% (43 errors per 796 opportunities). Errors involving the MedStation included seven unordered drug errors (ketorolac, Percocet, Unipen, and furosemide), and 14 wrong dose errors involving promethazine (12), lorazepam, and meperidine.

The effect of limiting the nurse to medications in the Medstation ordered for each patient on medication administration errors was studied by Borel and Rascati [26]. It was possible to override the system and obtain selected medications (for example, in emergencies) without prior pharmacist approval of the order. Overrides occurred during the study but it wasn't clear if any were associated with medication administration errors. The observation-based error rate was 16.9% (148 errors for 873 opportunities) before, and 10.4% (97 errors in 929 opportunities) after implementation of the MedStation-Rx system—a statistically significant difference ($\chi^2 = 16.24$, d.f. = 1, $p = 0.001$). The new system decreased the relative frequency of omission errors, but unauthorized drug errors and wrong dose errors were not affected. The rate of wrong time errors increased following the implementation of the system. These results suggest that having the medications available on the nursing unit affects omission errors, but there must be other reasons for different error types unrelated to automation.

Skibinski and colleagues [27] conducted a comprehensive evaluation of a series of technology implementations in a rehabilitation hospital, including a pharmacy computer system, automated dispensing cabinets (ADCs), and point-of-care products. The observation-based administration error rates decreased significantly for wrong time, wrong dose, and wrong route errors, but no change was detected for wrong drug and wrong form errors. No explanation was provided regarding this lack of effect of the bar code medication verification system on two types of errors that should have been prevented. The number of doses observed before and after implementation was not reported, limiting the ability to fully evaluate the analysis.

Paoletti and colleagues [1] evaluated the effect of changing from a five-day handwritten medication administration record (MAR) to an electronic, pharmacy-controlled MAR with bar code verification of medication administration on medication errors in a 521-bed general hospital. The new system decreased the error rate significantly on one of two study units when excluding wrong technique and wrong time errors, and there was no change in error rate on the control patient care unit (bar code system was not implemented). The lack of change on the patient care units that implemented the use of bar codes was attributed to differences in nurse practices. This study also reported the use of electronic surveillance data available from the bar code medication administration system. For example, over an 18-month period, medication scanning compliance ranged from 87% to 92% and prevented errors averaged 5 per 1,000 doses scanned.

Franklin and colleagues [28] conducted a before-and-after study of the effects of three technological implementations: computer prescriber order entry, an automated dispensing system on the ward, and bar code medication administration verification on medication safety. The study was performed in the United Kingdom with a new automated device. The nurse would select a patient on the device's screen which then presented her with a list of that patient's medications due within the next 2 hours. The machine would then release each of the medications to the nurse, who would remove the dose from the open drawer. All doses for each patient were placed in a bar code controlled drawer on an electric drug trolley. During medication administration, the drawer could only be opened by scanning the patient's bar coded wristband identification. The new technology resulted in a decrease in medication administration errors from 7.0% (1,473 doses) to 4.3% (1,139 doses), a significant decrease ($p = .005, \chi^2$).

Barker [29] described 10 evidence-based pharmacy automation features designed to reduce medication administration errors that should be used to evaluate systems:

1. *System controls are comprehensive.* System control over the drug delivery process should start at the point of computer order entry and continue to the point of dispensing or administration. System control should be integrated with the pharmacy information system.
2. *Scope.* All drug doses are accommodated by the system, including unique or error-prone medications.
3. *Dispenses unit drug doses.* Drug doses are provided to the nurse in ready-to-administer form and do not require any further manipulation (e.g., measuring a dose of an oral liquid).
4. *Alerts nurse when drug doses are due.* Reminders can help decrease wrong time errors and omission errors.
5. *Electronic identification (e.g., bar coding).* Each part of the medication system including the drug, patient, and person dispensing are identified in a unique and accurate manner to promote accurate use of the item.
6. *Access to medications is limited and controlled.* Medications are accessible only when needed, and only by authorized personnel.
7. *Dispensing and administration are documented.* Medication use information is recorded automatically and completely.
8. *Drug use information is provided.* Access must be immediate (e.g., at the bedside) and preferably on the medication package or on a labeled outer wrap. Examples of such information include warnings not to crush a dose and safe intravenous administration rates.
9. *Labeling machine printed and affixed.* Any additional labeling of the dose should be generated by a machine (not handwritten except in emergencies).
10. *Controls are not easily compromised.* Overrides of system warnings regarding errors are signaled visibly and audibly at the time of the incident, and documented by the system's software.

Despite the fact that most systems evaluated in these reports incorporate many if not all of these recommendations, error rates still result from the human intervention part of the drug administration process. Pursuit of the ideal system design to minimize the opportunities for adverse human interaction continues.

9.6 Technology and Automation: Community Pharmacy

As the pharmacist shortage continues while patients expect their prescriptions to be filled quickly, a number of semiautomated machines offer some relief. A wide range of products are available to do everything from counting tablets to fill and label prescription vials. Baker cells and Kirby-Lester counting systems have been available to count oral solids for over 30 years. ScriptPro offers an automated prescription filling system that presents the pharmacist with a filled and labeled vial for inspection as well as a photograph of the drug that should be in the vial. The Innovation PharmASSIST line of electronic products includes a scale for counting (SmartScale), machines that count oral solid doses (SmartCabinet), a device that fills and labels vials (ROBOTx), and a robotic arm that retrieves vials into which oral solids have been counted (RDS ROBOTx). Other devices include the AutoMed R800, and Parata RDS.

9.7 Effect of Automation on Dispensing Errors: Community Pharmacy

Dispensing errors are generally defined as delivery of a medication from the pharmacy with one or more deviations from the prescriber's order, and can occur within a hospital or at a community pharmacy when a prescription is delivered to a patient [30].

Flynn and Barker[31] measured dispensing error rates using observation in two pharmacies before and after implementation of a bar code based automated prescription filling and inspection system. Observation was performed by a pharmacist in one chain and one independent pharmacy for 12 to 13 days in order to inspect approximately 3,300 prescriptions for dispensing errors before and after system implementation (13,000 prescriptions total). The new system included the SP Central Pharmacy Dispensing Management System, SP Station, SP Checkpoint, and the ScriptPro 200 Automated Prescription Filling Device. At Pharmacy 1, the dispensing error rate was 2.8% before (99 errors on 3,427 prescriptions) and 2.1% after the new system (68 errors on 3,241 prescriptions). This was a significant decrease in error rates ($p = .099$). At Pharmacy 2, the dispensing error rate was 1.9% before (64 errors on 3,424 prescriptions) and 2.4% after the new system (74 errors on 3,028 prescriptions). This was not a significant difference ($p = .225$). The authors concluded that a comprehensive automated prescription filling system is capable of significant decreases in dispensing error rates, but requires staff to use built-in safety controls.

9.8 Examples of Problems with Automation That Can Affect Medication Safety

Overrides of automated system warnings should be monitored and followed up on to ensure that the benefits of technology are realized (e.g., if a wrong dose error warning is ignored when the nurse believes she or he has the correct dose). A recent national survey of hospitals found that the number of medications dispensed subsequent to an override decreased from 22.8% in 2002 to 13.3% in 2006 [5], which indicates there is still a high potential for error. Kowiatek and colleagues [32] described the use of an expert panel at their hospital to develop criteria for override access to a limited list of medications stored in automated devices. The application of these criteria and an override monitoring tool to randomly audit compliance with override practices significantly decreased opioid override rates.

Automation may give staff a false sense of security, leading them to be overconfident that the device always provides them with the correct drug and strength. Pharmacy-replenished unit-based drug dispensing devices require monitoring to ensure medication dispensing quality. Klibanov and Eckel found that 2.3% of device drawers audited contained incorrect unit dose medications—most were in drawers that allowed access to more than one medication upon opening [33].

Patterson and colleagues [34] observed nurses as they used bar code technology during medication administration and described the workarounds witnessed. The workaround strategies employed were typically performed because they were considered faster than the bar code system. The authors concluded that workarounds were particularly troublesome because they were not detectable by reviewing electronic documentation—they could only be witnessed by observation. When compliance with safety procedures associated with new technology differs significantly from actual practice, problems (e.g., medication errors) will occur. Dr. Patterson and colleagues have published best-practice recommendations for bar code medication administration based on studies in the Veterans Health Administration using observers with a human factors background [35].

9.9 Conclusion

Amalberti and colleagues described one of five system barriers to achieving ultrasafe healthcare as the need for simplification [36]. When technology leads to a more complex system, there is the increased potential for new kinds of errors that are more difficult to detect, and thus, the intended goal of safer care cannot be achieved.

Pharmacy automation has made an important impact on medication distribution in hospitals and community pharmacy settings, but healthcare providers must continue to be vigilant in testing each new system to ensure benefits are realized and safety is not compromised.

References

[1] Paoletti, R. D., T. M. Suess, and M. G. Lesko, et al., "Using Bar-Code Technology and Medication Observation Methodology for Safer Medication Administration," *Am J Health Syst Pharm,* Vol. 64, No. 5, 2007, pp. 536–543.

[2] "Neonatal Heparin Overdose Reveals Problems of Automated Drug Dispensing Without a Safety Net: Isolated Case or System Failure?" *J Clinical Engineering,* Vol. 31, No. 3, 2006, pp. 170–171.

[3] Barker, K. N., B. G. Felkey, and E. A. Flynn, et al., "White Paper on Automation in Pharmacy. Consultant Pharmacist," March 23, 1998, pp. 256, 261, 265–266, 268, 274–276, 279, 283–284, 286, 289–290, 293.

[4] Rough, S., J. Temple, "Automation in Practice," in *Handbook of Institutional Pharmacy Practice, 4th Editon,* pp. 329–352, T. R. Brown (ed.), Bethesda, MD: American Society of Health-System Pharmacists, 2006.

[5] Pedersen, C. A., P. J. Schneider, and D. J. Scheckelhoff, "ASHP National Survey of Pharmacy Practice in Hospital Settings: Dispensing and Administration—2005," *Am J Health Syst Pharm,* Vol. 63, No. 4, 2006, pp. 327–345.

[6] Eckel, S. F., and F. M. Eckel, "Medication Distribution Systems," in *Handbook of Institutional Pharmacy Practice, 4th Edition,* T. R. Brown (ed.), Bethesda, MD: American Society of Health-System Pharmacists, 2006, pp. 383–392.

[7] http://www.mckesson.com/en_us/McKesson.com/For%2bPharmacies/Inpatient/Pharmacy%2bAutomation/ROBOT-Rx.html.

[8] http://talyst.com/Products/Hardware/AutoCarousel.

[9] Guerrero, R. M,. N. A. Nickman, and J. A. Jorgenson, "Work Activities Before and After Implementation of an Automated Dispensing System," *American Journal of Health-System Pharmacy,* Vol. 53, No. 5, 1996, pp. 548–54.

[10] Shirley, K. L., "Effect of An Automated Dispensing System on Medication Administration Time," *Am J Health Syst Pharm,* Vol. 56, No. 15, 1999, pp. 1542–1545.

[11] Douglas, J. B., and DS Wheeler, "Evaluation of Trained Pharmacy Technicians in Identifying Dispensing Errors," ASHP Midyear Clinical Meeting, December 29, 1994, P-244(E).

[12] Woller, T. W., J. Stuart, and R. Vrabel, et al., "Checking of Unit Dose Cassettes by Pharmacy Technicians at Three Minnesota Hospitals: Pilot Project," *American Journal of Hospital Pharmacy,* Vol. 48, September 1991, pp. 1952–1956.

[13] Becker, M. D., M. H. Johnson, and R. L. Longe, "Errors Remaining in Unit-Dose Carts after Checking by Pharmacists Versus Pharmacy Technicians," *American Journal of Hospital Pharmacy,* Vol. 35, April 1978, pp. 432–434.

[14] Mayo, C. E., R.G. Kitchens, and L. Reese, et al., "Distribution Accuracy of a Decentralized Unit-Dose System," *American Journal of Hospital Pharmacy,* Vol. 32, November 1975, pp. 1124–1126.

[15] Taylor, J., and M. Gaucher, "Medication Selection Errors Made by Pharmacy Technicians in Filling Unit Dose Orders," *Canadian Journal of Hospital Pharmacy,* Vol. 39, February 1986, pp. 9–12.

[16] Hassall, T. H., and C. E. Daniels, "Evaluation of Three Types of Control Chart Methods in Unit Dose Error Monitoring," American Journal of Hospital Pharmacy, Vol. 40, June 1983, pp. 970–975.

[17] Hoffmann, R. P., K. H. Bartt, and L. Berlin, et al., "Multidisciplinary Quality Assessment of a Unit Dose Drug Distribution System," *Hospital Pharmacy,* Vol. 19, March 1984, pp. 167–169, 173–174.

[18] Cina, J., M. McCrea, and P. Mitton, et al., "Detection of Medication Error Rates," ASHP Midyear Clinical Meeting 2003, Vol. 38, p. MED-08.

[19] Cina, J. L., T. K. Gandhi, and W. Churchill, et al., "How Many Hospital Pharmacy Medication Dispensing Errors Go Undetected?" *Joint Commission Journal on Quality & Patient Safety,* Vol. 32, No. 2, 2006, pp. 73–80.

[20] Poon, E. G., J. L. Cina, and W. Churchill, et al., "Medication Dispensing Errors and Potential Adverse Drug Events Before and After Implementing Bar Code Technology in the Pharmacy," *Annals of Internal Medicine,* Vol. 145, No. 6, 2006, pp. 26–34.

[21] Oswald, S., and R. Caldwell, "Dispensing Error Rate After Implementation of an Automated Pharmacy Carousel System," *Am J Health Syst Pharm,* Vol. 64, No. 13, 2007, pp. 1427–1431.

[22] Barker, K. N., R. E. Pearson, and C. D. Hepler, et al., "Effect of an Automated Bedside Dispensing Machine on Medication Errors," *American Journal of Hospital Pharmacy,* Vol. 41, July 1984, pp. 1352–1358.

[23] Cooper, S., D. Zaske, and R. Hadsall, et al., "Automated Medication Packaging for Long-Term Care Facilities: Evaluation," *Consultant Pharmacist,* Vol. 9, January 1994, pp. 58–70.

[24] Dean, B. S., E. L. Allan, and N. D. Barber, et al., "Comparison of Medication Errors in an American and a British Hospital," *American Journal of Health-System Pharmacy,* Vol. 52, No. 22, 1995, pp. 2543–2549.

[25] Barker, K. N, and E.L. Allan, "Research on Drug-Use-System Errors," *American Journal of Health-System Pharmacy,* Vol. 52, No. 4, 1995, pp. 400–403.

[26] Borel, J. M., and K.L. Rascati, "Effect of an Automated, Nursing Unit-Based Drag-Dispensing Device on Medication Errors," *American Journal of Health-System Pharmacy,* Vol. 52, No. 17, 1995, pp. 1875–1879.

[27] Skibinski, K. A., B. A. White, and L. I. K. Lin, et al., "Effects of Technological Interventions on the Safety of a Medication-Use System," *Am J Health Syst Pharm,* Vol. 64, No. 1, 2007, pp. 90–96.

[28] Franklin, B. D., K. O'Grady, and P. Donyai, et al., "The Impact of a Closed-Loop Electronic Prescribing and Administration System on Prescribing Errors, Administration Errors and Staff Time: A Before-and-After Study," *Quality and Safety in Health Care,* 2007, Vol. 16, pp. 279–284.

[29] Barker, K. N., "Ensuring Safety in the Use of Automated Medication Dispensing Systems," *American Journal of Health-System Pharmacy,* Vol. 52, No. 21, 1995, 2445–2447.

[30] Flynn EA, Barker KN: Medication error research. In *Medication Errors, 2nd ed.,* M. R. Cohen (ed.), Washington, DC: American Pharmaceutical Association, 2006.

[31] Flynn, E. A., and K. N. Barker, "Effect of an Automated Dispensing System on Errors in Two Pharmacies," *Journal of the American Pharmacists Association,* Vol. 46, No. 5, 2006, pp. 613015.

[32] Kowiatek, J. G., R. J. Weber, and S. J. Skledar, et al., "Assessing and Monitoring Override Medications in Automated Dispensing Devices," *Joint Commission Journal on Quality & Patient Safety,* Vol. 32, No. 6, 2006, pp. 309–317.

[33] Klibanov, O. M., and S. F. Eckel, "Effects of Automated Dispensing on Inventory Control, Billing, Workload, and Potential for Medication Errors," *American Journal of Health-System Pharmacy,* Vol. 60, No. 6, 2003, pp. 569–72.

[34] Patterson, E. S., M. L. Rogers, and R. J. Chapman, et al., "Compliance with Intended Use of Bar Code Medication Administration in Acute and Long-Term Care: An Observational Study," *Human Factors,* Vol. 48, No. 1, 2006, pp. 15–22.

[35] Patterson, E. S., M. L. Rogers, and M. L., Render, "Fifteen Best Practice Recommendations for Bar-Code Medication Administration in the Veterans Health Administration," *Joint Commission Journal on Quality and Patient Safety,* Vol. 30, July 2004, pp. 355–365.

[36] Amalberti, R, Y. Auroy, and D. Berwick, et al., "Five System Barriers to Achieving Ultrasafe Health Care," *Annals of Internal Medicine,* Vol. 142, 2005, pp. 756–764.

CHAPTER 10

Automation Technologies in the Operating Room

Jonathan M. Sackier and Yulun Wang

10.1 Introduction

A Czechoslavakian playwright, a collection of Greek mythology and a surgeon's hand reaching into the future for stability and accuracy—what could these have in common?

In 1923, Karel Capek wrote a play about a Utopian society where automatons undertook the mindless work, thereby freeing humans to achieve a more meaningful existence. *Rossums Universal Robots* gave the world a new word, and as technology advanced, robots found their way into many spheres of activity, seeking to, but not quite achieving Capek's dream. Automation on the industrial production line saw robots assemble and spray paint cars and handle fragile and delicate components, the advent of urban terrorism had robots investigating suspicious objects, and space exploration deployed robotic arms to undertake complex, demanding, and repetitive maneuvers.

Creative spirits grasped the potential of robotics with passion, and from Isaac Asimov's three laws of robotics to the characters of popular media fame, a long line of intriguing personalities entered our lexicon. Indeed, the lovable but rather frenetic C-3PO from *Star Wars* was based on the robot Maria from Fritz Lang's *Metropolis* and the smaller, incomprehensible "droid," R2-D2, was (according to movie legend) rather ignominiously modeled on George Lucas's vacuum cleaner!

The prophetic author Isaac Asimov prescribed the "three laws of robotics," which state:

1. A robot may not injure a human being or, through inaction, allow a human being to come to harm;
2. A robot must obey orders given it by human beings except where such orders would conflict with the First Law;
3. A robot must protect its own existence as long as such protection does not conflict with the First or Second Law.

The generation that brought this combination of art and science to the operating room was influenced by the images they saw on their television screens, in the movie theaters, and in their imaginations.

10.2 Prosthetic Hip Surgery

For many years, degenerative joint disease has been treated with reconstructive surgery whereby the diseased and worn bone is removed and a prosthetic multipart joint is inserted. One of the most common procedures is to excise the head of the femur (the thigh bone) and to place a cup of high impact material in the acetabulum (the socket in the pelvis that receives the head of the femur) and a metal component into the femur itself. This second component consists of a smooth ball that fits into the socket and a long stem that fits into the interior of the femur. It has been usual practice to cement this stem in place, but in the 1990s there was a movement to make a closer-fitting prosthesis that would be cementless. This led to the work of Dr. Paul in 1992 where a robotic surgical device, "Robodoc," was used to fashion a cementless hip prosthesis procedure whereby a precise placement was achieved. This technology did not find its way into popular practice since achieving commercial viability a medical technology has to add perceived value to the physician, improve outcomes, reduce cost, and preferably do all three. Robodoc could not pass muster using these filters and thus did not proceed.

10.3 AESOP

Before launching into a description of the first commercially viable operating room robot, AESOP, it is important to understand the landscape it was designed for; the rather tortologous term "minimally invasive surgery." To understand the utility this robot provided, we need to describe the historical perspective.

The term *laparoscopy* literally means to "look into the loins" but was adopted to refer to the approach first utilized at the turn of the nineteenth century in rather tentative fashion by the German surgeon, Kelling. He used room air to inflate the abdominal cavity and then inserted a cystoscope, originally designed to look into the bladder, to visualize the contents of the abdomen. He correctly called this technique *"coeliscopy."* In Sweden, Jacobaeus reported 119 cases and then a physician from Johns Hopkins in Baltimore utilized a mirror designed for ENT work to gaze into the same space.

Another German, Heinz Kalk, wrote a monograph on the technique and reported on his experience with 2,000 liver biopsies. Herr Kelling took exception to a publication by Jacobaues on the topic of primacy of thought and responded with a rather vitriolic attack and then Dr. Kalk entered the fray over ownership of the "intellectual property." However, despite this passion, the technique and potential value of "peritoneoscopy," "coelioscopy," or "laparoscopy" was ignored for decades.

In the 1960s, gynecologists rediscovered the technique and used it to examine the pelvic organs and then developed the methodology to perform sterilization of women by ligating the tubes that carry eggs from ovary to uterus. Additionally, surgeons were using laparoscopic examination to diagnose causes of abdominal pain, triage trauma, stage cancer, and for a host of other applications. At this point in history, visualization was based on a rigid glass scope through which the operating physician would peer to the exclusion of everyone else in the room. In 1981, yet another German, Kurt Semm, published a technique to remove the appendix and the era of operative laparoscopy was born [1].

A few years later, Mûhe from Germany and others from France, America, and Britain reported techniques and results of laparoscopic cholecystectomy, removal of the gallbladder under guidance of the laparoscope and with small instruments placed via tubes (cannulae) into the abdominal cavity. The impact was massive and rapid, for now there was an operation that was performed by every general surgeon and that likely contributed to 25% to 35% of their incomes. Performing laparoscopic surgery was no longer an issue of whether this might be a "nice" or "interesting" technique, but was now elevated to a technique that was necessary to maintain one's business. The stampede to learn started and grew exponentially [2].

One of the developments that facilitated this relatively bloodless revolution was the attachment of a charge-coupled device (CCD) camera to the laparoscope that allowed the image of the surgical field to be displayed on one or more television screens in the operating room, thereby allowing others to see and participate in the procedure. Indeed, it was critical that others be able to see, for the surgeon needed three additional instruments to allow safe and expeditious extraction of the diseased organ; one to hold the gallbladder up and over the liver, one to pull it from side to side, and another to dissect it free from its surroundings; the fourth port provided access for the 10-mm diameter laparoscope.

This meant that an assistant was required to hold the scope and move it under orders from the operating surgeon and this became a source of frustration for all concerned. First, holding a laparoscope is fairly mindless and induces boredom and loss of concentration so the unfortunate individual would tend to "wander" off target, leaving the surgeon staring at anything other than the operative field. Due to the inherent magnification provided by the scope, camera, and large television screen, the smallest tremor induced by consumption of alcohol the night before, a morning cup of coffee, or even the excursions caused by one's heartbeat or a sneeze or cough could induce motion sickness in even the most robust individual.

In order to affect control, a set of verbal orders needed to be given that were confusing and frustrating for all concerned during this operation. The surgeon would stand to the patient's left, the camera operator to the patients' right and the TV screen(s) would be at the head of the operating table. If the surgeon asked the assistant to "move right," did this mean the surgeon's right, the assistant's right, the patient's right, or to screen right, for all are in different directions. Furthermore, sometimes an angled view laparoscope was used where the tip showed a view at 30° or 45° to the axis of the scope. If the CCD camera "twisted" along that axis relative to the laparoscope, "right" (whoever's right that might be!) could actually turn into "up!" Finally, even the most astute and committed laparoscope assistant would never be able to anticipate the surgeon's next move with any degree of reliability and this also caused stress and delays to the prosecution of the procedure. Of course, despite the shortcomings of the unfortunate individual condemned to hold the laparoscope, that person still commanded fixed costs to be borne by the healthcare system and could not be more profitably used for the duration of the operation.

A number of rigid and semirigid systems were developed to hold the laparoscope in a stable configuration but demanded that the surgeon repeatedly adjust the position of the scope.

This set the stage for the automated endoscope system for optimal positioning (AESOP) to be devised—as we have seen, robots are ideal at undertaking predictable, repetitive, and uninspiring work and anyone who has held a laparoscope would definitely characterize the job in that manner [3] (see Figure 10.1).

The device was developed by Computer Motion (Goleta, Calfornia) and was based on concepts devised for defense purposes, and the intellectual property was transferred to the healthcare arena as a result of a productive relationship between surgeons and engineers. The device consisted of several components:

- A robotic arm, whose base was rigidly attached to the operating table rail and which was covered with a sterile drape, thereby preserving the sterility of the operating environment. The arm provided 7 degrees of freedom with four active joints, two passive joints, and one that could be adjusted statically.
- A sterilized collar that attached to the laparoscope, pierced the sterile drape, and thereby connected to the arm.
- A controller unit that included the processor or "intelligence" of the device.
- A foot pedal that was operated by the surgeon and controlled the movement of the arm. Moving the pedal to the left moved the arm to the left, moving to the right moved the arm to the right, pressing the toe down moved the tip of the laparoscope down, and pressing with the heel moved the tip up. A combi-

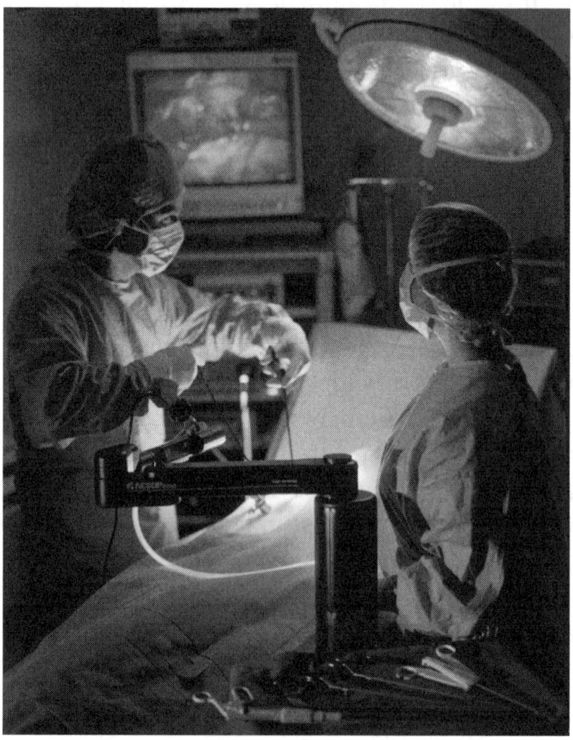

Figure 10.1 The AESOP 3000 was a voice-controlled robotic arm used to direct a laparoscope in minimally invasive surgery.

nation of movements of the foot pedal led to a similar combination of movements of the laparoscope and the greater the pressure applied to the foot pad, the faster the movements. Additionally, two buttons to the left and right of the foot pedal would facilitate zooming in and out and the device also had buttons to recall programmed positions—a very useful facility that allowed the surgeon to move from a panoramic view of the entire abdomen and a closeup of the operative field.

Obviously, patient safety was a concern and, in keeping with the precepts of Asimov's laws, the device was designed to be fail-safe. For instance, any undue pressure on the laparoscope that might be reflective of impending trauma to the patient would cause the collar to disengage and if a more radical issue arose, the robotic arm would suspend activity. The device obtained clearance under the 510K provision of the Food and Drug Administration in 1993 and the first clinical case at the University of California, San Diego, was attended by a great deal of media attention.

Subsequently, the device was widely adopted around the world; furthermore, laparoscopic surgery training courses were not complete unless they included an introduction to medical robotics. Research demonstrated the device led to a smoother and less frustrating procedure and that time for neophytes to learn how to control the laparoscope to perform reproducible tasks robotically was equal to manual control [4].

It became rapidly apparent that although foot control with AESOP was a vast improvement on third-party manual control, a more intuitive method was required and a voice-activated interface was necessary [5]. This required the surgeon to learn a series of specific commands such as

"AESOP right"
"AESOP, remember position 1"
"AESOP return position 1"

Having learned these commands, the surgeon would don a microphone-equipped headset and record them several times onto a chip that would then be inserted when that individual was operating so the robot would only respond to that doctor's commands and not to extraneous noises in the operating room. After regulatory approval had been obtained from the FDA in 1996, the voice-controlled robot was released to a receptive and appreciative surgical audience [6].

10.4 HERMES OR Control Center

Named for the mythical messenger, the HERMES OR Control Center is a computer server that networks and provides centralized control of all of the electronic devices in the operating room. HERMES enables the surgeon to use simple voice commands or a hand-held touch-screen pendant to control a network of smart medical devices that may need adjustment during a surgical procedure. The system was designed with an open architecture software platform, making it easy for medical device manufacturers to create HERMES-ready devices. Stryker Endoscopy partnered with Computer Motion to create the first array of networked devices that could be voice controlled by the surgeon (Figure 10.2).

The networked devices included video cameras, insufflators, recording equipment, surgical tables, operating room lights, and even robotic surgical tools like AESOP. HERMES also provided the surgical team with visual (via video monitor) and voice feedback on the status of each networked device. Increasing surgeon control and providing a useful means of conveying important device information feedback improved safety, operating room efficiency, communication, and augmented the quality of care delivered to the patient. It also enabled operating room support staff to focus more directly on patient care, as opposed to performing mundane tasks for the surgeon [7].

10.5 Zeus

The next generation device, Zeus, named for the mightiest of the ancient Greek deities, came to fruition at the very height of new developments in laparoscopy in the mid-1990s. The initial intent was to facilitate complex microsurgical procedures such as the repetitive but precise suturing required to complete a coronary artery bypass, one of the most commonly performed surgical operations [8].

This facilitation was predicated on the concept of "robotically enhanced surgery" whereby a stable two- or three-dimensional view of the operative field was provided by AESOP and gross human muscle movement could be accurately scaled down and filtered of extraneous movements such as tremor to allow exact manipulations. The device would be defined as a "master-slave" telepresence construct whereby the machine mirrored human movement, although not in a mathematically linear fashion. Additionally, the thinking was to take the surgeon from the

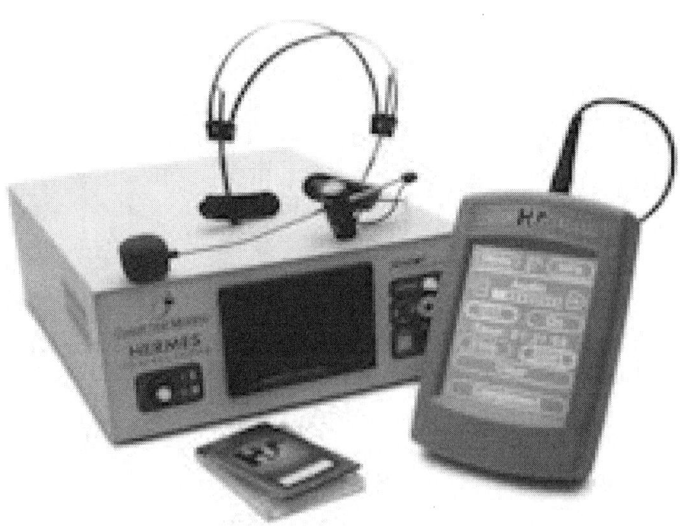

Hermes system

Figure 10.2 HERMES OR Control Center. The HERMES system was a voice-controlled computer that networked all of the electronic devices in the OR enabling centralized control.

strenuous, back-pain-inducing and ergonomically unsatisfying standing position and seat the operator in a comfortable chair with remote controls for the robot and an appropriate orientation to the patient [9] (Figure 10.3).

The construct of the robot was centered on the AESOP platform to provide stable endoscopic imagery and then two additional AESOP positioning robots to each of which was attached an instrument driver that was configured to hold a range of new instruments. The surgeon had a console with two handles, one for each hand, which controlled the respective manipulator and resembled the familiar laparoscopic instrument handle. Control of the imaging robot was through the now popular voice-recognition system controlled by the surgeon. The laparoscopic proceedings were either viewed on a conventional television screen for the 2-D version or with the aid of goggles for 3-D imagery.

A further development followed in 2001, the MicroWrist System, which added another degree of freedom and a new interface whereby there was direct correlation between surgeon finger motion and motion of the tip of the laparoscopic instrument [10].

To prevent the potentially dire consequences of inadvertent activation and to remain true to Asimov's principles, certain safety features were incorporated such as handles that needed to be gripped and a pedal to be depressed to ensure that any movement of a control was intentional; if a handle was manipulated without the

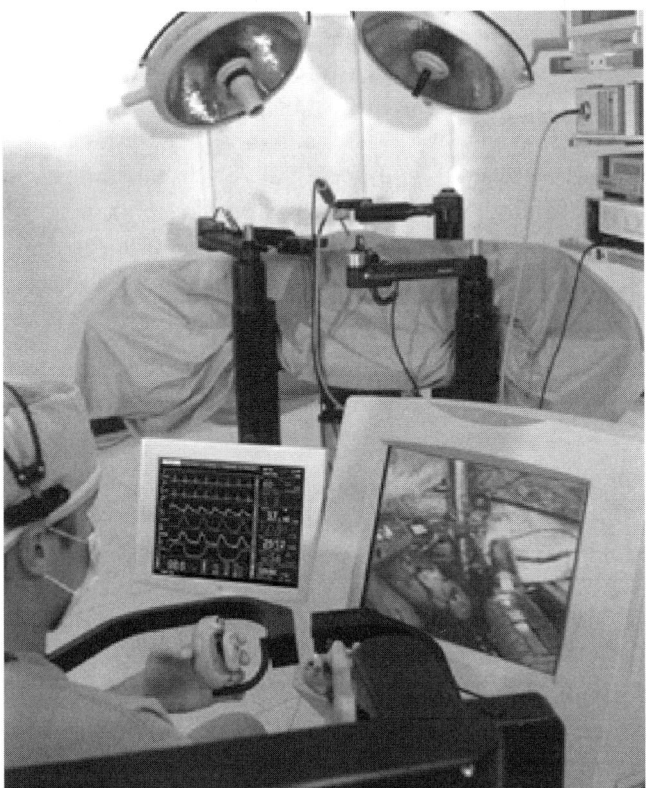

Figure 10.3 The ZEUS system was a trio of robotic arms controlled by the surgeon in a master-slave configuration enabling precise micro-surgical operations in minimally invasive surgical procedures.

pedal being depressed, the robot would ignore the command. If released during a procedure, the pedal would cause the robot to enter "hold" mode so the manipulators could be released, freeing the surgeon to view radiographs, discuss the case, or take brief respite from the proceedings. This "indexing" is similar to when a computer mouse is removed from the pad and then returned, and the screen pointer starts off again from where it was when the mouse was picked up rather than demonstrating any reference to the movement. Additional safety features included fail-safe modalities and error detection capabilities.

The robot also offered the ability for the surgeon to adapt a variety of customizable items such as the velocity with which the robot would respond, scaling of the relationship between master and slave, forces applied by the grippers, and so on.

The device demanded the design and construction of new instruments and tools as conventional laparoscopic instruments could not simply be attached. This demanded new thinking as well, for many conventional laparoscopic tools use wires to transmit the force from handle to tip and this presented some potential risks; therefore a push-rod design was chosen. In 1999, Dr. Douglas Boyd performed the first documented coronary artery bypass on a beating heart without the aid of a heart-lung machine and completed through minimally invasive means with the Zeus system. This demanded a system to stabilize the beating heart and this consisted of a tool that could be inserted through a port and then opened up inside the chest into a fan shape [11].

True remote telepresence was realized with a modified Zeus system in "Operation Lindbergh," whereby a surgeon in New York City removed the gallbladder of a patient in the French city of Strasbourg just days before the tragic events of September 11, 2001. Despite the great distance involved, the latent period caused by time for the signal to transmit did not cause any difficulties for the operating team.

This procedure led to many to dream of circumstances whereby this technology would allow highly skilled surgeons in one locale to offer lifesaving or enhancing procedures (or elements of procedures) to patients across the globe, to astronauts in orbit, to those "imprisoned" by an outbreak of a highly contagious disease or other complex environments [12].

10.6 da Vinci

The next progression in robotically enhanced surgery was named for one of the greatest visionaries, Leonardo da Vinci, and the technology and computer code required to achieve this milestone was surely in proportionate scale to the international acclaim that accompanied a published novel shortly thereafter!

In 1997, Cadiere performed a laparoscopic cholecystectomy with da Vinci. Other procedures followed swiftly, including gastric bypass for obesity and Nissen fundoplication for gastroesophageal reflux disease. The merger of two companies, Computer Motion with Intuitive Surgical, enabled all of their technology resources to focus on continuing to advance the da Vinci robot. By 2007, the da Vinci system has reached a critical mass in the marketplace where robotically assisted radical prostatectomy was becoming the standard of care as it allows for a more precise operation, and other gynecologic and cardiac procedures are rapidly gaining adoption. In general surgery, Heller's cardiomyotomy (a procedure to treat esophageal disease whereby muscle fibers are divided), removal of the adrenal gland, fixation of the rectum for prolapse, and many other procedures are driving the visibility of robotically assisted surgery into a wider audience (Figure 10.4).

10.7 RP-7

The RP-7 (In Touch Technologies, Santa Barbara, California) provides healthcare providers with the opportunity to be in two places at one time; that is, to remotely communicate and interact with patients and personnel when they cannot be there in person [12]. There are two components to RP-7 ("Remote Presence, 7th Generation"); a robot and a remote control station. The former consists of a mobile base unit that utilizes a holonomic drive topped by a video monitor that displays the face of the remote operator, and a video camera that allows the remote operator to see the view where the robot is located. The robot can be driven from one locale to another within the healthcare environment, is capable of movement in all directions, and the "head" can also move by panning and tilting. Additionally, the body is equipped with sensors that allow the remote driver to "sense" objects or people when they are not in view (Figure 10.5).

The remote operator sits at the control station (a computer with a broadband connection) and by means of a joystick can control the RP-7's position. The same control can be enabled with a wireless laptop computer. The computer screen displays not only the view from the robot but also what the distant audience is seeing

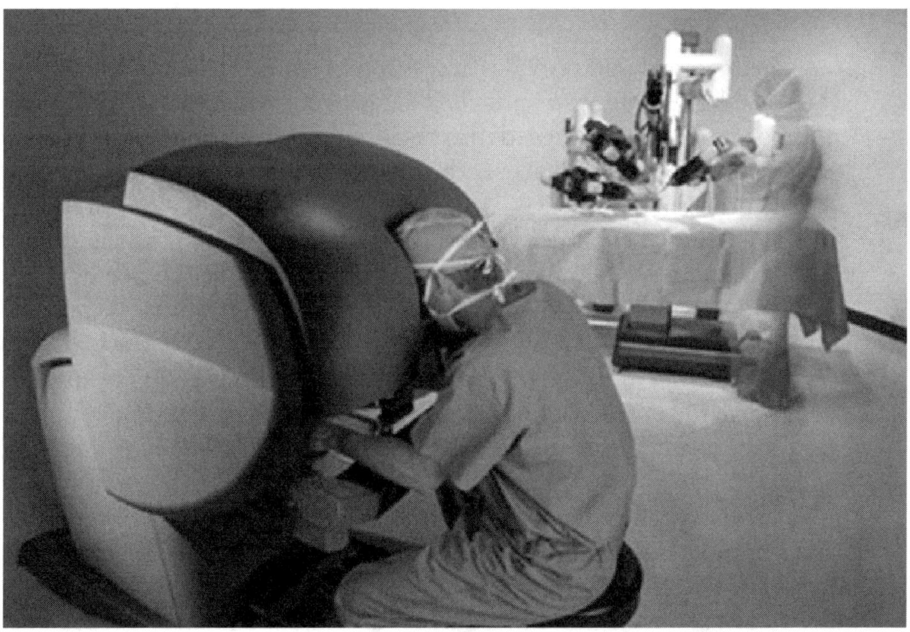

Figure 10.4 The da Vinci robot (da Vinci SurgeryT, Intuitive Surgical, Inc., Sunnyvale, CA), (pictured in the background) can operate through three quarter sized incisions in the patient's abdominal wall, while the surgeon (pictured in the foreground) controls the robotic manipulators remotely.

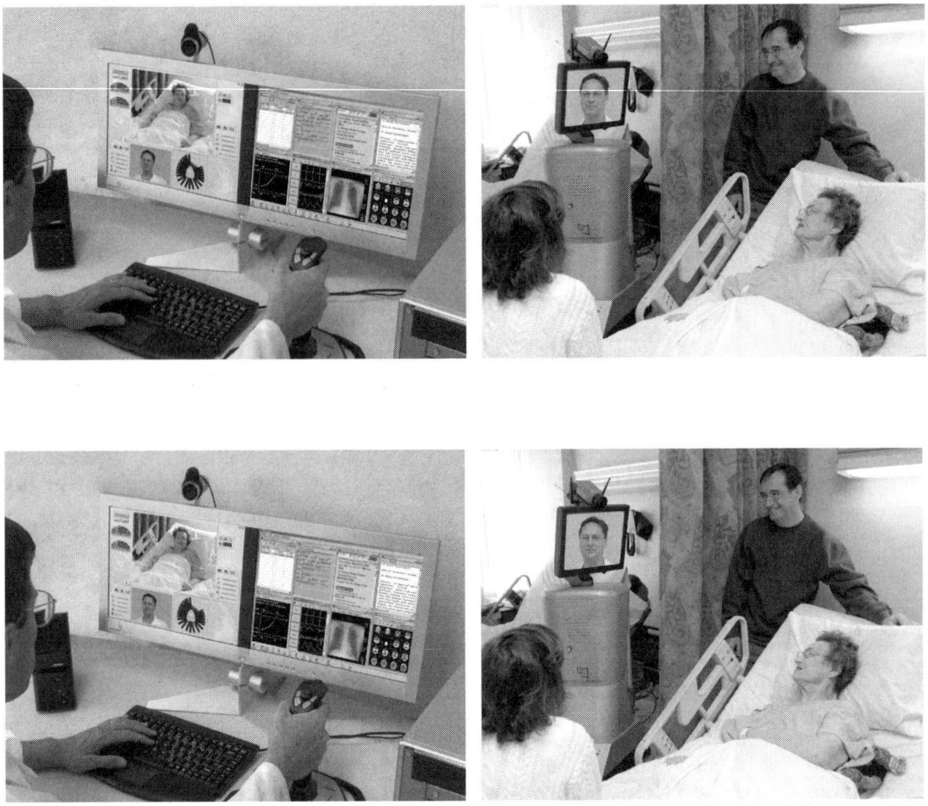

Figure 10.5 RP-7. The RP-7 is a robotic telepresence system enabling mobile, remote-controlled, 2-way audio/video communication.

(i.e., the operator) and provides access to electronic medical records. It can also be set up to take readings from attached medical devices such as a stethoscope or video endoscope that can be plugged directly into RP-7 by a local assistant and the distant physician can hear and interpret the physical signs [13].

The problem that the RP-7 addresses is that it is difficult, or impossible, for the proper healthcare expertise to be provided to the patient at the opportune time. Given the aging population, the growing shortage of healthcare professionals, and the continuing subspecialization of medicine, this problem is continuing to get worse. The RP-7 was designed to provide healthcare professionals "in-person" access to patients from geographically diverse locations [14]. The areas of most frequent usage today are in applications where patients may be critically ill and require rapid expert attention, such as in a hospital's intensive care unit or emergency department [15].

Several surgeons, including Drs. Petelin, Gandsas, Kavoussi, and Su, have begun to apply this technology to the operative environment whereby a surgeon can attend and even "proctor" another surgeon in performing a procedure. The learning curve of some surgical procedures is difficult and long, and it becomes impractical from many perspectives to have an expert surgeon in the operating room with the learning surgeon for a large number of procedures. This remote mentoring capability may be an important tool that might help enable a surgeon become proficient in a new procedure more efficiently and safely.

10.8 The Future

One of the most irresponsible and potentially embarrassing things a human might do is to predict the future, as history tells us again and again:

- In the 19th century, the president of the Royal Society predicted that no human could travel faster than 100 mph in a train due to lack of oxygen;
- In the 1940s, the chief executive of IBM predicted that there was probably a world market for four computers.

However, at the risk of future humiliation, we suggest the following; the achievements we might reach with robotics are limited only by our imagination, for as computational power increases, materials science provides us with the ability to make new components, and nanotechnology opens our minds to where smaller robots might go [16].

Robotics might help surgeons to do current operations faster and with more precision, facilitate new approaches, and may even render some current approaches totally redundant.

Might we see the day when an army or nanorobots enter the abdominal cavity, mobilize a diseased segment of bowel, invert and excise it, and then secure the separated ends together, bringing their spoils of war out through a natural orifice? Might we see "smart robots" act upon a surgeon's prompting but effect a procedure based on their greater skill, devoid of stress or emotion?

Whatever the future does hold, it is likely that the future referenced by Capek and framed by Asimov's rules will be more likely and will promise better, kinder, and more cost-effective healthcare for us and for our patients.

References

[1] Litynski, G. S., *Highlights in the History of Laparoscopy: The Development of Laparoscopic Techniques—A Cumulative Effort of Internists, Gynecologists and Surgeons*, Frankfurt, Germany: Barbara Bernert Nerlag, 1996.

[2] Sackier, J. M., "Protocols for Training in Minimally Invasive Surgery," *Surgery*, Vol. 15, 1997, pp. 273–276.

[3] Sackier, J., and Y. Wang, "Robotically Assisted Laparoscopic Surgery: From Concept to Development," *Surg. Endosc.*, Vol. 8, 1994, pp. 63–66.

[4] Jacobs, L. K., V. Shayani, and J. M. Sackier, "Determination of the Learning Curve of the AESOP Robot," *Surg. Endosc.*, Vol. 11, 1997, pp. 54–55.

[5] Sackier, J. M, C. Wooters, and L. K. Jacobs, et al., "Voice Activation of a Surgical Robotic Assistant," *Am. J. Surg.*, Vol. 174, 1997, pp. 406–409.

[6] Wang, W. F., D. Uecker, and Y. Wang, "Choreographed Scope Maneuvering in Robotically-Assisted Laparoscopy with Active Vision Guidance," *Proc. Third IEEE Workshop on Applications of Computer Vision*, 1996, p. 187.

[7] Wang, Y., and D. Roe, "A Voice Controlled Network for Universal Control of Devices in the Operating Room," *Minimally Invasive Therapy & Allied Technologies*, 2000.

[8] Reichenspurner, H., R. Damiano, and M. Mack, et al., "Use of the Voice-Controlled and Computer Assisted Surgical System ZEUS for Endoscopic Coronary Artery Bypass Grafting," *J. Thorac. Cardiovasc. Surgery,* Vol. 118, No. 1, 1999, pp. 11–16.

[9] Butner, S., and M. Ghodoussi, "A Real-Time System for Tele-Surgery," *Proc. IEEE International Conference on Distributed Computing Systems,* 2001, pp. 236–243.

[10] Ewing, D., A. Pigazzi, and Y. Wang, et al., "Robots in the Operating Room—The History," *Sem. Laparo. Surg.*, Vol. 11, 2004, p. 2.

[11] Argenziano, M., *Robotic Cardiothoracic Surgery*, New Jersey: Humana Press, 2007.

[12] Ellison, L., P. Pinto, and F. Kim, et al., "Telerounding and Patient Satisfaction after Surgery," *J. Am. Coll. Surg.*, Vol. 199, No. 4, 2004, pp. 523–530.

[13] Thacker, P., "Physician-Robot Makes the Rounds." *JAMA,* Vol. 293, 2005, pp. 2.

[14] Vespa, P., "Robotic Telepresence in the Intensive Care Unit," *Critical Care,* Vol. 9, 2005, pp. 319–320.

[15] Vespa, P., et al., "Intensive Care Unit Robotic Telepresence Facilitates Rapid Physician Response to Unstable Patients and Decreased Cost in Neurointensive Care," *Surg. Neurol.,* Vol. 67, 2007, pp. 331–337.

[16] Wang, Y., S. Butner, and A. Darzi, "The Developing Market for Medical Robotics," *Proc. IEEE,* Vol. 94, No. 9, 2006.

CHAPTER 11

Health Care Supply Chain Automation

Yue Xie

11.1 Introduction

A quick Internet search on the topic "healthcare supply chain automation" yields surprisingly few entries with relatively narrow coverage, while by comparison a similar search of "supply chain automation" yields a wide variety of articles expanding across the breadth of the topic. Is this just an example of how unremarkably healthcare as a whole has progressed in comparison to other industries in the development and application of supply chain knowledge and techniques that have evolved over the past few decades? The answer unfortunately is no, and much of the problems lie with the inherent complexity of the healthcare industry, with its myriad specialty products and the countless number of stakeholders that influence the supply chain decisions of the healthcare delivery organization. As a result, most healthcare providers are still struggling with internal supply chain functions such as product selection, demand forecasting, inventory management, information technologies, charge capturing, and supplier relationships. While healthcare struggles, other industries have found success in adopting the supply chain as a fundamental competitive strategy through expanding focus to include all suppliers and customers up and down the entire product supply-demand chain. Their success is based on the concept that cost savings and better customer service levels can both be achieved if suppliers and customers shared real-time demand information. Some examples of successful adopters of such supply chain strategies are Dell, Wal-Mart, and Toyota. By extending their respective supply chains upstream to include suppliers and downstream to include customers, these companies optimized supply chain strategies that transformed the competitive landscapes within their own industries, and as a result gained significant financial rewards. This is not to say that healthcare has not made any progress in supply chain. In fact, one area of healthcare supply chain that has managed to keep pace with other industries is the use of supply automation technologies such as the automated point of use (APU) systems and related information technology infrastructures that promote the transfer of supply information between the healthcare provider and its suppliers.

11.2 Rationale for Healthcare Supply Chain Automation

Walking through the hallways of a hospital, it is easy to observe the nature of supply disarray on a daily basis. Many hospitals, often because of lack of storage space and of the human tendency for hoarding under perceived scarcity, store supplies in hallways or in spaces where people often forget that the supplies are even there. In some extreme cases, supplies were discovered years later hidden above ceiling tiles just because staff members were wary of the consequences of supply shortages. Such waste, although understandable, only adds to the already high cost of healthcare supplies, which according to the Healthcare Financial Management Association (HFMA) is around 25% of the operating budget excluding supply chain labor and logistics costs [1]. If such costs were to be included, the overall supply chain costs may reach anywhere from 35% to 45% of an organization's operating budget [1].

Over the past 15 years, the near elimination of cost-plus reimbursement and the concurrent rise of managed care across the country along with the aforementioned high supply costs have continued to pressure provider operating margins. According to Solucient, the average operating margin has steadily declined to the low to mid single-digit range [2]. After adjusting for inflation, it seems that most hospitals achieve minimal return on investments (ROI). To remain financially viable for the long run, more and more providers are focusing on decreasing operating costs to counter the decline in reimbursement rates. Being one of the larger cost components, supply chain is naturally a focus of cost reduction and many providers are increasingly pursuing supply chain strategies that include the expanded use of information technology and automation solutions to create cost savings [3].

Along with margin pressure, healthcare providers face another related problem that is far more challenging to deal with than optimizing supply chains—nursing shortages. According to the U.S. Department of Health and Human Services' Bureau of Health Professions (BHPr), the nursing workforce incurred a 6% shortage, or 110,000 of the total number of nurses needed, in the year 2000. What's more alarming, is that the same report also shows that this number may even reach 29% by 2020 [4] (Figure 11.1).

One of the reasons why there is increasing nursing shortage is the growth and aging of the population. The U.S. Census Bureau projected in 2004 that the total U.S. population would increase from the 2000 census of 282,125,000 to 308,936,000 by 2010, and again to 363,584,000 by 2030 (Table 11.1), an increase of 9.5% and 28.9%, respectively, over the year [5]. At the same time, the percentage

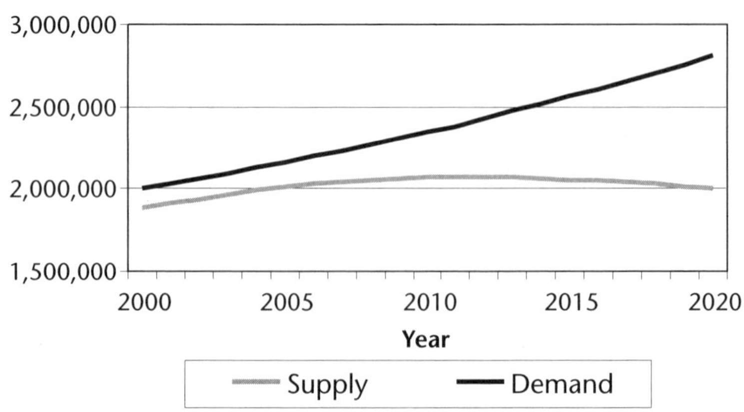

Figure 11.1 Projected shortage of nurses 2000–2020. (Source: U.S. Department of Health and Human Services 2002; ftp://ftp.hrsa.gov/bhpr/nationalcenter/rnproject.pdf.)

Table 11.1 Projected U.S. Population Growth

	2000	*2010*	*2030*
Population (thousands)	282,125	308,936	363,584
Increase (%)	—	9.5	28.9

Source: U.S. Census Bureau, International Data Base, Table 094.

of population over 65 would also increase from 12.4% in 2000 to 13% by 2010 and 19.7% by 2030 (Table 11.2).

This continued strong growth in U.S. population is a problem that the rapidly aging Europeans would love to have [6], but to the U.S. healthcare system, already at 16% of GDP (Figure 11.2) and projected to increase to 20% of GDP by 2015, this is a significant problem, not only in terms of cost (Figure 11.3), but also in the ability to supply the nursing workforce to support the growing and aging population.

Another contributing factor to the nursing shortage, according to research, may be the increased job stress levels that bedside nurses now face [7]. One unfortunate side effect of medical technology advancement and payor pressure on length-of-stay reduction is that people now live longer, but are generally sicker by the time they are admitted to hospitals. In addition, providers are increasingly experiencing higher patient-to-nurse ratios as population growth outpaces nursing workforce growth. As a result, nurses are currently experiencing high levels of job dissatisfaction and burnout, which again adds to the problem of nursing shortage. In a cross-national study involving more than 43,000 nurses, researchers reported that job dissatisfaction and burnout for U.S. nurses are the highest among five countries surveyed,

Table 11.2 Projected U.S. Percent of Total Population Over Age 65

Age	*2000 (%)*	*2010 (%)*	*2030 (%)*
65–84	10.9	11.0	17.0
85+	1.5	2.0	2.6
>65	12.4	13.0	19.7

Source: U.S. Census Bureau, International Data Base, Table 094.

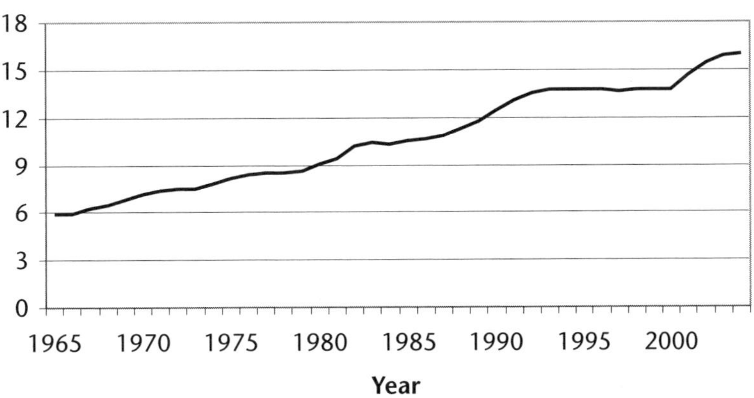

Figure 11.2 National health expenditures as percent of GDP 1965–2004. (*Source:* CMS, http://www.cms.hhs.gov/NationalHealthExpendData/downloads/nhegdp04.zip.)

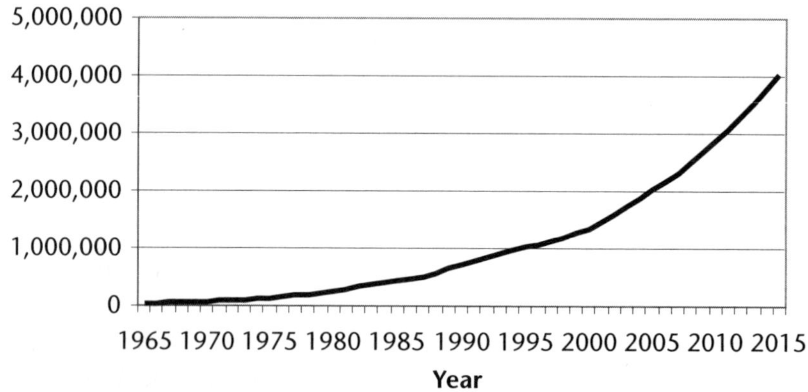

Figure 11.3 National health expenditures 1965–2015 ($ million). (Source: CMS, http://www.cms.hhs.gov/NationalHealthExpendData/downloads/nhegdp04.zip.)

and that one-third of nurses under the age 30 are considering leaving the profession within one year [8]. This high percentage of burnout, however, seems to reflect the overall troubling state of the nursing labor pool. In 2002 the U.S. Department of Health and Human Services' Bureau of Health Professions (BHPr) reported that nearly 500,000, or more than 25%, of the 1.89 million licensed RNs are employed outside of nursing, and the remaining nursing population is aging rapidly [4]. Figure 11.4 shows that the average age of nurses has gotten older over the years and it is now approaching 45. It also shows that the number of nurses entering the profession may not be sufficient to replace those that are leaving the workforce as they reach retirement age.

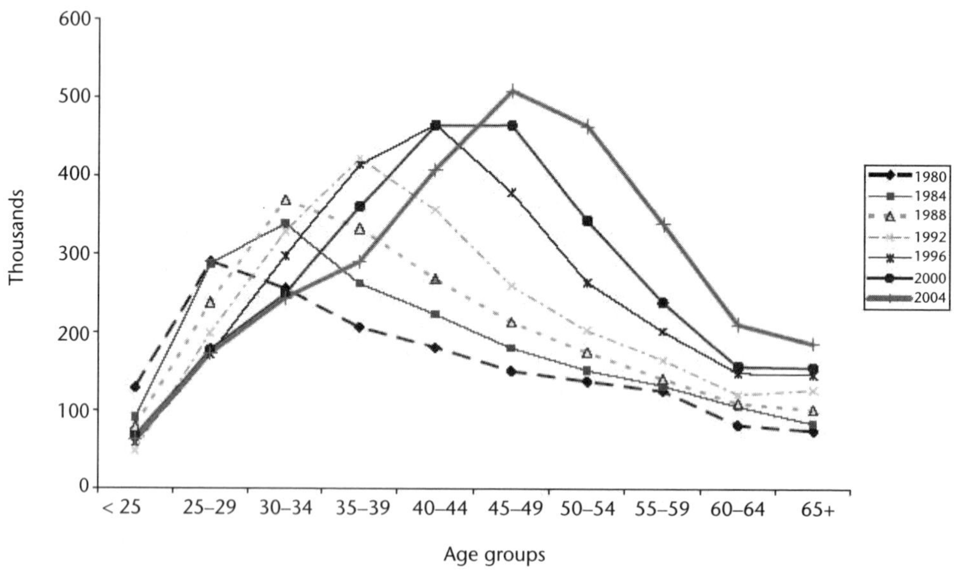

Figure 11.4 Age distribution of the registered nurse population, 1980– 2004. (Source: DHHS; http://bhpr.hrsa.gov/healthworkforce/reports/rnpopulation/preliminaryfindings.htm.)

As most industry experts would agree that there is a serious shortage of bedside nurses in the United States and that this trend would not change any time soon, many in the healthcare industry have started looking at technology not only as a way to improve quality, productivity, and nursing job satisfaction, but also to help recruit younger nurses [9, 10]. Given the amount of time nurses spend on handling supplies, it is logical that supply chain automation options are being actively explored by the healthcare industry to decrease both the cost of supplies and the amount of time nurses spend on nonclinical activities.

11.3 History of Supply Chain Automation Technologies in Health Care

Supply chain automation in the healthcare industry can be traced back to the American Hospital Supply Corporation's (AHSC) Analytic Systems Automatic Purchasing (ASAP) computer-based remote ordering, tracking, and supply management system, which began operations in 1957 as an internal solution [11]. In 1963, ASAP (by then called Tel-American), using proprietary punch card based IBM Dataphones, was finally introduced to the client hospitals and ushered in the "prime vendor" phase of the vendor-customer relationship in the healthcare industry. AHSC continued to improve the ASAP system through the early 1970s with its focus on high inventory customer service levels. By the mid 1970s, however, driven by customers' increasing need to improve internal supply management and cost, AHSC began to offer select customers value-added services, such as "customer purchase analysis" reports, and for a limited number of large customers, supply-chain consulting services. This new strategy provided AHSC (acquired by Baxter) dominant market share in the late 1980s with approximately 5,500 out of 6,900 hospitals nationwide using the ASAP system.

In the mid to late 1980s, however, Johnson & Johnson and Abbott Laboratories both introduced their own versions of computer-based supply chain systems to compete with Baxter. In addition, they no longer allowed Baxter to distribute the supplies they produced. In response, Baxter introduced ASAP Express in 1988, a multivendor system that sold eight other vendors' products in addition to Baxter's own.

In 1990, Baxter again proved its ability to provide a leadership position in innovation by advancing the supplier-customer relationship with the introduction of the Valuelink program. Valuelink was sold as an integrated, strategic partnership based logistics system that synchronized the flow of products and information between the customer and Baxter via consolidated purchases and multiple deliveries to the point-of-use inside the hospital seven days a week [11]. In other words, this was the beginning of vendor-managed-inventory (VMI), a form of a just-in-time (JIT) inventory system that allows the vendor to manage supplies within the hospital walls and benefits the hospital in significantly decreased onsite inventory requirements. The concept of the Valuelink program, however, has its roots in the stockless program that Baxter developed in the mid 1980s. The underlying business rationale for Valuelink is that the overall efficiency within the manufacturing-distribution-customer chain increases as the distributor assumes inventory and distribution functions in exchange for long-term customer purchase commitment [11]. Byrnes et al.

(1991) describes the strategic alliance between Baxter and the hospital as a "partnership of specialists" [12]. In this case, Baxter specializes in materials management and the hospital, in patient care. Byrnes, however, cautions that for this strategic partnership to reap the potential benefits, a company must change the fundamental way it does business: "Managers must act strategically, developing inter-functional capabilities, and link resources both inside and outside their companies." Unfortunately, achieving complete supplier-customer integration in the healthcare industry has so far been elusive.

A more recent advancement in hospital supply chain technology has been the use of automated point of use (APU) systems. A couple of examples are shown in Figure 11.5. These supply stations are specifically designed for either medical supplies or medications. Both systems can be configured to use with bar code scanner technology, which is the standard for many hospitals in tracking supplies, instruments, equipments, medication dispensing, and even patients. These APU stations are usually strategically placed in the various clinical units of the hospital. For many organizations that have older facilities, it is often difficult to find space on the unit floors for the APU stations, which could make deployments more challenging. To access these stations to pull supplies, authorized users usually have to be authenticated based on fingerprints and/or passwords. In addition, to accurately track supply usages, the APU stations require that the supply pull be accompanied by a user action that registers the item(s) taken, such as pressing the "take" button on the appropriate bin.

The benefits of APU for managing inventories are many. First, it has the ability to track items taken and automatically send orders to the supply distributor electronically at preestablished frequencies and up to preestablished inventory levels. Also, APU has the ability to provide accurate charge capture information by tracking supply usage against the patient that it is used for. Finally, APU has the ability to provide inventory visibility across the hospital through the linking of individual supply stations to a centralized computer that administers the system. In the event that an item is stocked-out at one station, this networked system has the ability to

Figure 11.5 Pyxis Supply Station and Omnicell OmniRX Medication Station.

cross-check other stations within the facility to see if the same item is stocked at any other location. In concept, this provides the benefit of the inventory pooling effect that allows the facility to decrease inventory stock levels, also known as par levels. In practice however, nursing attitude towards supply sharing may be a key factor in nurses' satisfaction rating of the APU system, which will be further explored later in this chapter [13].

In addition to the APU systems, the latest trend in supply chain automation has been the increased adoption of radio frequency identification (RFID) technology for inventory tracking and stock level monitoring (Figure 11.6). RFID utilizes radio frequency waves to send and receive unique identification information from the inventory that has been tagged to the receiver system that centrally stores the information, which may optionally include details such as item location, type, age, condition, and so forth. There are currently three types of RFID tags: active, passive, and semipassive, and they differ mainly in power, function, and cost.

The active tags, as the name may suggest, are self-powered and are able to broadcast their own identification information over extended distances continuously for the designated readers to retrieve until the power is consumed. An example of such a tag is the automobile toll tag that many drivers use to go through automated highway toll booths (Figure 11.6). Because these tags are self-powered, the active variety usually has more capability to store and broadcast identification information than the passive ones, but the downside is that it is also much more expensive at ten dollars or more per tag, and therefore limited to more expensive or mission critical assets and inventories.

On the other end of the power-function-cost spectrum are the passive tags (Figure 11.6). As opposed to the active tags, passive tags carry no individual power source, but are normally quite inexpensive and small enough to be concealed. In operation, the passive tag absorbs the necessary energy to power its information broadcasting function from strategically positioned readers that send out radio waves to find these passive tags. Because of its limited power source, the broadcast range on the passive tags is much more limiting than that of the active tags. However, compared to the active tag's ten dollars and up price, a passive tag usually costs below ten cents and therefore can be more widely deployed in inventory tracking. As the adoption rate has increased in recent years, some tags are now reaching five cents, which according to the estimation of the AUTO-ID Labs of the Massa-

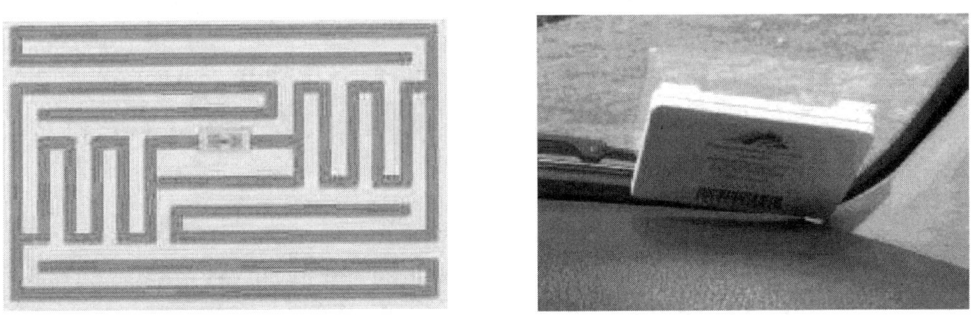

Figure 11.6 Examples of a passive RFID tag (left) and an active RFID tag (right). (Source: http://en.wikipedia.org/wiki/RFID.)

chusetts Institute of Technology (MIT), the thought leader on current RFID technology, is where the price point needs to be for wide-scale adoption.

As for the semipassive tag, also called the battery-assisted tag, it is a hybrid of both the active and the passive tags in design and in cost. Although powered, a semipassive tag does not actively broadcast information like that of an active tag; instead, to conserve battery life, it just powers the microchip circuitry and only wakes up to broadcast after receiving a signal from a compatible reader. The advantage of a semipassive tag is that it has the broadcast range of an active tag, while providing longer battery life. Also, in terms of cost, the semipassive tag is currently priced at a dollar and up as compared to the active tag's ten dollars or more and passive tag's five cents per tag prices.

Although RFID technology has been around for decades, the cost associated with using this technology has only recently been lowered to the point where it is now feasible to be adopted commercially. Wal-Mart for instance, had achieved early success in requiring all of its suppliers to place RFID tags on all pallets by the end of 2006, and by early 2008 Wal-Mart started penalizing suppliers for noncompliance by the pallet. As for healthcare, RFID development has so far been limited to a few startups, but the potential is there for RFID to prescribe order to the hectic healthcare work environment. Just suppose that if all supplies used in a surgical procedure could be automatically tracked and logged, it would not only decrease the amount of time spent on clinical documentation and suture and instrument counts, but it would also improve healthcare quality and patient safety. Financially, the providers would also benefit as all supplies used are properly accounted for and accurately charged to the patient. As for inventory management, all items used are tracked and orders are placed automatically with suppliers to replenish the supplies used during the procedure. Even though there are many technical hurdles remaining in the widespread adoption of RFID in healthcare, the potential benefits of this technology definitely warrants its further exploration.

In summary, the evolution of healthcare supply chain management has pushed both suppliers and providers into a more integrated relationship. To this end, technology advancement in healthcare supply chain management has enabled some suppliers to initiate partnerships with hospitals that are based on the successful integration of information that is necessary for a mutually beneficial and lasting relationship. As new technologies continue to emerge, such as the weight-based shelving systems similar to those used at supermarket self-pay stations, healthcare will ultimately benefit as automation increases clinician productivity and satisfaction and improves patient care quality while providing real-time information integration between providers and suppliers to allow increased supply chain optimization and cost reduction.

11.4 Supply Chain Automation Implementation

In a traditional hospital that has yet to adopt supply chain automation technologies, a typical supply process flow may start with the distributor delivering the ordered supplies to the hospital's loading dock at a predetermined time interval. From there the hospital's materials management team would then move the supplies to either the pharmacy or the general store. The pharmacy, however, may maintain its own distributions within the hospital, but the overall distribution processes remain similar. For supply replenishment, usually on a daily basis, members of the materials management team would deliver supplies to each clinical unit and also perform inventory audits to determine what and how much to order for restocking. Some hospitals, however, may require that orders for specialty supplies, such as those used by the Cardiac Cathiterization Lab or Endoscopy, be initiated by a clinical staff member rather than the materials management team member whose primary focus is commodity items. In either case, once the order is placed with the distributor or vendor, the hospital's supply chain process cycle is complete. Figure 11.7 shows a generic supply flow diagram that should be similar to most traditional hospital supply chain processes [13].

To be fair, there is absolutely nothing wrong with the traditional hospital supply process flow. Instead, what supply chain technologies are trying to accomplish is to automate much of the hospital's internal supply management processes and therefore save costs and increase customer-service levels. A service scope of a sample supply chain automation implementation agreement between a fictitious national wholesale supplier/distributor (Primera) and a fictitious hospital (BBC) is shown in Figure 11.8. In the following sections, this particular agreement will be used to illustrate potential processes and considerations that may be required to successfully implement a supply chain automation solution.

Under the scope of this sample agreement, Primera, the wholesale supplier and distributor, and BBC, the hospital, agreed to implement a suite of solutions centered on the adoption of APU technology for both medication and medical/surgical supplies. The breath of the implementation scope encompassed four components and is detailed as follows:

- *Consultation:* A 15-month process with the goal of identifying opportunities for standardizing supplies and reducing costs;

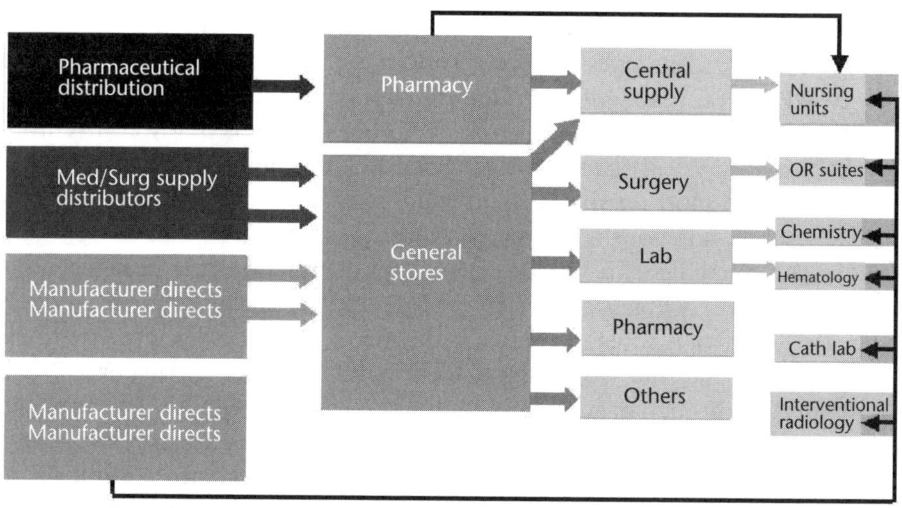

Figure 11.7 Traditional supply chain example.

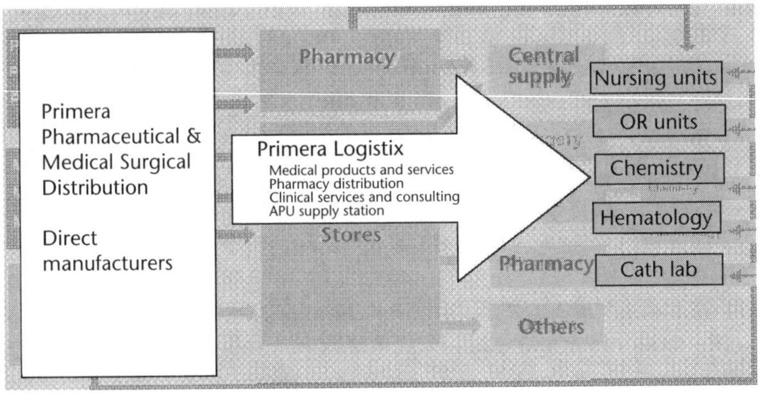

Figure 11.8 Primera Logistix functions.

- *Supply automation*: Automated supply distribution and reordering through the use of APU stations that allow ordering based on need, reducing waste, and providing better control of supplies;
- *Pharmacy distribution:* Direct medication distribution from Primera and reordering through the use of the medication APU;
- *Medical products and services:* Tote specific product picking and procedure-based kitting (PBK) at supplier warehouses based on automated APU ordering.

In addition to the above-mentioned services, the agreement also called for Primera to become the sole supply distributor for all items that Primera carries, therefore BBC can only order specialty items from other suppliers for categories that Primera does not carry. The benefit to BBC is the automation of supply chain functions and a guarantee of lower supply prices while leaving the distributor, Primera, responsible for periodic inventory tracking, warehousing, ordering, and delivering—essentially the adoption of the vendor managed inventory (VMI) model. Unlike what is shown in Figure 11.8, however, in this particular case BBC does ultimately retain control of in-house inventory replenishment of the floor APU stations. One added advantages for BBC to adopt the VMI model is that the supplier, in this case Primera, is actually carrying the supply inventory until BBC is ready to use it; this means that BBC benefits from VMI both in terms of lower inventory costs and in actual physical space savings. For BBC, not unlike most hospitals that always seem to be short on space, the freeing up of precious real estate for other revenue generating functions is definitely a plus.

Are there benefits to this business model for Primera? In reality, the wholesale medical supply business is a fiercely competitive, low margin operation. As a wholesaler, Primera is squeezed between the manufacturers, who are oftentimes large oligopolies, and the hospitals, which band together to negotiate better-termed contracts through group purchasing organizations (GPOs). To supplement the low margin supply distribution business, the wholesalers are all extending into higher margin value-added businesses such as supply chain consulting and procedure-based kitting, both of which are offered in the scope of this sample agreement. In addition, by becoming a prime vendor and locking the client hospital into a multiyear VMI agreement, over time even a low margin business can expect to generate sizable profits through volume.

As for the project, from the start the project plan called for a phased approach to implementation with different BBC units going live in a staggered fashion. For phase I, the plan was for a three-month implementation involving the operating room (OR), inpatient units (In Pt), emergency department (ED), and endoscopy (Endo) departments of BBC's satellite (Satellite 1) facility along with the main facility's cardiac OR, urology OR, plastics OR, GYN, vascular, Endo, intensive care unit (ICU), cardiac catheterization lab (Cath Lab), and electrophysiology lab (EP Lab) departments. For phase II, the implementation would focus on additional OR units along with nursing units, ED, and anesthesia at the main facility. Finally, for phase III, the initial plan was to roll out the APU stations for BBC's clinic areas as well [13] (Figure 11.9).

One of the most important ingredients of successful technology implementations is the level of attention that senior leadership pays to an active project. Because of the extended financial scope of this particular project, senior executives of BBC placed exceptional importance on the successful outcome of this supply chain automation project. In addition to frequent project steering committee meetings that included BBC's C-level executives, periodically the hospital's senior management team, including the CEO, would also meet with Primera's regional senior executives to assess progress using previously agreed metrics. From Primera's side, they were also committed to the success of the project. Besides frequent visits by regional executives, Primera selected an experienced project manager to be located full-time onsite at BBC to lead the implementation.

Another important but often overlooked ingredient for a successful implementation is the training of end users prior to the project go-live. In this case, because of prior implementation experiences, Primera was quite innovative and experienced in

Phase I	
Satellite 1 complete	**Main Facility**
OR	OR Cardiac
In Pt	OR Urology
ED	OR Plastics, GYN, Vascular
Endo	Endo
	ICU's
	CL
	EP Lab
Phase II remaining Main Facility	
OR completion	
CVIR	
CNIR	
Nursing Units	
ED	
Anesthesia	
Clinics - PhaseIII	

Figure 11.9 BBC APU implementation phases.

getting clinicians trained on APU station operations. Immediately prior to go-live, Primera introduced a working APU station filled with candies for a period of time to make it fun for the nurses to use the stations. In addition, as implementation moved from unit to unit, Primera trained unit "super users" prior to unit go-live so that each unit would have its own system experts to provide guidance. As for BBC, its employees also took training seriously. Trained BBC system administrators not only created reference manuals, but also symbolic APU user certification exams to assist the clinicians in learning the APU systems.

On average the rollout of the APU stations to each unit took about three days to complete, depending on the level of unit involvement in APU layout design. The go-live date was usually set on a Wednesday to allow for operational adjustments in the two days leading to the weekend when supplies were not scheduled for delivery from Primera warehouses. BBC system administrators, acting as project "super users," usually configured the APU stations for operation with help from unit nurse managers and Primera consultants. For some units, however, bedside nurses were also involved to make the implementation successful. In addition, BBC administrators were also helpful in serving as conduits for nursing feedback post-unit go-live, which would provide the basis for continuous improvements.

The restocking of supplies would usually take place early in the morning from Monday through Friday, based on the previous day's electronic data feeds from the APU stations to Primera warehouses (passing through both BBC and Primera information systems). Overnight, Primera would pick the orders and deliver the supplies separated by APU-assigned totes from its warehouses to BBC Central Supply, and from there BBC refill technicians would then take the totes to the units and restock the APU stations. While restocking, the refill technician would also perform a physical inventory of the stations to check for discrepancies between system recorded inventory levels versus actual inventory levels. Figure 11.10 shows the actual supply chain process flow from Primera to BBC post phase I implementation.

Although all clinicians who handled supplies anticipated some degree of process changes in their patient care routines prior to the implementation of the APU stations [13], some were quick to embrace the changes while others did not. At first glance, the changes that were embraced were usually the ones that had benefited the clinicians, such as nurses in some units were no longer required to stay after their shift to stock the shelves, while the changes that were rejected were usually the ones that had increased task complexity or changed task routine for the clinicians, such as changing the item take procedure or changing the location of an item needed for a procedure.

To evaluate the overall effectiveness of the Primera/BBC supply chain automation project and provide lessons learned insights for similar future projects, a study expanding over six months was conducted during this implementation [13]. Because nurses are the primary caregivers and also the primary users of supply chain automation technologies, the focus of this study was to use their satisfaction levels as related to the use of APU stations as gauges in measuring automation and implementation success. Although one may argue that financial success would be a better gauge, it should not be difficult to understand that without implementation success, there cannot be any financial success.

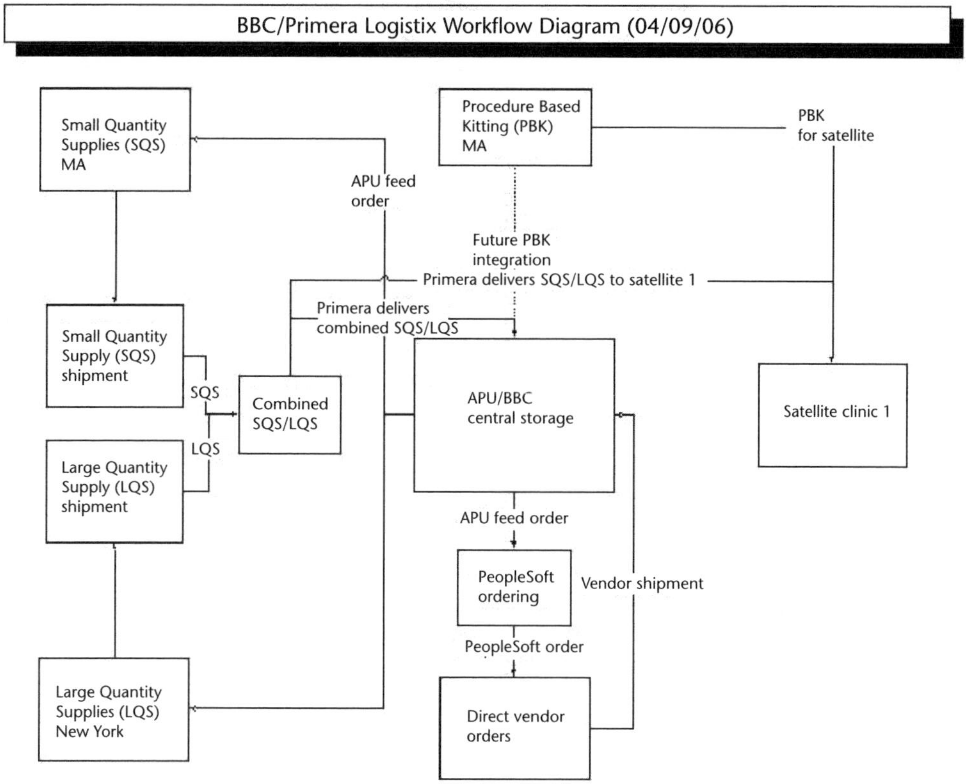

Figure 11.10 BBC/Primera supply chain process flow.

The study concluded that supply chain automation implementation projects can be successful as measured by nursing satisfaction if certain values that are important to nursing are incorporated. One original assumption of the study was that supply chain automation frees up nurses' time in managing supplies. What the study found was that, while timesaving is achievable through supply chain automation, it is only achieved when implementation initiated process changes were designed with the involvement and support of the nurses who are the frontline users of the automated supply stations. In other words, success of the implementation depends on first getting buy-in from nursing staff by adding tangible value that helps the nurses. More important, however, it was the central finding of this study that although timesaving is an important element in perceived value, it is only one of four elemental values that nurses perceive as important measures of overall supply automation satisfaction. Together, these four elements of clinician perceived supply automation values that contribute to overall APU satisfaction are (Figure 11.11):

1. Timesaving in managing supply (time savings);
2. Perceived quality of the products (product quality);
3. The availability of supplies (supply availability);
4. The accessibility to supplies (product accessibility).

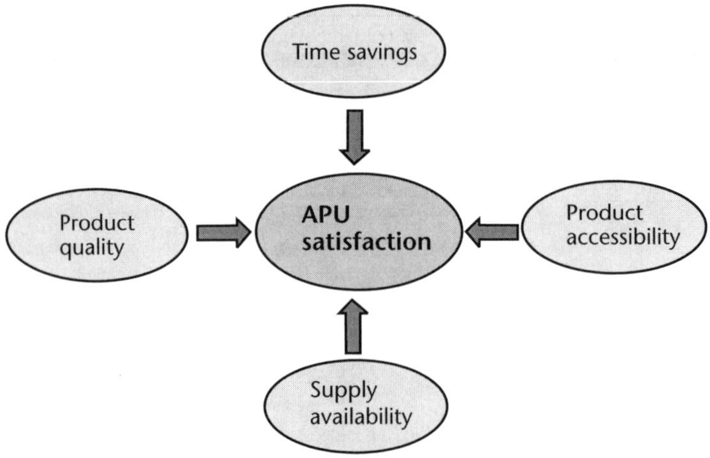

Figure 11.11 Four contributing elements of clinician perceived APU value.

Therefore, in order for supply chain automation to fully succeed in hospitals, it is the task of the implementation team to design process changes that maximize these four elemental values. In addition, it is also the job of the implementation team to tailor these values to the individual units by maximizing what is most important to the unit; for example, product accessibility, which is how quickly can nurses access the supplies from the APU, may be the most important value for emergency room nurses.

Based on this finding, the study also made the following recommendations for supply chain automation implementations:

1. Both the vendor and the healthcare provider must work closely together to find process changes that add tangible value based on the four elements; some examples that should be carried out early in the implementation process wherever possible are:
   - *Accessibility:* Design the supply layout of the critical care units' APU stations to be based on procedures so that nurses could quickly access the supplies, and therefore decrease concerns for accessibility.
   - *Product quality:* Showcase supply quality through product demonstrations, and at the same time, provide a degree of choice for the nurses to pick supplies, such as gloves, among the different brands that the vendor carries.
   - *Supply availability:* Readjust the par levels to adequately supply weekends, and educate users to the functional value of cross-facility visibility advantage of the APU system.
2. In the process of creating value, the implementation should:
   - Always involve nurses in decision-making to get buy-in and create trust.
   - Be proactive in searching out resistance and demonstrate effort in providing timely solutions for problems causing resistance.
   - Improve organizational communication others who are interested in the implementation through frequent progress updates such as monthly implementation newsletters, periodic cross-unit "super user" meetings to share ideas , and physician/nurse/materials management roundtable discussions.

- Emphasize that implementations are team efforts involving and affecting the organization from senior management, to physicians, to nurses, to material management personnel, and others in between; senior management should lead the way in bringing the organization together as a team.
- Provide more training to more people in a well-timed manner both prior to the implementation as well as after the implementation to increase user knowledge and realize additional value (e.g., sponsoring periodic friendly postimplementation competitions that showcase how a skilled nurse can quickly and accurately take supplies).
- Empower a system administrator who is energetic and motivated to learn the new system and not necessarily encumbered by the experience of the old system.
- Create a knowledge base for APU implementation and usage and allow everyone to both access and contribute to the knowledge base, therefore leveraging the collective knowledge of the entire organization (or across multiple organizations if the knowledge base is vendor sponsored) to achieve additional value.
- Learn to fail or fail to learn—do not be afraid of initial failure; as long as both the vendor and the provider work together as a cohesive team to learn from initial setbacks and continuously improve the overall process, ultimate success would follow.

As for the implementation project, by the end of the study most units that had implemented APU stations were doing well, especially those that had time to work out the initial kinks. One measurement that was tracked, activity compliance, which is performing proper procedures when taking an item, was mostly in the 90-plus percent range for the units that had been live for four months or more. In general, nurses adapted to the APU system over time. In addition, some units that had great difficulties in using the APU system underwent station redesigns to better accommodate the ways the nursing units worked. One such unit completely changed its view of the APU stations after the redesign and started to routinely give the automation technology the highest marks among all the units surveyed.

11.5 Future Possibilities

As more vendors and customers start to integrate both in process and in technology, the next step in supply chain automation in healthcare should go beyond the traditional realm of materials management related systems and services. With the advancement of information technology, the time is right for the healthcare industry to leapfrog into the era of real-time information exchange of compatible systems that allow everyone, vendors and customers, executives and clinicians, to feel the pulse of the organization and allow strategic actions to be taken based on real-time key measurements that drive forward the goal of achieving high-quality patient care with highly satisfied staffs and excellent financial return.

Increasingly, hospitals are turning to real-time dashboards to inform them of clinical measures. However, it is likely that hospitals are also interested in knowing their key inventory levels and how satisfied their employees and patients are in real time. Large healthcare supply distributors are naturally well suited to extend the APU supply stations into central hubs of hospital data that both collect and disseminate information so that real-time issue management becomes a realistic possibility. In exchange, suppliers can add this capability as another value-added feature in the partnership formula and deepen the relationship between the supplier and the hospital in the long run.

There are challenges that must be overcome before the information integration becomes a reality for both the supplier and the partnering providers. Currently, there are myriad technology platforms that coexist in the healthcare environment that offer little or no true integration capabilities with other platforms. For information integration to succeed, an industry-leading supplier must take the thought leadership position and leverage the unique relationships that it has built with its customers and create a common information integration platform similar to that of the ASAP program in concept. Perhaps, also similar to ASAP, this new initiative would usher in a new paradigm of truly integrated suppliers and providers in the healthcare industry that may rival that of the more heralded supply chains of Toyota, Wal-Mart, or Dell.

References

[1] "Resource Management: The Healthcare Supply Chain 2002 Survey Results," *Healthcare Financial Management Association*, 2002.

[2] "The Health of Our Nation's Hospitals 1997–2004," *Solucient*, 2005.

[3] "Supply Chain Benchmarking Survey: Managing Resources to Achieve Improved Economic Outcomes and High-Quality Care," *Healthcare Financial Management Association*, 2005.

[4] "Projected Supply, Demand, and Shortages of Registered Nurses: 2000–2020," *BHPr*, 2002.

[5] "International Data Base," Table 094, *U.S. Census Bureau*, 2004.

[6] Geddes, A., "Europe's Ageing Workforce," *BBC News*, June 20, 2002, http://news.bbc.co.uk/1/hi/world/europe/2053581.stm, 2002.

[7] Aiken, L. H., et al., "Hospital Nurse Staffing and Patient Mortality, Nurse Burnout, and Job Dissatisfaction," *JAMA*, Vol. 288, No. 16, 2002.

[8] Aiken, L. H., et al., "Nurses' Reports on Hospital Care in Five Countries," *Health Affairs*, Vol. 20, No. 3, 2001, pp. 43–53.

[9] Janiszewski, G. H., "The Nursing Shortage in the United States of America: An Integrative Review of the Literature," *J. Adv. Nurs.*, Vol. 43, No. 4, 2003, pp. 335–350.

[10] Case, J., et al., "Can Technology Help," California Health Care Foundation Study, 2002.

[11] Short, J. E., and N. Venkatraman, "Beyond Business Process Redesign: Redefining Baxter's Business Network," *Sloan Manage. Rev.*, Vol. 34, No. 1, 1992, pp. 7–21.

[12] Byrnes, J. L. S., and R. D. Shapiro, "Intercompany Operating Ties: Unlocking the Value in Channel Restructuring," Harvard Business School Working Paper, No 92-058, 1991.

[13] Xie, Y., "Supply Chain Automation and the Effect on Clinician Satisfaction and Patient Care Quality in the Hospital Setting," MIT masters thesis, Massachusetts Institute of Technology, 2006.

CHAPTER 12

Process Management Using Information Systems: Principles and Systems

Norbert Stoll and Kerstin Thurow

12.1 Introduction

Medical process flows are usually highly complex distributed systems consisting of several subsystems capturing various different parameters. This combination places particular demands on process administration, control, and regulation and thus on process management.

Flexible information management that can handle the unrestricted planning of project-specific tests is always a requirement for medical automation projects. These tests can be extremely varied, with a spectrum ranging from high-throughput in vitro tests to all different kinds of in vivo diagnostics and therapy. The data collected during a test can range from a multitude of material measurements and physiological parameters and readings to subjective assessments and objective evaluations.

A universally valid procedure to support this diverseness of data has not been provided to date by project-specific information management systems. The data that is to be captured is usually fixed in advance. There is a lot of work associated with adding new information classes. Deficits particularly exist with regard to how the available IT and communications technologies are utilized. Because of this, open, Web-based mapping systems facilitating the flexible capture of medical data in variable patient and test subject files that can then be subsequently archived in a suitable and retrievable manner are needed.

Manual and automated data capture is currently supported by various different interfaces. Distributed data in external systems (e.g., parameters from stationary or mobile data capture systems) can be automatically imported via communication frameworks that support typical interfaces such as text, CSV and Excel files, XML, process databases, and Web services. By using syntax and semantic converters, the imported records can be flexibly assigned to the test parameters.

Technologically and conceptually speaking, powerful database systems (in terms of their structure and data volumes) are almost always behind these solutions nowadays. The issue of flexible system integration, process optimization, and specific data selection has not yet been comprehensively or satisfactorily solved. In hierarchical medical automation, information management systems (IMS) occupy a central position for integrated process mapping and data and workflow management and present an initial basis for the development of medical information systems for this interesting field of application. Research to date has concentrated on the deficits of universal system integration and flexibility that have been clear since 1997 and have not yet been sorted out despite their increasing importance. This is even more remarkable given that there has been a significant improvement in previous years in the technology for implementing system integration, especially Web engineering tools and methods.

Medical and telemedical information management systems are at an initial stage in their development. However, the hierarchical systems currently available for the problem areas mentioned above do offer solution approaches right now that are interesting for the handling of medical process information due to numerous parallels. As with laboratory automation, resolved comprehensive system integration has a crucial effect on the efficiency of medical tests and interactions. IMS concepts can be used as personalized medical application environments and integration platforms with flexible process communication.

In the area of "large" information systems for medical applications, work on electronic patient files should be mentioned. However, this philosophy as it is currently understood is not sufficient for medical information management in a general sense. There are, for instance, advanced concepts from INTEL and GE Healthcare that also facilitate the inclusion of diagnostic data, whereby a solution suggested by IBM (the M-Health system) contains the most wide-ranging concept with regard to systems engineering.

12.2 Process Management and Control: Definitions

12.2.1 Process and Business Process

ISO 12207 defines a *process* as a set of interrelated tools and activities that turn input states into output states. The medical automation processes studied at the observed abstraction level of the process planning, process implementation, and process optimization with the inclusion of resource management are discontinuous (discrete). A *business process* pursues the goal of product or service generation (synonyms: flow, procedure, and process). It is assumed that business processes can be modeled and automated.

12.2.2 Process Management

Process management, also called *business process management*, deals with the discovery, design, documentation, and improvement of business processes. The goal of process management is to use existing information to improve and optimize processes. In addition to this, exact process knowledge is as necessary as process flow documentation. Clear interfaces must be defined between different processes so that process chains and nesting can be mapped simply. Essentially, process management consists of the following three aspects:

- Process planning and modeling;
- Carrying out work in accordance with processes;
- Process monitoring.

Ideally, the findings from the monitoring should flow back into the process planning.

The planning phase is about identifying processes. As such, either existing processes can be discovered and documented or new processes planned. One possibility is to simply define rules for recording the process flow in a first step. The data can then be recorded. The discovery of the actual processes can then be attempted through process discovery and/or process mining. These identified processes can then be analyzed and used as the basis for further planning.

Business processes can be illustrated by using flow charts, business rules, and the like.

Planning influences process implementation. The conventional tools for organizing processes can be used. Depending on the type of business, the implementation can be supported electronically to a greater or lesser degree by workflow management systems, business rule engines, and special software such as CRM systems.

Business process monitoring involves both *short-term activities* and also *longer-term activities such as the generation of key data* that can then be used again to influence the planning. The discovery of a company's own *actual processes* is also part of this—the data collected from executing the process can be combined with modern methods based on BPMN in order to graphically illustrate the actual processes. The process data is stored in a *process warehouse*. This is a data warehouse in which the process data plays a strong role. This data can also be analyzed from the beginning if it is first implemented in a data warehouse.

12.2.3 Workflow, Workflow Management, and Workflow Management System

A *workflow* (WF) can be characterized as a predefined sequence of activities relating to operational and technical processes. Operational business process factors in the narrow sense (costs, benefits) are not included in the workflow documentation/modeling. Automation or semiautomation is the purpose and goal of a workflow. A workflow is a business process or part thereof.

Workflow management (WM) is used to identify, model, specify, and simulate workflows. It builds on the administration of sequences of individual activities as atomic execution units. Besides needing human resources, medical automation activities also require mechanical resources at differentiated automation levels and can have a sophisticated temporal dependency, among others, in the form of sequences, a minimum dwell period, or a time-out condition.

A *workflow management system* (WFMS) is an IT system for supporting the design and implementation of workflows. The overlapping documentation and coordination tasks of a WFMS (roles, tasks, process, and environment) are specified according to the needs of the complex medical automation processes.

12.2.4 Resources

Resources can be defined for the automated workflow in order to identify the operating tools that must be provided. Resources are abstractly understood as a black box with I/O interfaces and states to be defined. Resources may have a hierarchical organizational structure. Resources are used to implement process stages with defined segmentation or block formation. Resources with the same or similar characteristics in relation to the correct execution of tasks can be bundled into resource classes. As a result, a parallel-machine model can be used as a generalization of the one-machine model. Machines with the same abilities can take on identical WF stages in parallel in order to achieve throughput goals.

12.3 System Architecture

When creating automation concepts for medical automation, various degrees of automation can be assumed.

The conventional approach to manual process management is generally known. All subprocesses are executed manually by the operator, independently of one another. All of the parameters are entered using manually operated devices and the data that is then collected is manually entered into datasheets or computer-based archiving systems. As the work is mostly carried out in sequence, the throughput is very low. This puts a lot of pressure on personnel and is associated with a high likelihood of errors.

A first step towards automation, which is also often the final one nowadays, is an island of automation approach. In this, the automation of special systems involving regularly repeated, time-critical subprocesses or those that are dangerous for the operator is to the fore. Individual subprocesses are automated. The data collected is recorded by the computer in the respective process control systems and is available there for visualization and—in individual cases—processing. Sample transfer between substations is carried out manually by an operator. Throughput can thus be increased in comparison with manual approaches and there is a reduced likelihood of errors.

The term "island of automation" was coined in the 1980s and nowadays refers to either a nonintegrated automation process (an automated process which runs in a system that is not completely automated) or a separate information technology or automation system that is usually not integrated or compatible with other systems.

The last step is full automation that has already been realized in certain sectors, such as the automotive industry. However, there is still a huge need for development with regard to medical automation processes involving complete automation of all subprocesses and sample transfers. All partial processes should be networked with one another in terms of the transport of material and information and the data systems should be compatible with one another so that all of the data collected can be made available via a central data management system to users and operators. Manual intervention would only take place if there were errors or maintenance was being carried out on the systems.

Crucial for the development of automation concepts are, among other things, the issue of the number of measurements to be collected and the necessary flexibility of the systems as these will affect how the systems are structured in terms of hardware and the process control system and information processing requirements.

12.4 Scheduling Workflow in Real Time

12.4.1 Introduction

Distributed hierarchical automation for medical applications is placing new demands on workflow management. Process planning, scheduling, and resource management are becoming enormously complex due to the need to integrate process flows and the changing granularities of strongly networked process stages with highly variable runtimes (from minutes to weeks). Therefore, new approaches to system structuring for coping in the future with largely automated, computer-based, optimized, controlled, and integrated (business) processes are the focus of scientific and technical interest. Method-oriented process planning and scheduling rely on an approach involving networked islands of automation that can best execute complex sequences in a process stage, for example, thereby facilitating powerful I/O interfaces between communicating process mappings. These can be consolidated at the level of integrated workflow management. Concepts and solutions for distributed scheduling at a workflow management level under the constraints of medical automation requirements and the current heterogeneous system environments of automation components must be found. The IT approach can be seen as a network of distributed process mappings on distributed computer nodes, including several de facto standardized service-oriented communication protocols and dedicated communication servers.

Process mappings of medical applications are characterized by a high degree of flexibility and short lifecycles. Integrated process mapping can be carried out by applying a new approach using laboratory information management systems (LIMS) as an integration platform, for example. In such an instance, the LIMS represents the workflow level in the hierarchical operation automation. The goal of resource management is the flexible application of hierarchical scheduling methods for integrated research-oriented medical automation.

Process mappings can be symbolically identified by using three quantities. However, these must be considerably more structured for distributed medical automation processes:

- Set of resources SOR $R = \{R_1, \ldots, R_r\}$;
- Set of processes SOP $P = \{P_1, \ldots, P_p\}$;
- Set of hard constraints SOC $C = \{C_1, \ldots, C_c\}$.

The holistic process view of medical automation integrates, for example, material and substance allocation processes, high-quality storage, checking, resource control, quality assurance and evidence of quality, process development, and validation, taking all process stages, including information processing (IT validation), into account.

12.4.2 Scheduling

Scheduling is the temporal execution of parallel workflows or, under certain circumstances, workflows competing for resources in accordance with their activities and conditions. Flexible medical automation with short application process lifecycles calls for optimization compromises between resource economics and the application input. When scheduling, the schedule links structural elements of the process with resources in the sense of an optional alternative assignment that is called the allocation. Back coupling the process flow to the scheduling leads to the possibility of dynamic scheduling (i.e., one that accords with runtimes).

Scheduling the workflow level optimizes processes as opposed to the specific informatics approach of scheduling in relation to distributed calculation processes on distributed processors, memories, and I/O resources. Nevertheless, there are analogies in relation to the scheduling models and optimization.

Method scheduling is related to workflow stages in the sense of mastering complex dependent workflows. Description definitions for any indivisible workflow stages (sequences, sequence structures) that correspond to the typical process scenarios of medical automation applications must be developed at the WF level in order to implement distributed scheduling.

A *scheduler* is a suitable IT solution for the temporal execution of tasks under given operating conditions (operating systems, business processes/resource cases, etc.). The application development of schedulers under the conditions of coping with complicated state and event dependencies at a WF level are crucial for the project goal. Therefore, hierarchical scheduler configurations are favored in design terms.

12.5 Usable Software Languages

12.5.1 Platforms

12.5.1.1 The .NET Platform

The Microsoft.NET platform is a software component that is part of Microsoft Windows operating systems. It has a large library of precoded solutions for standard program requirements and manages the execution of programs written specifically for this platform. The .NET platform was developed under the premise of making a general platform available for use by most new Windows applications. The library covers a large range of programming needs, including user interfaces, data access, database connectivity, cryptography, Web application development, numeric algorithms, and network communications. The library is used by programmers who combine it with their own software code to produce applications. Programs based on .NET execute in a software environment that manages the program runtime requirements. This runtime environment, which is part of .NET, is known as the Common Language Runtime (CLR). The CLR provides an application virtual machine so that the capabilities of the CPU executing the program do not have to be taken into consideration when programming. The CLR also provides other important services such as security management, memory management, and exception handling. The library and the CLR together make up the .NET application platform.

The .NET platform is included with Windows Server 2003, Windows Server 2008, and Windows Vista operating systems and can also be installed on most older versions of Windows.

12.5.1.2 Other Programming Environments

Besides .NET, there are a number of other software platforms that can be used to program process management and process control applications. These include:

- *ASP.NET* is a Web-based development platform that provides programmers with the services they need to develop commercial Web-based applications. ASP.NET allows conventional programming languages such as C# and VB.NET to be used for the simple development of Web applications. ASP.NET integrates different programming languages without compatibility issues. In doing so, the code and implementation are kept separate.
- *C#* is a modern, object-oriented programming language derived from the C and C++ programming languages. Due to this similarity, C# (pronounced "C sharp") can be easily used by programmers who are already familiar with C or C++. C# was specially developed for programming the MICROSOFT.NET framework. C# combines the high productivity of Visual Basic with the raw power of C++. C# is a component of Microsoft Visual Studio 7.0, which also supports Visual Basic, Visual C++ ,and the VBScript and JScript scripting languages. C# does not have its own library. C# provides a Web-oriented environment that is harmonized with HTML, XML, and SOAP.
- *VS.NET* is a tool for the quick development and integration of XML Web services and applications. It significantly increases programming speed and provides new business opportunities. VS.NET is the only development environment that has been specifically developed for XML service applications. Through common data usage over the Internet, XML Web services facilitate application development that is independent of the platform and programming language.
- *VB.NET* facilitates the time-saving development of large Web-based applications for Windows platforms, the integration of data access for different database scenarios, and the development of components with minimal code. VB.NET has a number of new and improved language features, such as inheritance, interfaces, and overloading, which make it a strongly object-oriented programming language. Furthermore, Visual Basic developers can create multithreaded, scalable applications.
- *XML web services* are the fundamental components for using distributed computing on the Internet. Open standards and a focus on communication and collaboration between people and applications have created an environment in which XML web services are the platform for application integration. XML web services use a standardized web protocol, usually SOAP. They provide the means of describing your interfaces with a sufficient level of detail to allow the user to create a client application. The description is usually made available in a so-called web service description language (WSDL) document. XML web services are registered so that they can be found easily by potential users.
- *JScript.NET* is Microsoft's implementation of JavaScript. JScript.NET extends JScript with new features such as direct support for object-oriented programming techniques.

- *SOAP* is a specification protocol for exchanging information messages on servers, components, and objects in decentralized, distributed environments. SOAP was originally developed by Microsoft. SOAP provides a solution that facilitates the linking of Web sites and applications.
- *UDDI* (Universal Discovery, Description, and Integration) is a public registry for structured information about businesses and their services. The data can be classified according to standard taxonomies so that information can be found on the basis of categories. UDDI contains information about the technical interfaces of business services.
- *Visual.NET* offers newcomers to programming with a little bit of experience in C++ or Java, a modern programming language and robust development environment for creating XML Web services and applications for Microsoft Windows operating systems based on Microsoft.NET.

12.5.2 Databases

12.5.2.1 Introduction

Database systems (DBS) are systems for electronic data management. The main duty of a DBS is to efficiently and consistently store large quantities of data for a lengthy period of time and to present partial amounts needed in different presentation forms according to the needs of users and application programs. Database systems consist of the actual database (which represents the quantity of data to be managed) and the management software, the database management system (DBMS). The management software organizes the structured storage of data internally in accordance with a prespecified database model and controls all read and write access to the database. It provides a database language as an external interface for formulating queries, inputting and changing data, and for administrative commands. In addition to the actual data, the database also contains a description of the data, the so-called data catalog.

The database model specified by the DBMS manufacturer forms the basis for how the data and its interrelationships are structured. Depending on the database model, the database scheme must be adjusted to the particular structuring possibilities. Databases can be structured in the following ways:

- *Hierarchically*: the data objects exist exclusively as parent/child relationships.
- *Network-based:* the data objects are linked together in networks.
- *Relational:* the data is maintained as rows in tables. There can be any kind of relationships between the data. They are specified according to table columns determined according to values.
- *Object-oriented*: the relationships between data objects are administered by the database system itself and characteristics and data can be inherited from other objects.

There are many other kinds of mixed forms and subvarieties, such as the object-relational model.

Database languages are a subclass of computer languages that have been developed for use in database systems. A user or application program can use the database language to communicate with the database system. As an important part of working with database systems is formulating queries, a (database) query language is also usually part of the language scope.

There are a wide range of database languages, from call interfaces that closely resemble machine code to types of formal English. The most commonly used database language is SQL for relational database systems.

A common way to categorize database languages and their elements is according to DML, DDL, and DCL:

- *Data Manipulation Language* (*DML*): language or parts of language for querying, adding, changing, and deleting reference data;
- *Data Definition Language* (*DDL*): language or parts of language for adding, changing, and deleting data structures;
- *Data Control Language* (*DCL*): language or parts of language for access control;
- *Transaction Control Language* (*TCL*): language for the final storage or resetting of the results from DML commands.

12.5.2.2 SQL Database Language

Structured Query Language (*SQL*) is a database language for defining, querying, and manipulating data in relational databases. SQL is both an ANSI and ISO standard and is supported by almost all current database systems.

In SQL, all elements are unified in one language through various different instructions. In the case of the historical database system IMS, DML, and DDL have their own languages (DL/I and Assembler macros) and the DCL is realized using operating system tools.

SQL has a relatively simple syntax and is semantically aligned with English as it is commonly spoken. SQL provides a range of commands for defining data structures according to relational algebra, for manipulating data (adding, editing, and deleting records) and for querying data. SQL has great importance through its role as the quasi-standard, as this makes it possible to remain largely independent of the software used. However, most SQL implementations also offer manufacturer-specific expansions that do not correspond with the standard language scope, with the consequence that the same functions developed in parallel by manufacturers may use different language elements.

Most database systems used today, such as DB2, Informix, Microsoft SQL Server, MaxDB, MySQL, Oracle, PostgreSQL, 4D SQL, Borland InterBase, Firebird, Sybase, SQLite, Lotus Approach, and Microsoft Access, implement parts of the SQL language standard. As a result, applications that are independent of the database system used can be created.

The first SQL standard was passed by ANSI in 1986 (and then ratified by ISO in 1987). The standard was considerably revised in 1992 and published as SQL-92 (or also SQL2). All current database systems essentially follow this standard version. The newer version SQL:1990 (ISO/IEC 9075:1999, also called SQL3) has not yet been implemented in all database systems. SQL:2003 has not yet been extensively implemented.

12.5.2.3 MySQL Database System Software

MySQL is a relational database management system from the Swedish company MySQL AB. MySQL is available as open-source software for different operating systems and forms the basis for dynamic websites.

MySQL is free software available under General Public License (GPL). It can also be used with a commercial license (dual licensing system). MySQL is being used as an imbedded database system in hundreds of products. With more than 6 million installations and over 35,000 downloads per day, it is the most popular open-source database management system in the world.

Besides many UNIX versions, Mac OS X, and Linux, MySQL also runs on Windows OS/2 and i5/OS (formerly OS/400). There has also been a Symbian version since the beginning of 2008.

While earlier MySQL versions only supported parts of the SQL3 language (e.g., view definitions were not possible), version 5.0 offers a significantly expanded language scope that largely corresponds with the SQL3 standard.

A replication system has been available since version 3.23.xx. It is designed for use in a computer cluster, whereby several databases on different computer nodes are allocated to the database management system (DBMS). One of the databases functions as the master and the database contents are changed there. The replication system then distributes the data-changing SQL commands to the other databases, which implement these changes locally on their tables. Thus, this is considered asynchronous replication of the SQL commands.

One popular area where MySQL is used is in Web service data storage. MySQL is often used in connection with the Apache Web server and PHP. This combination is described as LAMP, MAPM, or WAMP (XAMPP), according to the initial letters of the software involved (whether the Linux, Mac OS, or Windows operating systems are being used).

Some Web services avail of this architecture. They run several hundred MySQL servers from which access from the Net can be handled.

MySQL is often used as a DBMS in a Web environment. In such a case it is used to ensure that many read accesses can be executed quickly. SQL's cluster concept in version 5.1 is designed for a computer cluster in a shared nothing architecture. This means that each computer node has its own hard drive and memory. If the individual computer nodes have sufficient memory, all data can be held in the memory. MySQL has proven to have better access times than Oracle and DB2 for such requirements in several tests.

12.5.2.4 Oracle Database System Software

Oracle Database (also Oracle Database Server, Oracle RDBMS, and the like) is a database system software. Together with DB2 and MS SQL server, Oracle is one of the market leaders in the RDBMS segment. In the mainframe area, Sun Fire machines with the UNIX system Solaris and IBM machines are the commonly used platforms. In the mid-range segment, almost all UNIX systems are supported and used, along with OpenVMS. Along with Solaris, Linux has been favored as the main strategic platform for a long time and is very widely used. Due to its high proliferation, Windows is also very strongly supported for strategic reasons.

An Oracle database system can store both relational and object-relational data. It also has the ability to process and relationally transform XML data structures (XMLDB, XDK). It implements ACID properties, provides good scalability, and has a very large functional scope.

Due to its architecture and scalability, an Oracle database management system (DBMS, RDBMS, ORDBMS) is also a very good choice when there are a large number of simultaneously active users. Oracle database software is available for almost all of the operating systems in use today and the Express Edition (XE) is free to use. However, the Express Edition is more limited than the free DB2 version as it does not support Java.

The Oracle database server in its current versions supports 8 exabytes in one database and therefore has a storage capacity that any individual organization would currently find it difficult to exhaust (2006).

Other important basic features are cross-platform support for distributed databases, data warehouse functionality, messaging, OLAP and data mining, intelligent backups (electively with block change tracking), Java stored procedures (not in the XE version), support for regular expressions in queries, table versioning (long-term transactions), virtual private database functionality, and numerous mechanisms for data protection.

12.5.3 Audit Trails

The term "audit" (from Latin for hearing) refers to general inspection proceedings that can be used to evaluate process flows with regard to how well they meet requirements and guidelines. These are often part of a quality management system.

Depending on the area, the actual situation is analyzed in an audit or the original objectives are compared with the goals that have actually been achieved. An audit should often also serve to detect general problems or a need for improvement so that they can be dealt with.

When setting up, certifying, and maintaining management systems, audits play an important role. The audit types can be differentiated between according to various criteria:

- Compliance audit (checking whether there is compliance with a body of rules and regulations, list of questions);
- Performance audit ((also known as a legitimacy check) is an objective and systematic check of goal achievement (effectiveness) and of whether the resources involved have been economically and efficiently used);
- Process audit (looks at individual processes);
- System audit (looks at the management system);
- Product audit (looks at the product on the basis of customer expectations);
- Project audit (looks at the progress of projects).

Audits relating to the certification of management systems include the following aspects:

- Preaudit for establishing certifiability, also known as a "friendly audit";
- Certification audit with a check on the documents and fulfillment of the set of rules to be certified on the basis of a list of questions;
- Monitoring audit (usually carried out annually) for monitoring the further development of the management system;
- Repeated audit or recertification is carried out on most management systems every 3 years.

In information technology, the term audit is used for various internal checks on software projects.

- Regular and occasional checking of project components for adherence to internal rules (the use of specific templates, conformity with the overall project, suitability for implementing the module requirements, etc.);
- The systematic checking of source code (e.g., for unclean implementations, quality of the source text formatting, and the completeness of documentation (code audit)};
- The systematic searching for potential security gaps in programs and IT infrastructure (security audit);
- Checking whether a company has a sufficient number of licenses for the software used.

Selected Bibliography

Alapati, S. R., and C. Kim, *Oracle Database 11g: New Features for DBAs and Developers,* Apress, 2007.

Alonso, G., and F. Casati, *Web Services—Concepts, Architectures and Applications,* Berlin, Heidelberg, New York: Springer Publishing Company, 2003.

Brush, M., "LIMS Unlimited," *The Scientist,* Vol. 11, No. 15, 2001, pp. 22–27.

Dishman, E., "Measuring and Monitoring Social Health in a Wireless World: An Intel Case Study with Cognitive Decline Households," *Proc. 1st World Conference Medical Automation,* June 29, 2005, Helsinki.

Kettley, P., "M-Health System," Keynote Speech, *1st World Conference Medical Automation,* June 29, 2005, Helsinki.

McIntosh, R. L., and A. Yau, "A Flexible and Robust Peer-to-Peer Architecture with XML-Based Open Communication for Laboratory Automation," *JALA,* Vol. 1, No. 8, 2003, pp. 38–45.

Reese, G., R. J. Yarger, and T. King, *Managing and Using MySQL. A Database Optimized for Speed and Interactivity,* O'Reilly Media, 2002.

Särkka, P., "Technological Advances in Wireless Applications—Communication and Tracking," *Proc. 1st World Conference Medical Automation,* June 29, 2005, Helsinki.

Smith, C. A., and A. B. Corripio, *Principles and Practices of Automatic Process Control,* John Wiley & Sons, 2005.

Troelsen, A. W., *Pro C Sharp 2005 .NET 2 Platform,* Apress, 2005.

UNIX Systems Labs, *Audit Trail Administration,* Prentice Hall PTR, 1993.

Van der Aalst, W., and van Hee, K., "Workflow Management: Models, Methods, and Systems," *Cooperative Information Systems Series*, B&T, 2004.

Van der Laans, R. F., *Introduction to SQL: Mastering the Relational Database Language,* Amsterdam: Addison-Wesley Longman, 2006.

CHAPTER 13

Telehealth and Telemedicine Technologies: Overview, Benefits, and Implications

Karen S. Rheuban*

13.1 Introduction

Telemedicine is defined as the use of advanced technologies to facilitate the exchange of medical information and medical images via a host of communications services to facilitate the delivery clinical care. Such services can be delivered in live interactive formats via videoconferencing technologies (wherein both the patient and the health professional are present), or asynchronously through store-and-forward technologies such as teleradiology, teledermatology, teleophthalmology, and remote monitoring (Figures 13.1–13.4). Telehealth, more broadly defined than telemedicine, incorporates additional services that include health-related distance learning of patients and health professionals. Telehealth facilitated care can be delivered in a variety of venues, including hospital, clinic, school, workplace, and home settings.

Telehealth providers face continued challenges related to the interoperability of devices, electronic medical records, and the communications infrastructure over services provided. Standardization efforts range from videoconferencing protocols (such as IP, ISDN, and wireless), to devices and image acquisition formats (DICOM), to electronic medical records formats, and to clinical protocols vetted by the specialty societies. All will further drive utilization.

Clinical services delivered via telehealth technologies span the entire spectrum of healthcare, and across the continuum from prematurity to geriatric care, with applicability to more than 50 clinical specialties and subspecialties. Telecardiology, teledermatology, teleophthalmology, acute stroke intervention, pulmonary medicine, teleradiology, e-ICU, home telehealth, telemental health, and telepathology are but a few of the many applications in general use. Where local specialty care services are not available, particularly in rural and underserved regions and in health professional shortage areas, telehealth can offer timely access to care and spare patients the burden of long-distance travel for access to that care.

The societal integration of advanced technologies into everyday venues has profound implications for the development, support, and delivery of a new paradigm of healthcare services in the digital era. The powerful tools of health information technologies are critical to the transition from a culture in which health-related services

* Dr. Rheuban has no financial or personal conflicts to disclose in relationship to this chapter.

Figure 13.1 Example of live interactive videoconference connecting a patient with a pediatric plastic surgeon.

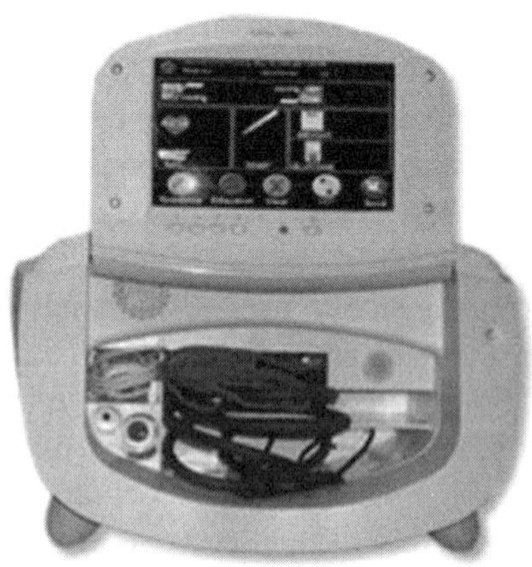

Figure 13.2 Example of a home monitoring device with pulse oximeter and blood glucose analyzer, which is capable of checking blood pressure, daily weight, and other measurements. (Courtesy of Vitelnet.)

are primarily delivered in a balkanized model on an episodic basis to an integrated systems approach focused on disease prevention, enhanced wellness, chronic disease management, decision support, quality, and patient safety. Through the incorporation of such tools and technologies, clinicians will be able to satisfactorily manage the exponentially expanding volumes of medical information, research, and decision support analytic tools.

The aging of our population has already created increased demand for specialty healthcare services to address both acute and chronic disease in the elderly. Such a demand, in the face of anticipated provider shortages, requires a fundamental shift from the model of physician-centered care to one focused on patient-centered care using interdisciplinary teams, evidence-based medicine, the use of informatics in

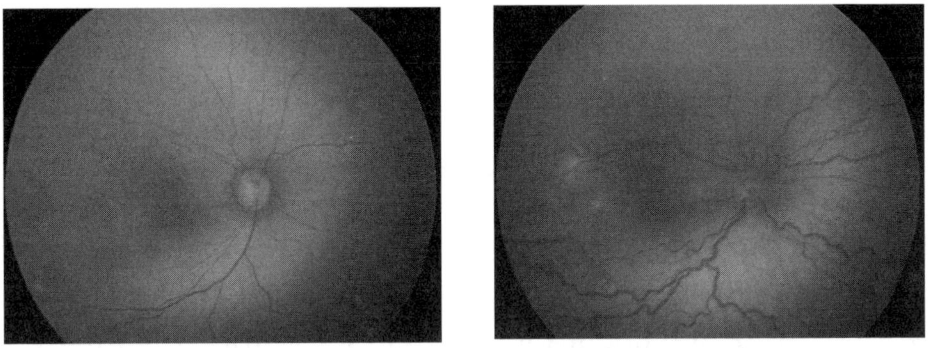

(a) (b)

Figure 13.3 (a) Wide-angle posterior retinal image from eye that was diagnosed by telemedicine as having no retinopathy of prematority. Vessels are thin and smooth, denoting absence of significant disease. (b) Wide-angle posterior retinal image from eye that was diagnosed by telemedicine as having treatment-requiring retinopathy of prematurity. Vessels are dilated and tortuous, denoting presence of severe disease. (Courtesy of Michael Chiang, M.D., Irving Assistant Professor of Ophthalmology and Biomedical Informatics, Columbia University.)

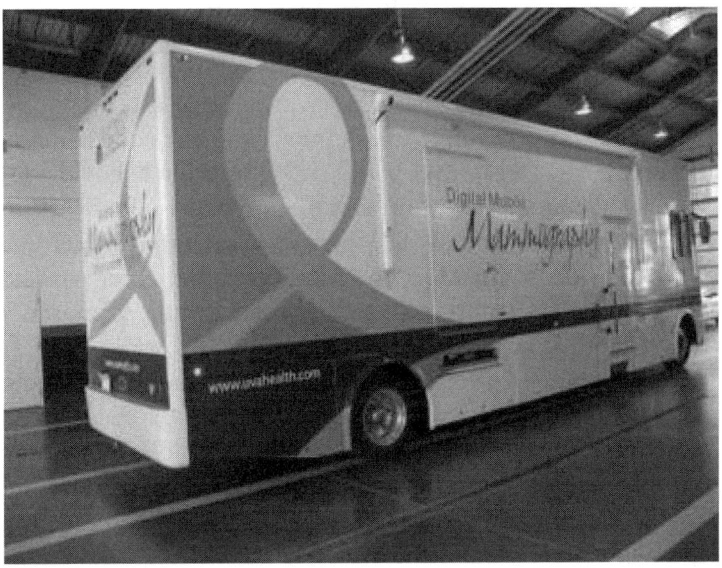

Figure 13.4 Example of mobile digital mammography van with mini-PACS and broadband image transfer capabilities.

decision support, and telehealth technologies where specialty care services are either not locally available or for other consultative needs. With the aging of the population and greater numbers of patients with chronic illness, home telehealth and home monitoring tools offer a cost effective mechanism to improve health and provide early intervention where appropriate.

Access to local specialty care remains inadequate for many Americans, attributable to a host of factors that range from geographic to economic to societal imposed barriers. It is widely accepted that the United States faces a shortage of physician providers, ranging from 85,000 to 200,000 physicians by 2020 [1, 2]. Lack of access in rural areas is exacerbated by the limited numbers of specialists who practice in rural communities and the limited resources generally available in those communities. The incorporation of telehealth technologies into integrated systems of healthcare offers tools with great potential to address the challenges of access, specialty shortages, and changing patient needs both in the rural and urban setting. Such tools support the goals of the federal Healthy People 2010 initiative.

13.2 Status of and Implications for Rural Healthcare in America

Rural America currently represents approximately 20% of the population, as defined by the Bureau of Census. These citizens reside in diverse communities that range from towns adjacent to suburban areas to remote and frontier communities with very low population densities [3].

Federal definitions of rural vary according to both the federal agency and the program. The United States Department of Agriculture has long maintained the most liberal definition of rurality, whereas HRSA and many of its agencies have chosen programmatic eligibility definitions based on countywide metro- and micropolitan statistical area designations and/or or rural–urban continuum codes or rural–urban commuting area codes [3]. These latter definitions have tended to limit eligibility for programmatic support for many communities that are otherwise rural in character. Definitions based on commuting behaviors fail to recognize that local economic underdevelopment is often the cause for work-related travel rather than geographic proximity. The Federal Communications Commission recently approved an order to liberalize the definition of rural for purposes of access to telecommunications discounts in the Rural Health Care Support Mechanism, as mandated by Congress in the Telecommunications Act of 1996 [4].

Demographically, rural populations tend to be older, with greater numbers of citizens who exhibit adverse health behaviors (smoking, lack of fitness, obesity) and the subsequent chronic diseases that result from those behaviors. Without adequate screening programs, education, and programs that mitigate unhealthy behaviors, rural patients are at a disadvantage from their urban counterparts.

Although rural communities face the same basic challenges in access, quality, and cost as their urban counterparts, they do so at far greater rates, attributable to a host of factors. "Core health care services" as defined by the Institute of Medicine as primary care, emergency medical services, long-term care, mental health and substance abuse services, oral health, and other services are considerably less accessible in rural communities [2].

Attracting health professionals to rural communities remains a daunting task and keeping those health professionals in practice in those rural communities is all the more difficult. Rural healthcare providers tend to work longer hours, see more patients, and have greater on-call demands because of lack of cross-coverage opportunities [5, 6]. Strategies to recruit and retain clinicians to practice in rural and frontier communities must also include innovative applications that reduce the chronic sense of isolation experienced by those practitioners by affording enhanced connectivity to colleagues and educational opportunities.

Rural reimbursement rates for health providers have traditionally been lower than those of their urban counterparts, although the 2003 Medicare Modernization Act holds promise to increase reimbursement rates to rural practitioners and to further incentivize clinicians to practice in rural areas.

Lack of access to specialty care services is an even greater challenge. Rural communities lack sufficient patient volumes to support specialty and subspecialty practices. Telehealth technologies offer ready access to such services when rural communities partner with tertiary and quaternary care facilities, through such applications as telemental health, teledermatology, HIV/AIDS care, and a host of other specialty applications [1, 7–13].

Acute stroke intervention, wherein thrombolytic therapies may be administered in the 3-hour window post-ischemic stroke, is an excellent application of subspecialty care otherwise unavailable to rural patients [14]. Similarly, when time is of the essence, services can be provided to support the provision of emergency cardiac care [15–18].

Despite the clear benefit of telehealth facilitated access to remote specialty and subspecialty care, potential exists for changing referral relationships and underutilization of existing local or regional community specialty providers.

Rural healthcare providers generally lag considerably behind urban healthcare providers in the deployment of advanced technologies. Paper-based medical records are still the norm in rural practices, although rural hospitals have begun to integrate health information technologies. In 2003, Brailer et al. reported that only 13% of private in-patient facilities utilized an electronic health record, and that the adoption of such technologies in urban areas exceeded those in rural areas by 150% [19]. Community health information networks that facilitate the sharing of medical information among otherwise unaffiliated practices and hospitals remain relatively underdeveloped in rural regions of the United States.

The integration of clinical information systems, picture archiving, and communications systems (PACS) for image storage and sharing, community health data exchange, decision support analytic tools, and telehealth facilitated consultative and educational capabilities promise to bring state-of-the-art healthcare to rural regions of the nation. Teleradiology and cardiology PACS applications allow for remote interpretation of digitally acquired images. Screening for diabetic retinopathy using a nonmydriatric retinoscope allows remotely located healthcare providers to capture digital retinal images and transfer those images to an ophthalmologist for determination of risk for diabetic retinopathy, and the need for intervention [20]. Similarly, retinoscopic images may be acquired and transmitted to evaluate premature infants for retinopathy of prematurity, an increasing challenge in the face of greater numbers of infants at risk and a limited number of pediatric ophthalmologists to screen those infants [21].

New mobile digital mammography equipment developed in platforms with integrated two-way satellite capabilities as have been deployed for the Indian Health Service by Healthcare Anywhere allows for immediate transmission and interpretation of mammograms and timely (<30-minute) feedback to the patient should the acquisition of other images or studies be indicated.

Remote monitoring, hospice, and telehomecare initiatives that incorporate technology in the management of chronic diseases such as diabetes, asthma, congestive heart failure, and wound care offer effective tools that ameliorate chronic illness, reduce hospitalizations, and offer the potential of lowering healthcare costs. However, nationwide, fewer than 200 home health programs have deployed such technologies and the majority are located in urban areas [3, 22–24].

Statutory and regulatory burdens that must be overcome include continued limited reimbursement from many third-party payers, improved but still limited reimbursement from Medicare, and inconsistent reimbursement nationally by state Medicaid programs, licensure, and acceptance of technology by providers and patients alike. As of 2008, twenty-eight Medicaid states cover telehealth facilitated encounters. In addition, the Medicare legislation passed by Congress in July 2008 has increased the eligible consult origination sites to include skilled nursing facilities, hospital dialysis facilities, and community mental health centers.

Coverage for store-and-forward applications remains limited to teleradiology, telepathology, teleophthalmology, and second opinion (generally self-pay) services in teledermatology. By statute however, store-and-forward applications are considered a covered service in Alaska and Hawaii.

The recent expansion of the Rural Health Care Support Mechanism of the Universal Services Administrative Corporation of the Federal Communications Commission offers hope for continued enhanced deployment of sustainable telehealth solutions. The broadening of eligible providers to include certain rural for-profit hospitals (with an emergency room), and the expansion of acceptable benchmarks for comparable bandwidth anywhere in the state, through the November 2003 order has enhanced utilization of that program [25]. The December 2004 order broadening the FCC's definition of rural, and allowing telecommunications discounts for mobile satellite services to levels of comparable terrestrial rates promises to further stimulate innovative applications for rural telehealth [4].

In 2006, the FCC issued an order creating a new $50 million, 2-year pilot program with a goal of facilitating the creation of statewide and regional networks to increase the utilization of the Rural Health Care Support Mechanism.

Telehealth technologies should be viewed as integral to rural development. Communities not yet familiar with the value of broadband need to be advised about the other educational and economic benefits of telemedicine to include provisions for shared utilization of bandwidth by other businesses in the community. Data from some telehealth providers have indicated that more than 85% of patients seen via telehealth technologies remain within their community healthcare environment, resulting in a reduction in unnecessary transfers, less hospital lost revenue, as might occur with patient transfers, and the potential of enhanced economic viability of the community hospital [26]. A viable community healthcare environment and workforce ultimately provides incentives for the relocation of industry, thereby enhancing community economic development.

13.3 Implications for Urban Healthcare

Although the challenges of unfavorable geography and distance are uniquely rural, the socioeconomic barriers to access to quality healthcare in urban areas remain equally compelling. Poverty, unhealthy behaviors, and adverse health status indicators are as prevalent in the shadow of our finest urban academic health centers as they are in rural communities.

Healthcare services, while abundant in most urban areas, are less accessible to uninsured or underinsured Americans. For nation's 44 million uninsured citizens, fear of medical debt often prevents people from seeking medical care, and a lack of access to timely and/or preventive care results in overutilization of the country's emergency rooms. The balkanization of healthcare delivery systems, persistence of paper-based medical records, and the paucity of community health information networks that offer the sharing of patient data and images across proprietary health systems often results in the delivery of lesser quality care, costly duplication of effort, and a lack of coordination of care [3, 27].

Telehealth services have been successfully integrated into urban settings, and include robust initiatives that include telepsychiatry projects, pediatric telemedicine services, teledermatology applications, tele-ICU support, teleophthalmology initiatives, telecardiology, HIV/AIDS care, home monitoring, and telehomecare for patients with chronic illness [20, 28].

While the barriers of limited broadband deployment and costly telecommunications services are not as applicable as is the case in rural settings, the challenge of limited reimbursement by Medicare, Medicaid, and private payers in the urban environment create additional roadblocks to implementation. Currently, only a handful of Medicaid states support urban telemedicine projects for their enrollees.

13.4 Correctional Populations

Correctional organizations, often significantly lacking in on-site specialty care services, serve a litigious population with both specific and general healthcare needs. These agencies with large transportation and security budgets are ideal venues for the implementation of telehealth initiatives. Support of emergency, acute, and chronic illnesses within the inmate populations is easily facilitated as long as the clinical challenge of mutually agreed upon protocol development, therapies and follow-up care and technical challenge of firewalls, and secure transfer of medical data between otherwise unaffiliated entities can be resolved.

HIV infection is one of the most significant healthcare issues affecting U.S. inmate populations [29–34]. Telehealth tools have increasingly been applied to the care of incarcerated HIV-infected persons, since most correctional systems already have designated healthcare professionals on site to help coordinate referral care and to enact specific recommendations. As with care provided to noncorrectional patients, specific blood or radiologic studies can be done in advance of the telemedicine consultation, allowing the best advice to be given in an immediate, informed manner. Most important, routine chemistries and CD4 counts, HIV viral load, and HIV resistance studies in advance of the visit allow more rapid decisions regarding appropriate antiretroviral therapy than can usually be made in the standard outpatient setting. Directly observed administration of medications is an additional tool that enhances the outcome in the correctional environment. Additionally, these patients can be seen as often as necessary. Transportation needs are decreased and security issues are lessened, although cost savings may be less than anticipated if the cost of implementation of antiretroviral therapies are incrementally increased by virtue of enhanced access to HIV specialists.

Preliminary data from the UVA experience has shown that care delivered via telehealth technologies in the prison setting has correlated with excellent therapeutic success. In a chart review of 126 incarcerated patients receiving telemedicine facilitated care, 77% of patients previously untreated with pharmacologic agents attained undetectable viral loads as did 59% of all patients, including those with previous antiretroviral experience and AIDS-defining illnesses [34].

Challenges remain, however, with financial barriers to the provision and coordination of care exacerbated by fiscal constraints, and disparate healthcare systems. Although these issues are relevant to all populations, the transient nature of some HIV-infected individuals, especially those post-release from prison, makes the delivery of coordinated care a significant challenge.

13.5 Serving the Elderly

Services provided to elderly citizens using advanced technologies have the potential to greatly impact the cost and quality of care to those populations. The elderly have grown in numbers at double the rate of other populations and longevity has increased attributable to advances in healthcare. The U.S. Census Bureau reports an estimated 35 million people aged 65 and older, which represents approximately 14% of the total population. It has been estimated that chronic disease will impact nearly 50% of this population and that for every nursing home patient, there are as many as three to four times as many patients residing at home with similar needs [35, 36].

Migration of our younger populations to urban areas, and out-migration of retirees to rural areas has changed the demographics of our population in rural communities. Regardless of rural versus urban location, however, limited access to healthcare services can present a significant challenge for the elderly who are generally less mobile, visually and/or motor skills impaired, are remote from participatory family members, or suffer from chronic illness.

Telehealth technologies may be suitably applied to our elderly populations in the form of live interactive consultations in the inpatient or outpatient setting for acute or chronic disease management, interactive home telehealth, or monitoring of vital signs and/or disease-specific parameters integrated into the home, hospice, assisted living, or skilled nursing home setting [37]. Telemental health in the nursing home environment has been demonstrated to prevent hospital admissions in geriatric patients [38]. Wound monitoring and care is particularly applicable to both interactive and store-and-forward technologies [39, 40].

13.6 Challenges to Telehealth: Reimbursement

A significant expansion of Medicare coverage of telemedicine as originally authorized in the 1997 Balanced Budget Amendment occurred with the Medicare,

Medicaid, and SCHIP Benefits Improvement and Protection Act of 2000, effective October 1, 2001. Section 223 provided a framework for the revision of Medicare reimbursement for telehealth services and Section 504 addressed the use of telehealth in the delivery of home health services [41].

Current reimbursement for telehealth facilitated care provided to Medicare beneficiaries is limited by the program rules as below:

1. System requirement:
   a. Interactive telecommunications video system
   b. The use of store-and-forward technologies is permitted only in Alaska and Hawaii
2. Eligible entities:
   a. The physician and facility fee payments are only available for services provided to an "eligible telehealth individual" for a telehealth service furnished at an eligible originating site
   b. The Act sets forth the following definition of the originating site:
   (i)The term "originating site" means only those sites at which the eligible telehealth individual is located at the time the service is furnished via a telecommunications system and only if such site is located in
      a. An area that is designated as a rural health professional shortage area under section 332(a)(1)(A) of the Public Health Service Act (42 U.S.C. 254e(a)(1)(A));
      b. A county that is not included in a Metropolitan Statistical Area; or
      c. From an entity that participates in a Federal telemedicine demonstration project that has been approved by (or receives funding from) the Secretary of Health and Human Services as of December 31, 2000.
3. Eligible originating sites (site of the patient) include:
   a. Hospitals
   b. Critical access hospitals
   c. A physician office
   d. A rural health clinic
   e. A federally qualified health center
   f. A federally funded TeleHealth program in existence as of December 31, 2000, regardless of geographic location.
4. Eligible Providers – Medicare eligible practitioners who can bill for services provided via interactive telehealth encounters include:
   a. Physicians
   b. Nurse practitioners
   c. Physician assistants
   d. Certified nurse anesthetists
   e. Clinical psychologists
   f. Clinical nurse specialist
   g. Clinical social worker
5. Eligible CPT procedure/service codes within the ranges of:
   a. Office or other Outpatient Management—99201–99215

b. Consultations Codes—99241–99275
c. Individual Psychotherapy—90804-90809
d. Psychiatric Diagnostic Interview—90801
e. Pharmacological Management—90862
f. Dialysis Management—G0308, G0309, G0311, G0312, G0314, G0315, G0317, and G0318 are approved for January 1, 2006

6. Payment of a facility fee for originating sites of $21.68 for 2005.
7. Allows for payment of store-and-forward consultations in demonstration projects only in Alaska and Hawaii.
8. Payment for home health visits with Prospective Payment dollars. However, visits conducted via telecommunications may not count as a visit for the purposes of triggering a higher or lower level of payment.

Despite the forward momentum imparted by the 2000 statutory change, current regulations continue to foster an environment of unequal access for Medicare beneficiaries by limiting eligible services, locations, and types of eligible originating sites. Fears of fraud and abuse and skyrocketing Medicare expenditures related to telehealth services have not been realized. Until reimbursement of telehealth facilitated healthcare is provided for services originating in all Medicare eligible facilities, for all Medicare certified evaluation and management and procedure codes and by all Medicare eligible practitioners, regardless of geographic location, broader acceptance and deployment of telehealth technologies will fail to become a reality.

Medicaid reimbursement for telehealth facilitated consultations, ongoing care, and for home telehealth services for elderly Medicaid recipients varies markedly state by state. Although greater than 50% of Medicaid funding is federal in origin, there is, as yet, no federal mandate for coverage of traditional telehealth services or for home telehealth. The greatest financial burden to state Medicaid programs relates to services provided to residents of nursing homes and children with special healthcare needs. By supporting services provided through telehealth technologies, the greatest cost savings imparted to Medicaid programs could be expected in transportation costs, reduced emergency department visits, and a reduction in complications attributable to a delay in diagnosis and treatment.

13.7 Home Telehealth

Home telehealth is defined as the use of synchronous or asynchronous communications technologies by a home care provider to link patients to out-of-home sources of medical care, education, or other services. These services may incorporate interactive home telehealth technologies using plain old telephone service (POTS) or broadband telecommunications technologies. Home telehealth programs generally include monitoring devices such as pulse oximetry, blood pressure, EKG, blood sugar, weight, temperature, and passive monitoring of motion. It has been reported by the Food and Drug Administration, which regulates medical devices, that home care systems represent the fastest growing segment of the medical device industry.

Home telehealth can be utilized by traditional home health agencies, for the delivery of hospice care, or for case management by clinics or hospitals to facilitate chronic disease management for patients after discharge from the hospital [42, 43]. Home telehealth programs reduce readmission rates, visits to the emergency room, physician visits, and impart significant cost savings. As the labor pool of health professionals decreases, and with an ever-aging population, it is likely that models of home telehealth will play a greater role in the efficient day-to-day management of chronic illness in all age groups. The federal government supports major initiatives for aging in place such as the Program of All Inclusive Care for the Elderly (PACE), but does not pay for the technologies that will help keep people independent and living in their own homes.

For a cohort of patients with congestive heart failure monitored and treated with home telehealth technologies, Dimmick and Roth both reported a significant reduction in days of hospitalization from 6.2 to 1.2 days per year or 66%, respectively. Even when the cost of providing telehomecare was factored into the analysis, the total cost of care was 53% of national norms for the same diagnosis related group [44, 45].

Integration of home telehealth into rural models of healthcare is an efficient cost-effective choice when one considers the distances traveled by home health staff in rural areas. The average home telehealth visit approximates 18 minutes; the average in-home visit is 48 minutes, and yet patient satisfaction with home telehealth is excellent [42, 43]. Factoring in the time spent traveling to the home, significant cost savings accrues with the use of these technologies. Dimmick et al. report the cost savings associated with the supplementation of regular in-home visits by home telehealth visits at approximately $50/visit [46].

Reimbursement for home telehealth programs was authorized in October, 2000 with the Medicare Prospective Payment System (PPS), through which payment is supported through the model of "episodes of care." Providers may utilize home telehealth technologies as a component of those episodes of care in a capitated model to promote efficiencies. However, the home health agency may not count telehealth encounters for the purposes of billing or costing. In fact, the current strategy for the use of telehealth as outlined in PPS actually forces a specific number of visits before telehealth can be used. This position contradicts data that demonstrates improved clinical outcomes, patient safety and quality, and cost savings inherent in the use of telehealth technologies.

Hospice is a range of healthcare and comfort care services provided to patients nearing the end of life [47]. The goals of telehospice initiatives, like home telehealth, are to provide clinical care, physiologic monitoring, and other support services in the nonhospital setting. The applications, however, are focused on the terminally ill and their families and thus, vary from those of home telehealth. Telehospice services may include monitoring of vital signs, management and treatment of pain, monitoring of infusions, and support of the patient and his or her family [48, 49]. These technologies allow the provision of a broader array to hospice services in rural communities, where traditionally, geographic barriers have limited the number of services provided attributable to distance related inefficiencies.

13.8 Requirements for Telehealth Technologies

13.8.1 Acceptance and Usability of Advanced Technologies

Patient acceptance of the use of telehealth technologies for consultation and ongoing acute and chronic care has been remarkably positive, attributable in part to the obvious benefit of timely access to locally unavailable specialty healthcare that spares patients the burden and expense of travel to remote tertiary and quaternary healthcare facilities [50]. It has been reported that 93 million Americans use the Internet to access health information and that physician-patient relationships have been affected by such Internet use [51]. Electronic communication between physicians and patients has become more prevalent, since the 2001 survey reporting 14% of physicians communicated with patients via e-mail [52, 53]. Patient comfort with telehealth technologies has increased with greater use of technology in general and with consumer awareness campaigns.

Provider acceptance of advanced technologies and telehealth tools has been equally gratifying for patient consultation, for distance learning opportunities, for acquisition of timely information services, and for clinical decision support. In 2001, more than half of all physicians used e-mail to communicate with other physicians [52]. Utilization of Web-based educational tools has increased dramatically [54]. The exponential increase in broadband deployment has further encouraged use of such technologies [55]. Provider comfort and utilization of telehealth technologies also relates to ease/usability of the technology, frequency of use, and the relationship between the referring and consulting health professionals.

13.8.2 Standardization and Interoperability

The development of standards and guidelines, addressing both interoperability (such as HL7, DICOM, or H.320, 323, 324, compression standards for videoconferencing) and specialty specific applications (such as teleophthalmology or telepathology), will likely serve to increase functionality related to and acceptance of advanced technologies applied to healthcare. Specialty organizations that have published position statements include the American Telemedicine Association Special Interest Groups for Telepathology, Teledermatology, Teleophthalmology and Telehomecare, the American Academy of Dermatology, the Society of American Gastrointestinal Endoscopic Surgeons, the American College of Radiology, the American Psychiatric Association, and the American Nurses Association [56, 57].

13.8.3 Backend IT Systems (EMR, EHR, PHR)

Integration of electronic medical records, community health information networks, decision support tools, and telehealth technologies into a seamless platform/system of healthcare delivery will enhance access to healthcare for all Americans, improve quality, reduce costs, and increase patient safety. The crafting of integrated and compatible healthcare, education, and telecommunications policies by the relevant agencies to facilitate and incentivize innovation will smooth the transition to this new paradigm of healthcare in the digital era and support the goals of Healthy People 2010.

13.9 Conclusion

Advanced technologies have led to a new paradigm of healthcare services in the digital era and promise to aid our society in its transition to a coordinated systems approach focused on disease prevention, enhanced wellness, chronic disease management, decision support, quality, and patient safety. Through the incorporation of such tools and technologies, clinicians will be able to satisfactorily manage exponentially expanding volumes of medical information, research, and decision support analytic tools. The deployment of advanced technologies will minimize the barriers of distance and geography to enhance access, and foster the delivery of innovative integrated healthcare.

References

[1] Cooper, R. A., "Weighing the Evidence for Expanding Physician Supply," *Ann Intern Med*, Vol. 141, 2004, pp. 705–714.

[2] Blumenthal, D., "New Steam from an Old Cauldron: the Physician Supply Debate," *N Engl J Med*, Vol. 350, 2004, pp. 1780–1787.

[3] *Quality Through Collaboration, The Future of Rural Health,* Institute of Medicine, National Academies Press, 2004.

[4] *FCC Second Report and Order, Order on Reconsideration, and Further Notice of Proposed Rulemaking Federal Register,* Vol. 70, No. 24, February 7, 2005.

[5] Hart, L. G., et al., "Rural Healthcare Providers in the United States," *J Rural Health*, Vol. 18(S), 2002, pp. 211–232.

[6] Williams, J. M., et al., "Emergency Medical Care in Rural America," *Ann Emer Med,* Vol. 38, No. 3, 2001, pp. 323–327.

[7] Burgiss, S. G., et al., "Telemedicine for Dermatology Care in Rural Patients," *Telemed Journal,* Vol. 3, 1997, pp. 227–33.

[8] Pak, H. S., et al., "Teledermatology: An Intraobserver Diagnostic Correlation Study, Part II," *Cutis,* Vol. 71, No. 6, 2003, pp. 476–480.

[9] Krupinski, E., et al., "Diagnostic Accuracy and Image Quality Using a Digital Camera for Teledermatology," *Telemed Journal,* Vol. 5, 1999, pp. 257–263.

[10] Kvedar, J. C., et al., "The Substitution of Digital Images for Dermatologic Physical Examination," *Arch Dermatol,* Vol. 133, 1997, pp.161–167.

[11] Peterson, K., et al., *Arctic Investigations Program*, National Center for Infectious Diseases, CDC.

[12] Bergmo, T. S., "An Economic Analysis of Teleconsultation in Otorhinolaryngology," *J Telemed Telecare,* Vol. 3, 1997, pp. 194–197.

[13] Lopez, A., et al., "Pilot Data in Teleculposcopy in a Primary Care Women's Clinic," ATA presentation, Annual meeting, Orlando, FL, 2003.

[14] Schwamm, L. E., et al., "Virtual Telestroke Support for the Emergency Department Evaluation of Acute Stroke," *Acad Emer Med,* Vol. 11, No. 11, 2004, pp. 1193–1197.

[15] Scalvini, S., "Assessment of Prehospital Chest Pain Using Telecardiology," *J Telemed Telecare,* Vol. 8, No. 4, 2002, pp. 231–262.

[16] Srikanthan, V. S., "Use of Fax Facility Improves Decision Making Regarding Thrombolysis in Acute Myocardial Infarction," *Heart,* Vol. 78, 1997, pp. 198–200.

[17] Huffer, L., et al., "Feasibility of Remote Echocardiography with Satellite Transmission and Real Time Interpretation to Support Medical Activities in the Austere Medical Environment," *J Amer Soc Echocardiography,* 2004.

[18] Alboliras, E. T., et al., "Transmission of Full Length Echocardiographic Images Over ISDN for Diagnosing Congenital Heart Disease," *Telemedicine J*, Vol. 2, No. 4, 1996, pp. 251–258.

[19] Brailer, D., and E. Terasawa, "Use and Adoption of Computer Based Records," California Healthcare Foundation, 2003, pp. 1–42, available at www.chcf.org/documents/ihealth/UseAdoptionComputerizedPatientRecords.pdf.

[20] Flowers, C. W., et al., "Teleophthalmology: Rationale, Current Issues, Future Directions," *Telemed J* ,Vol. 3, No. 1, 1997, pp. 43–52.

[21] Chiang, M., L. Wang, and M. Busuioc, et al., "Telemedical Retinopathy of Prematurity: Diagnosis, Accuracy, Reliability, and Image Quality," *Arch Ophthalmol*, Vol. 125, 2007, pp. 1531–1538.

[22] Chetney, R., "Interactive Home Telehealth: Moving from Cost Savings to Reimbursement," *Telemed Today*, Vol. 9, No. 3, 2002, pp. 19–20.

[23] Field, M. J., and J. Grigsby, "Telemedicine and Remote Patient Monitoring," *JAMA*, Vol. 288, No. 4, 2002, pp. 423–425.

[24] Johnson, K. B., et al., "Hopkins Teen Central, Assessment of an Internet Based System for Children with Cystic Fibrosis," *Pediatrics*, Vol. 107, No. 2, 2001, pp. E24.

[25] *FCC Report and Order, Order on Reconsideration and Further Notice of Proposed Rule Making*, Federal Register, Vol. 69, No. 14, January 22, 2004.

[26] Meyer, K., "Can Telemedicine Deliver What It Promises," *Fam Prac Mgmt, AAFP*, March 1996.

[27] *Insuring America's Health: Principles and Recommendations*, National Academies Press, 2004.

[28] Breslow, M. J., "Effect of a Multiple Site Intensive Care Unit Telemedicine Program on Clinical and Economic Outcomes: An Alternative Paradigm for Intensivist Staffing," *Crit Care Med*, Vol. 32, No. 1, 2004, pp. 31–38.

[29] Crosland, C., M. Pohkus, and J. D. Rich, "Treating Prisoners with HIV/AIDS: The Importance Of Early Identification, Effective Treatment, And Community Follow-Up," *AIDS Clin Care*, Vol. 14, 2002, pp. 67–71.

[30] Maruschak, L. M., "HIV in Prisons, 2000," *Bureau of Statistics Bulletin*. U. S. Department of Justice, Office of Justice Programs, NCJ 196023, October 2002.

[31] Spaulding, A., et al., "Human Immunodeficiency Virus in Correctional Facilities: A Review," *Clin Infect Dis*, Vol. 35, 2002, pp. 1130–1136.

[32] McCue, M. J., P. E. Mazmanian, and C. Hampton, et al., "The Case of Powhatan Correctional Center/Virginia Department of Corrections and Virginia Commonwealth University/Medical College of Virginia," *Telemed J*, Vol. 3, 1997, pp. 11–17.

[33] Becker, C., "High-Tech Treatment: Federal Grant Allows HIV/AIDS Patients to Receive Treatment Through Telemedicine," *Mod Healthcare*, Vol. 32, No. 50, 2002, pp. 18–19.

[34] Rheuban, K., et al., *HIV/AIDS, Telemedicine Technical Assistance Documents*, University of Missouri Press, 2004, pp. 134–150.

[35] Pawlson, L. G., "Healthcare Impact of an Aging Population," in *Principles of Geriatric Medicine and Gerontology*, pp. 167–176, W. R. Hazzard, et al. (eds.), New York: McGraw Hill, 1994.

[36] Dimmock, S., et al., *Telehomecare, Telemedicine Technical Assistance Documents*, University of Missouri Press, 2004, pp. 151–187.

[37] Burgiss, S. G., et al., "Telehealth in Homehealth Care," in *Telecommunications for Nurses: Providing Successful Distance Education and Telehealth*, M.L. Armstrong and S. Frueh (eds.), New York: Springer Publishing Company, 2002.

[38] Hilty, D., et al., "Clinical and Educational Telepsychiatry Applications: A Review," *Can J Psych*, Vol. 49, 2004, pp. 12–23.

[39] Wirthlin, D. J., et al., "Telemedicine in Vascular Surgery: Feasibility of Digital Imaging for Remote Management of Wounds," *J Vasc Surg*, Vol. 27, No. 6, 1998, pp. 1089–1099.

[40] Braun, R. P., et al., "Telemedicial Wound Care Using a New Generation of Mobile Phones," *Arch Derm,* Vol. 141, 2005, pp. 2.

[41] *Medicare, Medicaid, and SCHIP Benefits Improvement and Protection Act of 2000*, 42 U.S.C. 1305 (HR 5661).

[42] Johnston, B., et al., "Kaiser Permanente Medical Center's Pilot Home Telehealth Project," *Telemedicine Today,* Vol. 4, No. 7, 1997, pp. 16–19.

[43] Johnston, B., L. Wheeler, and J. Deuser, et al., "Outcomes of the Kaiser Permanente Tele-Home Health Research Project," *Arch Fam Med,* Vol. 9, 2000, pp. 40–45.

[44] Burgiss, S., et al, http://gsm.utmck.edu/telehealth/presentations/Cost_Care_Reductions.pdf.

[45] Roth, A., "Telecardiology for Patients with Chronic Heart Failure: The SHL Experience in Israel," *Intl J Cardiol,* Vol. 97, No.1, 2004, pp. 49–55.

[46] Dimmick, S., et al., "A Case Study of Benefits and Potential Savings in Rural Home Telemedicine," Home Healthcare Nurse, Vol. 18, No. 2, 2000, pp. 124–135.

[47] Kinsella, A., "End-of-Life Care Ethics: Tailoring Telemedicine to End-of-Life Needs," *Virtual Mentor,* March 2001, p. E5.

[48] Kinsella, A., "Home Telecare in the United States," *J Telemed Telecare,* Vol. 4, No. 4, 1998, pp. 195–200.

[49] Doolittle, G. C., "A Cost Measurement Study for a Home-Based Telehospice Service," *J Telemed Telecare,* Vol. 6, No. 1, 2000, pp. S193–195.

[50] Demeris, G., et al., "Measuring Patient's Perceptions of Telemedicine's Role in Home Care," *Telemedicine J,* Vol. 6, 2000, p. 176.

[51] Madden, M., and L. Rainie, "America's On-Line Pursuits:The Changing Picture of Who's Online and What They Do," Pew Internet and American Life Project, December 23, 2002, available at: http://www.pewinternet.org/pdfs/PIP_Online_Pursuits_Final.PDF.

[52] Harris Interactive, "New Data Show Internet, Website and E-Mail Usage by Physicians All Increasing," aAvailable at: http://www.harrisinteractive.com.

[53] Borowitz, S. M., and J. C. Wyatt, "The Origin, Content and Workload of E-Mail Consultations," *JAMA,* Vol. 280, 1998, pp. 1321–1324.

[54] Bodenheimer, T., and K. Grumbach, "Electronic Technology: A Spark to Revitalize Primary Care?" *JAMA,* Vol. 290, 2003, pp. 259–264.

[55] Strode, S., et al., "Technical and Clinical Progress in Telemedicine, *JAMA,* Vol. 281, 1999, pp. 1066–1068.

[56] Hutarew, G., et al., "Two Year Evaluation of Telepathology," *J Telemed and Telecare,* Vol. 9, No. 4, 2003, pp. 194–192.

[57] Weinstein, R., et al., "An Array Microscope for Ultrarapid Virtual Slide Processing and Telepathology," *Human Pathology*, Vol. 35, No. 11, 2004, pp. 1303–1304.

CHAPTER 14

Automated In-Home Patient Monitoring: Geriatric Care Application

Majd Alwan

14.1 Introduction

The majority of the world's increasing older adult population requires some degree of formal and/or informal care due to loss of function as a result of failing health. In the United States, nearly three-quarters of older adults suffer from one or more chronic diseases according to the Centers for Disease Control (CDC); examples include arthritis, hypertension, and diabetes. The cost and burden of caring for older adults is steadily increasing [1]. Changes in the Medicare system in the United States have led to a shift in the responsibility of care from institutions (e.g., nursing homes) to the community (individuals and families). Meanwhile, the role of informal caregivers in providing care to the older adult population has greatly increased over the past two decades. Consequently, informal caregivers have come to be viewed as an unpaid extension of professional caregivers; providing most of the care to older adults requiring long-term care. In fact, national databases derived from independent sources have provided unequivocal evidence that family and friends are the sole care providers for about three-quarters of all community-dwelling older adults [2]. Informal caregivers have experienced increased physical burdens and emotional strains as a result of this shift in long-term care responsibilities. Furthermore, healthcare providers are faced with a shrinking professional caregiving work force at the same time [3]. On the other hand, the proportion of the world's population of individuals over the age of 60 is expected to double by 2030 to 20%. In the United States, the number of older adults is expected to grow to 108 million over the next 15 years, which represents 45% of the adult population. Older adults currently account for 60% of the overall healthcare spending in the United States. Appropriate management of chronic disease in older adults can reduce the U.S. healthcare bill by up to 50%. Furthermore, 92% of these older adults live alone in their own apartments, homes, independent living facilities, or assisted living facilities, including about 50% of those 75 and older. Such statistics clearly demonstrate an urgent need for innovative telehealth/telecare tools that enable older adults to live indepndently and maximize caregivers' efficacy by providing timely health information and delivering more effective care [4]. This change in the demographics, and its potential economic impact on industrialized nations has prompted active research in AI-based systems for automated functional and health status monitoring and assistance; a comprehensive review of research on the potential of exploiting automation technologies as caregivers is provided in [5]. In the meantime, modern sensor and communication technology, coupled with advances in data analysis and artificial intelligence techniques is causing a paradigm shift in remote management and monitoring of chronic conditions. In-home monitoring has the added benefit of measuring individualized health status and reporting it to the primary care provider and caregivers alike; allowing timelier and targeted preventive interventions [6].

Health monitoring in the home environment can be accomplished by (1) ambulatory monitors that utilize wearable sensors and devices to record physiological signals (reviewed in [7]), (2) sensors embedded in the home environment and furnishings to unobtrusively collect behavioral and physiological data, or (3) a combination of the two [7]. Passive monitoring has the inherent benefit of obviating the problems associated with device usability and incorrect use as well as potential patient noncompliance, thus in this chapter we will focus our attention on passive nonconstraining health monitoring.

Throughout the reminder of this chapter a model for geriatric care that is enabled by in-home monitoring and ambient intelligence technologies will be presented. We will also present system design criteria that are crucial to the success of this model. For the past few years, the Medical Automation Research Center (MARC) at the University of Virginia has been developing, validating, refining, and evaluating passive monitoring technologies that fit the technology-enabled care paradigm. The chapter will also briefly describe example systems, developed at MARC, that conform to the presented care delivery model and will present a summary of field pilot studies conducted with some of the described systems to demonstrate the validity of the model and the utility of the presented example systems.

14.2 Model for Care Enabled by Ambient Intelligence

It can be argued that a new paradigm for geriatric care can emerge with more integrative technologies. For example, the activities and selected physiological parameters of an older adult can be monitored in his or her own living setting through sensors embedded in the environment or the other objects, telehealth devices, and other technologies. The environment may be the person's house or apartment in the community, or a residence provided by an eldercare services provider (a continuing care retirement community, an independent living apartment, assisted living unit, etc.) Safety, activity, physiological, and health data can be analyzed, archived, and mined to detect indicators of early disease onset, deterioration, or improvement in health conditions at various levels. The care delivery diagram depicted in Figure 14.1 illustrates the process. Data analysis results, at various levels, can be made available to all stakeholders in the care process, including the monitored older adults, their professional caregivers, informal caregivers and primary health care providers, and integrated into an electronic medical or personal health record accessible to authorized caregivers whenever they need them.

The monitored individual can use the analysis results in personal health maintenance (e.g., diet, exercise). Informal caregivers will get an objective assessment of their loved ones' ability to remain independent, and peace of mind when everything is fine. This reassurance will eliminate interrogation, questioning, and role reversal

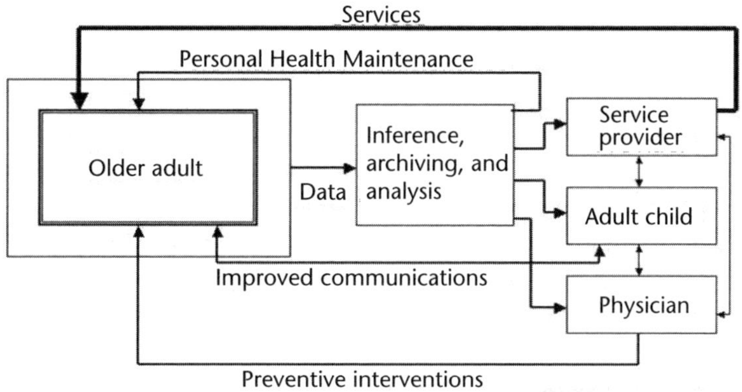

Figure 14.1 Model for the technology-enabled geriatric care paradigm.

between the older adult and their adult children and would increase the social content of their communications. This will improve the quality of life for both parties, as well as reduce unnecessary early institutionalization of older adults driven by the anxiety of their children.

When the older adult needs assistance in some of his or her activities of daily living (ADLs) or instrumental activities of daily living (IADLs), professional caregivers accessing the reports will have an objective assessment of their actual needs and can determine the appropriate care package. They can coordinate, dispatch, and track the delivery of care and services to the monitored older adults via home care agencies (e.g., meals on wheels, bathing) if they live in the community, or on-site direct care workers if they live in a continuum of care facility.

Primary health care providers can perform an evaluation of the monitored older adult's health that is more comprehensive than the "snapshot" assessment obtained during an annual physical examination. They may be able to detect the early onset of disease and prescribe appropriate interventions (including preventive interventions), and can monitor the efficacy of these interventions objectively and longitudinally.

Finally, access to the analysis of the same objective data by all authorized stakeholders is expected to improve the communication between them (e.g., the eldercare services provider and the adult child, when deciding on the most appropriate care package for the older adult).

This paradigm exploits the technical capabilities of embedded sensing, ambient intelligence, interoperability, and interconnectivity between different devices in the home, as well as other information and communication technologies, in automating the continuous assessment, documentation, and communication. It enables a network of professional and informal caregivers to coordinate and deliver high-touch care when needed. The paradigm is expected to prolong and enhance the independence of seniors, delay their transition to nursing facilities, and thereby reduce the overall cost of care.

This model, although presented in the context of geriatric care, is applicable to other populations with a variety of chronic conditions and in need of care and supportive services.

14.3 Requirements for Success

Acceptance of the technology (in terms of what is monitored and how) its usability, the security and privacy of the collected information and who it is shared with, cost and affordability, as well as the demonstrated benefits are key to the success of the paradigm presented above.

In addition, the ability to integrate the data from multiple sources and store the information in an electronic health record (EHR) or a personal health record (PHR), such that it can be shared with and between providers to allow synthesizing the data into knowledge to facilitate decision-making, plays a significant role in maximizing the benefits of in-home patient monitoring technologies.

In a recent study [8], we assessed the acceptance, perceived utility, and benefit of in-home monitoring technologies, using custom-designed survey instruments. Data were collected from 24 older adults, 20 informal caregivers, and 19 professional caregivers. The majority of surveyed participants (79% of older adults, 75% of informal caregivers, and 100% of professional caregiver) thought that it was a good idea to monitor the older adults' activities in their living settings. Some older adults (17%) and informal caregivers (10%) did not think the monitoring was a good idea, while the rest (4% of older adults and 15% of informal caregivers) were unsure. 71% of the surveyed older adults did not think they needed monitoring. Similarly, 80% of surveyed informal caregivers did not think that the person they care for needed monitoring. Interestingly, however, all surveyed professional caregivers thought that the older adults they provide care for needed monitoring. 16% of the participants (primarily professional caregivers) identified the data security and potential unauthorized access, as well as the private nature of older adults as possible privacy concerns that may hinder the acceptance of the monitoring technologies. Falls, walking abilities and balance, and taking medications were identified by all three participant groups as the most important parameters to monitor. The participants found that automatic Personal Emergency Response Systems (PERS), in-home fall detectors, wearable fall detectors, and motion detectors were the most acceptable monitoring technologies, while cameras, cameras that produce body outline images without details, and sound recorders were the least acceptable monitoring technologies.

The data showed that it was most acceptable to all parties to share the monitoring information with a hospital and emergency services, and it was not acceptable to share the information the health insurance company, Medicare, Medicaid, friends, neighbors, and relatives (other than the spouse, adult children, and siblings). The data showed that the upper threshold for the monthly monitoring fee was $100.

The majority of participants felt that the monitoring would enable older adults to live a safer life in their living setting, and would reflect positively on their quality of life. Moreover, the majority of professional caregivers felt that in-home monitoring could help them provide better care, while the majority of informal caregivers felt that the monitoring would help their peace of mind [8].

MARC's in-home monitoring system was implemented on an open source electronic medical record (EMR) system, OpenEMR, which was adapted to accept ADL, basic sleep, and gait data.

One of the barriers to the wide use of these technologies is the fact that interventions made on the basis of these technologies are not reimbursable, unlike some televisits or teleconsults with two-way video and most store-and-forward telemedicine technologies.

14.4 Example Passive Monitoring Systems

14.4.1 Activity Monitoring System

MARC's activity monitoring system is comprised of several off-the-shelf wireless motion sensors and a threshold temperature sensor above the kitchen range. All sensors transmit data wirelessly to a PC-based appliance as shown in Figure 14.2. This dedicated computer appliance collects all the sensor generated data and periodically dials into a secure remote data server, located in MARC's Robotics Lab, to write new data into the remote database. The in-home data collection appliance also runs routines to check possible emergency conditions. A Web-based inference application performs preprogrammed data analysis routines on the transferred data and then displays the inference and analysis results.

In order to develop the sensor suite and refine the activity inference algorithms, the system was tested for 18 months in a community home that served as a "living laboratory." In preliminary studies, the activity data of a normal healthy middle-aged participant was logged and analyzed using several data analysis techniques, including clustering, mixture models [9], and a rule-based approach, where spatial-temporal relationships among sensor events are exploited to infer the occurrence of activities with a high degree of confidence. The basic set of rules and algorithms were developed using knowledge acquisition principles to generate generic rules that would be applicable to almost anyone and can be used as a template that could be later refined and customized. The method is general, customizable, and is computationally inexpensive.

As an example, let us consider meal preparation. Meal preparation would generally entail at least motion in the kitchen and use of cabinets where food, plates, and/ or utensils are stored. Depending on the type of meal, it may also involve the use of appliances such as a stove, oven, or microwave oven. A rule describing meal preparation activity could be:

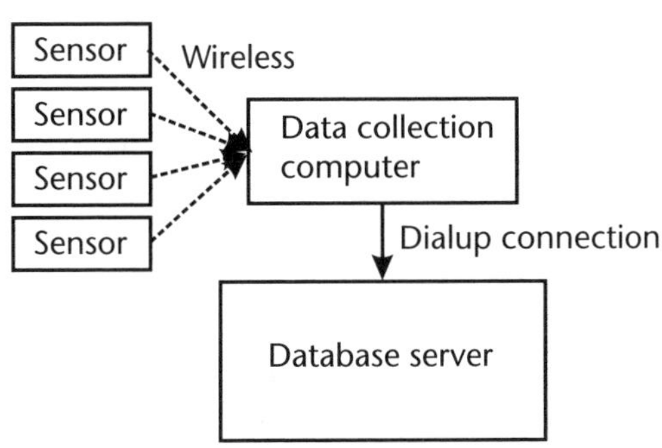

Figure 14.2 Monitoring system architecture.

IF the resident was in the kitchen AND (resident accessed meals ingredients cabinet AND resident accessed plates or utensils cabinet) OR resident used an appliance THEN a meal was prepared.

Presence in the kitchen can be indicated by motion in the kitchen persisting for a minimum period of time, whereas use of meals ingredients can be indicated by use of a food storage cupboard or the refrigerator. The rule-based approach was adopted for the inference of the activities of interest for its simplicity, computational efficiency, and scalability. Clustering and mixture models were useful in only identifying clusters representing activities but did not readily allow the automatic labeling of these clusters (i.e., assigning an activity name to the clusters). Moreover, both methods were computationally expensive, requiring significant learning, and thus may not scale easily. Nonetheless, these methods, together with the experimentation with the rule-based inferences, allowed limiting the number of sensors used while maintaining an acceptable level of confidence in our inferences. The system and rule-based inference methods were validated against 37 days of the subject's self-report, recorded in real-time using a personal digital assistant (PDA) based electronic diary developed specifically for the validation study. The validation results of the activity inference rules were reported in [10].

Later, the knowledge-engineered rules were compared to rules automatically generated by supervised learning using random forest. Results have shown an acceptable validity of knowledge-engineered rules compared to automatically generated rules. The kappa correlation coefficient between the classification results of the random forest model and the PDA record was 0.79, with 85% sensitivity and 93% specificity, compared to kappa=0.84, with 91% sensitivity and 100% specificity for the knowledge-engineered rules. The results showed that rule-based activity inference using knowledge-engineered rules is a viable approach to activity inference, and they highlighted the importance that insight knowledge engineering could have on activity inference results and their validity [11].

14.4.2 Passive In-Bed Vital Signs Monitor

MARC's noninvasive analysis of physiological signals (NAPS) system (see Figure 14.3) uses ballistocardiography (BCG) to detect minute forces generated during cardiac contraction and relaxation, and can also detect body movement from respiratory effort and postural changes. Preliminary data [12, 13] has shown strong correlations between the heart rate passively measured using the NAPS system and conventional clinical techniques, such as pulse oximetry. The NAPS system relies on a highly sensitive pressure transducer pneumatically connected to a compliant force-coupling pad installed on top of the mattress of any standard bed, on which the subject lies in order to acquire the data. Multiple pads can be used to acquire data from different parts of the body. The system is sensitive enough to gather data even when sheets and blankets are applied over the sensor, which rests on the mattress. The analog signal is filtered and amplified before being digitized by A-to-D converters. An algorithm has been developed to provide automatic scoring of the instantaneous heart rate and respiration data recorded by the NAPS system. Additionally, the passive nature of the NAPS system allows data to be recorded longitudinally and to establish personalized norms for individuals that can be

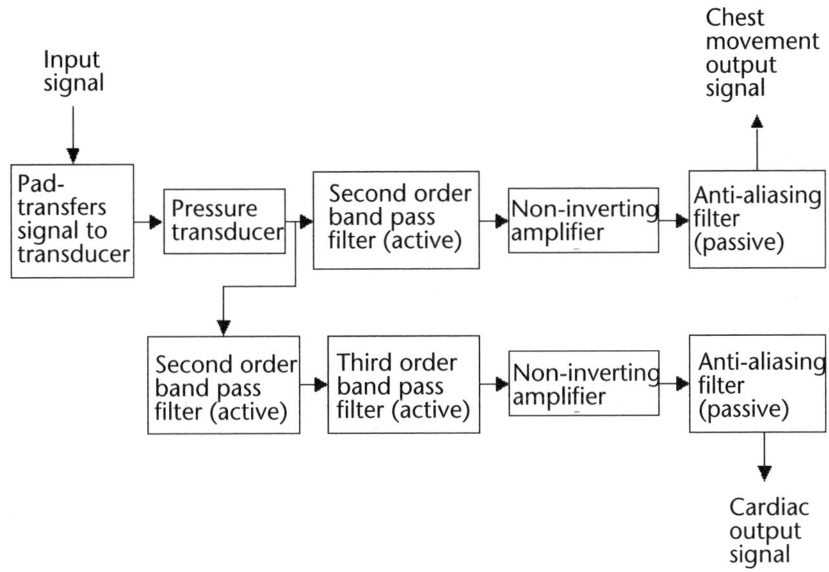

Figure 14.3 NAPS system block diagram.

used to detect changes in physiological parameters, and/or to assess the efficacy of interventions.

The ability of the NAPS system to measure heart rate, as compared to EKG, and its apnea and arousal detection capabilities, as compared to conventional polysomnography, were investigated. The NAPS average heart rate was highly correlated with the EKG average heart rate ($r = 0.972$, $p < 0.0001$). The NAPS system failed to produce an average heart rate for an epoch 10.6% of the time where the majority of those were due to movement artifacts. Of the detected heart rates, 80.0% were at or below the standard error of ±2.54 beats per minute (BPM). The standard error for the delta sleep portion of the data was ±2.20 BPM while the standard error for the segments with at least one apnea present was ±2.83 BPM. This is due to the increased movement and breathing anomalies that occur during and immediately following apneas and arousals, which can adversely affect the NAPS heart rate signal quality. Apneas and arousal events detected with the NAPS system were compared to those detected with polysomnography using two-way contingency comparisons, without discriminating between central and obstructive apneas (all hypopneas were included as well). Overall, the NAPS system achieved a sensitivity of 89.2% and specificity of 94.6% in the detection of apneas with a kappa correlation coefficient of 82.8%. The kappa correlation was chosen since the technician-scored polysomnography data is also subject to errors. Moreover, the NAPS system attained a sensitivity of 77.3% and a specificity of 96.2% in the detection of arousals with a kappa of 73.0%. In examining the aggregate detection of apneas by computing an apnea hypopnea index (AHI) for the 4 hours of data that were preliminarily analyzed, a strong agreement was found between the NAPS system (33.8) and polysomnography (32.5). Similarly, the computed arousal index also showed agreement between the NAPS system (16.8) and polysomnography

(16.5). Additionally, standard errors of ±9 seconds in detecting the start time of apneas and ±7 seconds for arousals were attained. Finally, the average length of apneas and arousals were closely matched to their respective polysomnography values (1% error in average apnea length and 4% error in average arousal length). The results presented here show that the NAPS system not only detected the right number of events, but also exhibited high timing alignment of events as compared to polysomnography [14].

A low-resolution low data rate version of the NAPS system, suited to longitudinal assessment of sleep quality in the natural living environment as well as verifying that the heart and breathing rates are within their normal ranges, is currently undergoing validation against actigraphy. This version has been integrated into the in-home health status monitoring systems that were pilot tested in the field in the studies summarized in Section 14.4.

14.4.3 Passive Floor Vibration-Based Fall Detector and Gait Monitor

MARC's floor vibration-based fall detector uses a special piezoelectric sensor, coupled to the floor surface by means of mass and spring, combined with preprocessing electronics to evaluate the vibrations of the floor. The size of the detector assembly is comparable to a small coffee can and it can be placed directly on the floor, typically one per large room. The detector continuously monitors the floor for vibrations that are produced by the common activities of the resident such as walking, sitting down, and tapping. Based on the experiments conducted, it was noted that there were significant differences in the patterns of vibrations induced on the floor by different activities. There was also a significant difference in the vibration pattern generated by a falling object as opposed to those generated by a falling anthropomorphic dummy or person. This difference in floor response for different activities was effectively used to detect falls with a higher degree of sensitivity and fewer false positives. The device detects a fall only when the vibration pattern (frequency, amplitude, duration, etc.) obtained from the floor over a small period of time matches the pattern induced when a person falls on the floor. The detector can then report the fall alert to the responder through an appropriate communications portal, such as the telephone line, to send a message to a radio pager or to a cellular phone.

The integrated fall detector was tested using anthropomorphic dummies (Hybrid-III seated crash test dummy, weighing 180 pounds, and a Rescue Randy, a 6-foot, 1-inch tall dummy weighing 165 pounds) that have mass distributions similar to those of humans to emulate various fall scenarios. The Hybrid-III dummy was used to emulate the scenario of a person falling when attempting to get out of a chair/wheelchair and the Rescue Randy was used to emulate tripping and falling from an upright position. The dummy falls were conducted at known distances from the sensor ranging from 2 to 20 feet. Experiments were repeated three times at each distance to ensure the repeatability of results. It is also important to establish the ability of the fall detector to distinguish a human fall from an object dropped within the detection range. The performance of the detector for nonfall events was tested by dropping objects weighing 5 and 15 pounds at various distances covering the entire range of fall detection, up to 20 feet. The fall tests, both for the dummy and objects, were repeated on different flooring treatments (carpet with and without foam padding) to evaluate the effect of floor treatments on the performance of the fall detector. Finally, the fall detector was placed in a room, while the dummy was dropped in an adjacent room, with a wall separating the two rooms, to test the possibility of falsely triggering the detector due to a fall in the neighboring apartment. This test was performed to determine the parameters governing the sensor's placement to limit such possibilities. The dummy was dropped less than 2 feet away from the wall, and the senor element was moved away from the wall in discrete steps until it no longer triggered for at least three successive falls.

The fall detection range for the sensor was found to be 20 feet in the case of mezzanine concrete floor covered with linoleum and 15 feet on concrete slab floor. A dummy fall induced higher amplitude vibration signals on mezzanine concrete floors than on concrete slab floors at equivalent distances. Thus, the fall detector circuitry required different preprocessing thresholds for these two different floor types. This can be achieved by means of a simple toggle switch that selects threshold levels and sensitivities appropriate to the floor type where the device is installed. Utilizing these modifications in settings, the fall detection function produced impressive repeatable performance on both mezzanine concrete and concrete slab floors. The fall detector attained 100% detection rate of falling dummies and 0% detection rate of dropped objects having a weight of 15 lb as close as 2 feet away from the sensor (100% sensitivity and 100% specificity from a total of 70 dummy falls and 53 object drops). The performance was not significantly affected by different floor treatments including linoleum, carpet, and foam padding. Thus, the different flooring treatments did not require any changes to the circuitry or sensor-sensitivity settings to maintain the detection ranges specified above. The vibration signal transmittance tests showed that the fall detector's sensor should be placed no closer than 5 feet to walls for mezzanine concrete (4 feet for concrete slab) during installation to minimize the chances of falsely triggering the fall detector as a result of a fall in a neighboring apartment [15].

The same sensor technology has shown a promising ability to detect step-induced floor vibrations and provide various basic gait parameters including step count, cadence, and step duration, in addition to the ability to distinguish between normal, limping, and shuffling gait modes. The gait monitor may be augmented with additional sensors to estimate the traversed distance and evaluate average walking velocity; this would enable the calculation of additional gait characteristics such as average step length and average stride length. These parameters can be used to detect various gait anomalies [16].

14.4.5 Instrumented Walker for the Passive Monitoring of Gait and Balance

Under the passive assessment paradigm, the author developed with collaborators a method that allows the passive assessment of basic walker-assisted gait characteristics, including heel strikes, toe-off events, as well as stride time, double support, and right-and-left single-support phases, using only force-moment measurements from the walker's handles. The passively derived gait characteristics of 22 subjects were validated against motion capture gait analysis. The force-moment based heel initial contact detection algorithm has produced a high level of concordance with heel initial contacts detected by a human inspecting the heel marker data sets of the Vicon® video motion capture system. The algorithm has demonstrated 97% sensitivity and 98% specificity with a narrow 95% confidence interval of ±1% during all experiments, which included five navigational scenarios. Temporal error in detecting the instances of heel initial contacts were within 5.27±3.66% of the overall stride time obtained from video motion capture when the subjects walked in a straight line, whereas the toe-off instance estimation were within 5.18±2.75% of the gait cycle. The errors in determining the duration of stride time, single support, and double support were within 5.86±2.49%, 5.24±2.29%, and 4.34±2.13% of the gait cycle, respectively. The stride time estimated, using the method presented here, correlated well with stride time computations based on visual inspection of motion capture data, Pearson correlation coefficient r=0.86 for straight-line segments. However, absolute errors were too high to estimate the single and double support phases with acceptable accuracy [17].

Similarly, resolving the forces and moments applied on the walker's frame to the center of gravity of the walker allows computing a stability measure that accounts for the vertical load distribution on the walker and the effect of angular loads particularly moments about the axes of the walker's left and right sides continuously. The value of the stability margin indicates the stability of the walker platform under the user's influence, which obviously impacts the stability of the walker user. In general, if the walker is unstable, the human user is very likely to also be unstable [18]. The potential application of the instrumented walker and the gait and stability assessment methods is longitudinal basic gait and stability assessment that can be performed outside of the conventional gait labs.

14.4.6 Summary of Field Pilot Results

Field pilots with MARC's passive in-home monitoring system comprising the activity and bed monitors revealed several benefits to the different stakeholders in the geriatric care process. In a Volunteers of America assisted-living facility, the system enabled rapid assessment and early detection of health conditions that might otherwise have been missed, and it proved to be an effective care coordination and management tool for professional caregivers. Professional caregivers used the information gathered to avert potential illnesses. For example, at the earliest signs of increased restlessness coupled with more frequent nightly bathroom visits, an indicator of a potential urinary tract infection, caregivers intervened by increasing fluid intake, which can clear up the infection without pharmaceutical or clinical intervention [19]. Moreover, monitored older adults experienced a significant increase in their perceived quality of life following three months of monitoring, possibly due to an increased sense of security or improved care quality, despite the short period of the pilot and the small sample size of 15 older adults [20].

In a controlled 3-month followup pilot, the technology resulted in a statistically significant reduction of billable interventions (aggregate physician visits, emergency department visits, and hospital days), and therefore a statistically significant reduction in overall cost of care to payer. The study showed that the initial cost of the monitoring system, even as a research prototype, would produce a return on investment in care cost savings in 6 weeks of deployment. Professional caregivers' workloads at the pilot sites were significantly lower than the workloads of peers at the control sites who were not using the technology, despite the fact that the former were serving a significantly larger number of care recipients. The difference in the ratio of caregivers to patients indicates that caregivers at the monitoring sites may have achieved higher levels of efficiency than their control site counterparts, which could be due to the effective utilization of the monitoring technology. The difference in efficiency-normalized workload scale scores between the control and monitoring sites was statistically significant [21].

In a pilot with a home health agency managed by the Evangelical Lutheran Good Samaritan Society in Windom, Minnesota, the technology resulted in a statistically significant increase in perceived quality of life of monitored older adults after 4 months of monitoring. Providing wellness and activity reports to professional and informal caregivers may have contributed to improvement in care, as evidenced by the decrease in the number of combined physically and mentally unhealthy days for the older adults after monitoring. Additionally, there was a statistically significant decrease in the measured indices of caregiver strain in informal caregivers [22].

Use of MARC's monitoring system in independent senior housing apartments in Waconia, Minnesota, managed by the Good Samaritan Society, increased informal caregivers' level of engagement in caring for their loved ones. Demonstrated by a statistically significant increase in informal care hours per week, the monitoring system and the availability of wellness and health status reports increased caregivers' participation without any corresponding increase in strain or burden. The study also showed that there was no appreciable impact on patients' quality of life, possibly because the reports were not provided to professional caregivers to provide any interventions [23].

14.5 Related Work

The project undertaken by Togawa and associates was one of the first projects to use passive sensing for everyday activities to monitor subjects both physiologically and behaviorally [24, 25]. One of the pioneering research projects in telehealth was conducted at the University of New South Wales, Australia, aiming to explore whether functional health status among the elderly could be accurately determined remotely by continuously monitoring relatively simple parameters that measured the interaction between participants and their environment [26]. The researchers reported a high level of acceptance by both the participants and their primary care providers, since the system was easy to use, effective, and potentially increased the efficiency of chronic disease management. In the United Kingdom, research and clinical trials examined the capabilities of intelligent monitoring systems to identify emergency situations based upon detected deviation from normal activity patterns; of the 61 alerts generated, 46 were classified as false alerts and the other 15 as genuine, although no real emergencies occurred during the study [27]. Currently, research is being conducted at University of Joseph-Fourrier in France [28, 29], focusing on data fusion of multisensor information and data mining to generate health alarm conditions. The data analysis methods were tested using simulated physiological data. Glascock and Kutzik [30] described a nonintrusive system. In the proof of concept phase, this system was validated in the activities of daily living (ADL) suite of an urban hospital where a video camera and recorder captured the actual activities carried out by participants. An in-home testing phase was sequentially conducted in 1998 in several homes, with the longest monitoring data collected representing 13 consecutive days [30]. In [31], Glascock and Kutzik described a commercial behavioral mentoring system, and the theory behind it. The article presented data, collected from a field test in over 150 homes, showing approximately 98% reliability in detecting motion activities by the systems. The validity of the activities of daily living inferred from the sensors, data, however, has not been formally evaluated. Seven case studies collected from a 12-month pilot study with 22 monitored clients, qualitatively showed that the system has been effective in detecting unusual behaviors indicating emergency conditions (possible falls) and change of behavior and health conditions; this allowed the home health nurses using the system to follow their response protocols and intervene to affect change to the client's care plan accordingly.

14.6 Conclusion

Passive in-home health status monitoring technologies may bring about a new technology-enabled care paradigm that would allow care providers to extend their services into the communities they serve. The approach has shown great promise for significant value propositions to all the stakeholders in the process of caring for older adults, when applied to geriatric care. More diverse patient populations, and better controlled field evaluations and demonstration projects are needed to further prove the utility and benefits in different care settings. Significant evidence of efficacy and cost effectiveness may drive changes to the reimbursment system to allow wide proliferation and adoption of these technologies into the standards of care.

References

[1] National Institute of Nursing Research, Informal Caregiving Research for Chronic Conditions RFA, 2001.

[2] National Institute of Nursing Research, Priority Expert Panel (PEP) Report, Vol. 3, Long-Term Care for Older Adults, Vol. 3, 1994.

[3] Felder, R. A., S. Kell, and D. Mack, et al., "Applying Technology to Aging Populations: An Overview of Projected Needs and Emerging Solutions," *Third Annual Symposium on Aging*, Keynote Address, Knoxville, TN, September 2002.

[4] American Institute of Aging, "A Profile of Older Americans: 2001," www.aoa.gov/aoa/stats/profile.

[5] Haigh, K. Z., and H. A. Yanco, "Automation as Caregiver: A Survey of Issues and Technologies," *Proceedings of AAAI 02 Workshop "Automation as Caregiver,"* Edmonton, Alberta, July 29, 2002, pp. 39–53.

[6] Celler, B. G., N. H. Lovell, and D.K. Chan, "The Potential Impact of Home Telecare on Clinical Practice," *MJA*, Vol. 171, 1999, pp. 518–521.

[7] Tamura, T., T. Togawa, and M. Murata, "A Bed Temperature Monitoring System for Assessing Body Movement during Sleep," *J Clinical Physics and Physiological Measurements*, Vol. 9, pp. 139–145, 1988.

[8] Alwan, M., B. Turner, and S. Kell, et al., "Acceptance and Perceived Utility of Monitoring Older Adults in Their Living Environment," *Proceedings of the 2nd International Conference on Technology and Aging (ICTA)*, Toronto, Canada, June 16–19, 2007.

[9] Barger, T., D. Brown, and M. Alwan, "Health Status Monitoring Through Analysis of Behavioral Patterns," *IEEE Transactions on Systems, Man and Cybernetics (Part A: Systems and Humans)*, Vol. 35, No. 1, 2005, pp. 22–27.

[10] Alwan, M., J. Leachtenauer, and S. Dalal, et al., "Validation of Rule-Based Inference of Selected Independent ADLs," *Journal of Telemedicine and E-Health*, Vol. 11, No. 5, October 2005, pp. 594–599.

[11] Dalal, S., M. Alwan, and R. Seirafi, et al., "A Rule-Based Approach to Analyzing Elders' Activity Data: Detection of Health and Emergency Conditions," *Proceedings of the AAAI's Fall Symposium on AI in Eldercare: Caring Machines,* Washington, DC, November 2005.

[12] Mack, D. et. al., "Non-Invasive Analysis of Physiological Signals (NAPS): A Low-Cost, Passive Monitor for Sleep Quality and Related Applications," Poster Paper, *Carilion Biomedical Institute Steps to Success Conference, Roanoke, VA, 2002.*

[13] Mack, D., S. Kell, and M. Alwan, et al., "Non-Invasive Analysis of Physiological Signals (NAPS): A Vibration Sensor That Passively Detects Heart and Respiration Rates as Part of a Sensor Suite For Medical Monitoring," *Summer Bioengineering Conference Proceedings,* Key Biscayne, FL, 2003.

[14] Mack, D. C., M. Alwan, and B. Turner, et al., "A Passive and Portable System for Monitoring Heart Rate and Detecting Sleep Apnea and Arousals: Preliminary Validation," *Proceedings of the Transdisciplinary Conference on Distributed Diagnosis and Home Healthcare (D2H2),* Arlington, VA, April 2–4, 2006.

[15] Alwan, M., P. J. Rajendran, and S. Kell, et al., "A Smart and Passive Floor-Vibration Based Fall Detector for Elderly," *Proceedings of the 2nd IEEE International Conference on Information and Communication Technologies: From Theory to Applications* (ICTTA'06), Damascus, Syria, April 23—28, 2006.

[16] Alwan, M., S. Dalal, and S. Kell, "Passive Unobtrusive Gait Monitor: An Economic Device to Detect Gait Anomalies and Falls," *Carilion Biomedical Institute Steps to Success Conference,* Poster Session, Roanoke, VA, October 2002.

[17] Alwan, M., A. Ledoux, and G. Wasson, et al., "Basic Walker-Assisted Gait Characteristics Derived from Forces and Moments Exerted on the Walker's Handles: Results on Normal Subjects," *Journal of Medical Engineering & Physics, Elsevier,* Vol. 29, No. 3, pp. 380–389.

[18] Alwan, M., P. J. Rajendran, and A. Ledoux, et al., "Stability Margin Monitoring in Steering-Controlled Intelligent Walkers for the Elderly," *Proceedings of the AAAI Fall Symposium on AI in Eldercare: Caring Machines,* Washington, DC, November 2005.

[19] Alwan, M., S. Dalal, and D. C. Mack, Kell et al., "Impact of Monitoring Technology in Assisted Living: Outcome Pilot," *IEEE Transactions on Information Technology in Medicine and Biology*, Vol. 10, No. 1, January 2006, pp. 192–198.

[20] Alwan, M., J. Leachtenauer, and S. Dalal, et al., "Psychosocial Impact of Passive Health Status Monitoring Technology in Assisted Living: A Pilot Study," *Proceedings of the 2nd IEEE International Conference on Information and Communication Technologies: From Theory to Applications* (ICTTA'06), Damascus, Syria, April 23—28, 2006.

[21] Alwan, M., E. B. Sifferlin, and B. Turner, et al., "Impact of Passive Health Status Monitoring to Care Providers and Payers in Assisted Living," *Journal of Telemedicine and E-Health,* 2007, 13(3): pp. 279–286.

[22] Alwan, M., D. Mack, and S. Dalal, et al., "Impact of Passive In-Home Health Status Monitoring Technology in Home Health: Outcome Pilot," *Proceedings of the Transdisciplinary Conference on Distributed Diagnosis and Home Healthcare (D2H2)*, Arlington, VA, April 2–4, 2006.

[23] Alwan, M., S. Kell, and B. Turner, et al., "Psychosocial Impact of Passive Health Status Monitoring on Informal Caregivers and Older Adults Living in Independent Senior

Housing," *Proceedings of the 2nd IEEE International Conference on Information and Communication Technologies: From Theory to Applications* (ICTTA'06), Damascus, Syria, April 23–28, 2006.

[24] Yamagushi, M., M. Ogawa, and T. Tamura, et al., "Monitoring Behavior in the Home Using Positioning Sensors," *Proceeding of the 20th Annual IEEE Conference on Engineering in Medicine and Biology,* 1998, pp. 1977–1979.

[25] Korhonen, I., J. Parkka, and M. Van Gils, "Health Monitoring in the Home of the Future—Infrastructure and Usage Models for Wearable Sensors that Measure Health Data in the Daily Environment of the Users," *IEEE Engineering in Medicine and Biology Magazine,* Vol. 22, No. 3, May/June 2003, pp. 6–73.

[26] Celler, B. G, W. Earnshaw, and E.D. Ilsar, et al., "Remote Monitoring of Health Status of the Elderly at Home: A Multidisciplinary Project on Aging at the University of South Wales," *International Journal on Biomedical Computing,* Vol. 40, 1995, pp. 147–155.

[27] Sixthsmith, A. J., "An Evaluation of an Intelligent Home Monitoring System," *Journal of Telemedicine and Telecare,* Vol. 6, No. 2, 2000, pp. 63–72.

[28] Demongeot, J., G. Virone, and F. Duchêne, et al., "Multisensors Aquisition, Data Fusion, Knowledge Mining and Alarm Triggering in Health Smart Homes for Elderly People," *Compte Rendus Biologies,* Vol. 325, No. 6, June 2002, pp. 673–682.

[29] Virone, G., N. Noury, and J. Demongeot, "A System for Automatic Measurement of Circadian Activity Deviation in Telemedicine," *IEEE Transactions on Biomedical Engineering,* Vol. 49, No. 12, December 2002, pp. 1463–1469.

[30] Glascock A. P., and D. M. Kutzik, "Behavioral Telemedicine: A New Approach to the Continuous Nonintrusive Monitoring of Activities of Daily Living," *Telemedicine Journal,* Vol. 6, No. 3, 2000, pp. 33–44.

[31] Glascock, A. P., and D. M. Kutzik, "The Impact of Behavioral Monitoring Technology on the Provision of Health Care in the Home," *Journal of Universal Computer Science,* Vol. 12, 2006, pp. 59–72.

CHAPTER 15

Connected Medicine

Timo R. Nyberg

15.1 Introduction

Connected medicine is a phrase that can, and does, have many meanings. It will be used in this chapter to describe the new and emerging application of Internet, mobile, and wireless technologies to:

1. Connect the patient to expert advice and information knowledge databases;
2. Connect patients to each other in self-help groups;
3. Connect the patient to monitoring devices for self-diagnosis;
4. Connect the patient physiological measurement data to the clinician.

In doing so, three powerful drivers will be identified that will change the healthcare landscape and the mindsets of patients, the public, and clinicians.

First, there is a growing recognition in the Western world that increasing affluence and improving medical technologies results in improved longevity, but at a price for the increased incidence of chronic conditions [1]. Societies face the challenge of delivering affordable high-quality healthcare to all their citizens. Policy makers have realized that the most affordable way to deal with chronic disease is through delaying the onset, or preventing disease, through the use of education and increased awareness that lifestyle plays an important role in the etiology of chronic disease.

Second, there is an evolution that is attributed to the Internet that is moving today's patients from those who are "informed" to those who are "expert." In the early days of the Internet, a patient walking into a doctor's surgery or consultant's office with a printout from a medical Web site was to some clinicians a source of amusement or irritation. Today, the clinical practitioner will often refer the patient to Web sites for information. Anyone who is sufficiently interested can research the Internet to become knowledgeable about most subjects and patients are often very motivated to find out more about their particular condition. Self-help groups are also increasingly using the Internet.

The Department of Health (U.K.) Self Care Support Summary of Work in Progress (2005–2007) states that, "Patients no longer want information provided only by medical professionals; people find that exchange of experiences with other patients and ex-patients is the most reassuring and efficient way of getting support...."

Third, there is such a market penetration of mobile phone technology that in most developed countries it is unusual for those under 45 not to have a mobile phone [2]. The prevalence of mobile phone technology is such that it provides an infrastructure and momentum for new innovative medical applications. Similarly, other wireless technologies are also becoming widespread. Two common wireless technologies are WiFi and Bluetooth. WiFi is used for broadband connectivity typically at home, office, or in a public building (e.g., airport hot spots use WiFi technology; the range is tens of meters). Bluetooth is often used for connecting small devices like hands-free earplugs to a phone or a keyboard to a computer; the range is a few meters.

These technologies are constantly improving their functionality and merging into devices such as the Nokia 800, Apple iPhone, and potentially a personal health assistant (PHA) device. The advantage of using mobile phones is that they are familiar devices to most people. Mobile healthcare systems are easy and quick to install and systems can be updated without the need to return the device to a central location. Scientists predict that one day handheld supercomputers will be available [3].

One of the observable issues currently taking place across Europe is the struggle to develop and implement national patient databases fed by primary, secondary, and community care systems. These are ambitious and challenging programs requiring an enormous investment of resources over many years. At the same time, technology developments are being harnessed in small but innovative ways to bring quick, beneficial results in improving care processes and the lives of patients. These developments require relatively small amounts of resources and increasingly employ current consumer equipment and the existing skills and knowledge of the patient. These devices will help shift part of the diagnosis, responsibility, and control from the clinical practitioner at the hospital to the patient at home.

Next, a few of these innovations will be described in detail, some of which are already in place and in use while others are on the near horizon. It is likely that some will remain in niche markets but others will become absorbed into mainstream life.

15.2 Applications

15.2.1 Expert Knowledge: Web Sites

There are an enormous number of health-related Web Sites from trusted sources offering expert help and advice. In the United Kingdom, NHS Choices [4] is a service that aims to put the individual at the center of their healthcare. It has been designed to help visitors make choices about their health through to the practical aspects of finding and using different health-related services, for example , assistance with smoking cessation or reduction in the consumption of alcohol while increasing exercise. Figure 15.1 shows the Body Map page, through which you can select to view or hear expert medical advice for a particular disease.

Also from the United Kingdom is another site—*Map of Medicine,* which produces "Healthguides" to help individuals prepare for discussions with clinical practitioners or for anyone who wishes to find out more information about a health issue (Figure 15.2).

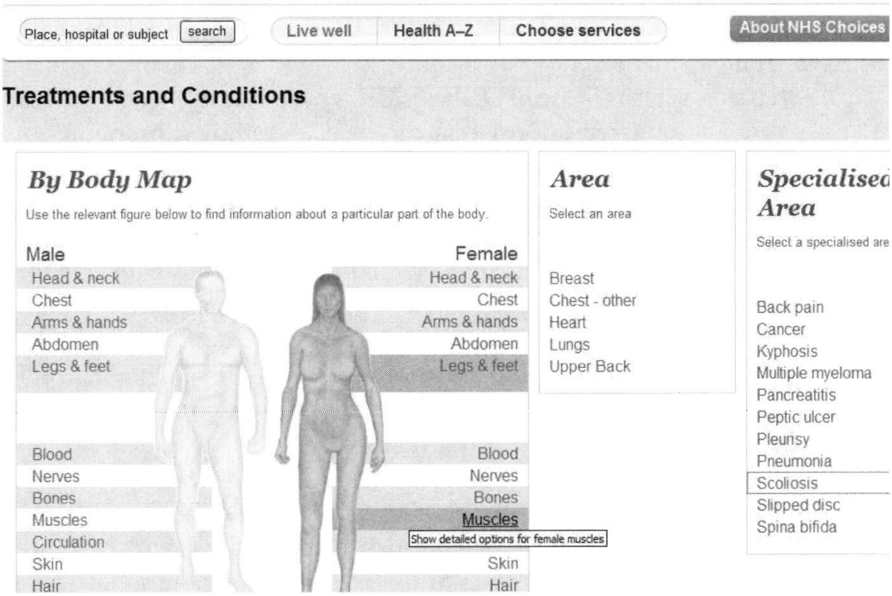

Figure 15.1 Starting page of www.mapofmedicine.com [5].

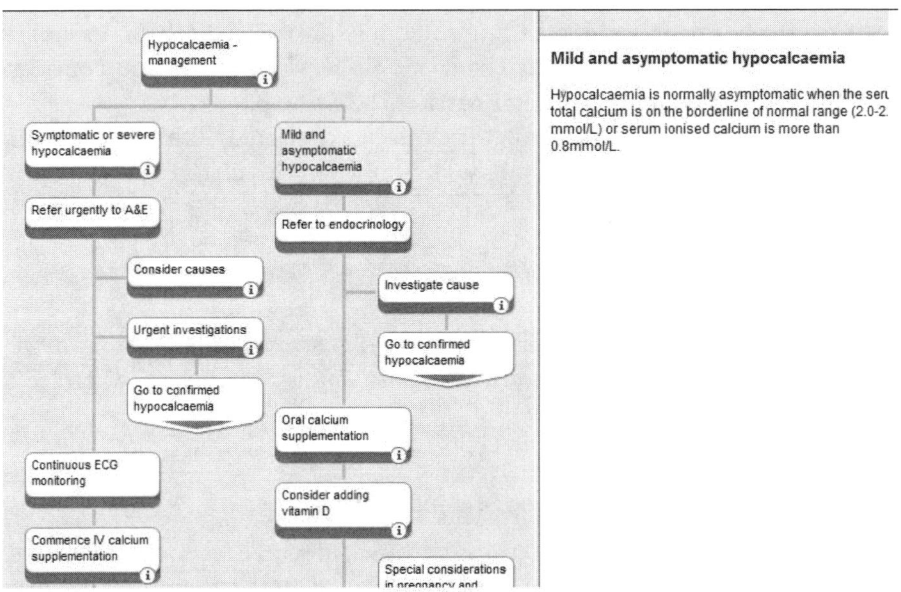

Figure 15.2 Healthguides has been developed by the Map of Medicine to give a clear understanding of the steps to better manage an individual's health. It gives the same trusted information available to healthcare professionals [5].

15.2.2 Self-Help

In addition to the sites run by national healthcare organizations are those run by charities and others that provide advice and support and discussion forums. One of the recent Internet phenomena is the growth in social networking Web sites such as Facebook. Since approximately 30% of people suffer from a long-term medical condition, inevitably social interaction on the Internet will include sharing of health issues and experiences.

Web sites such as those of Diabetes UK report over 4,000 visitors a day. Increasingly new and updated information can be acquired by subscribing to a Web feed such as RSS. In the typical scenario of using Web feeds, a site providing information will publish a link to where users can register within reader software on their own device. RSS on health information sites will enable patients to keep informed about the latest information relevant to their condition in an automated manner that's easier than checking sites manually. Receiving information like this has advantages over e-mail, which requires rules to sort incoming data into separate folders, and it is easy to remove the RSS link if the information is no longer wanted. St John Ambulance in the United Kingdom offers the facility to subscribe to medical advice feeds and download first-aid procedures to mobile devices.

Originally developed in America in the 1970s as a self-help program to aid people with arthritis, the Expert Patient Program has since been developed for a range of conditions across a number of countries to help patients with chronic problems to take more control over their health by understanding and managing their conditions. This has had great success in empowering people with chronic conditions, giving them control and a greater sense of independence, leading to an improved quality of life. Expert patients are said to have more effective conversations with health professionals, visit the doctor less, have less time off work, and are less likely to suffer acute episodes resulting in admission to hospital [6]. Participants report life-changing experiences acquiring the skills, knowledge, and confidence to deal with the mental and emotional barriers to self-help.

It is easy to see how the patient can be connected with self-monitoring devices linked to trusted medical databases described in this chapter to help with initiatives such as the Expert Patient.

15.2.3 SMS

There are a growing number of applications utilizing the fact that more and more people possess mobile phones. SMS messaging is showing an expanded use in various health applications.

15.2.3.1 Reminders for Appointments

Missed appointments are a significant waste of scarce resources. They are not only a waste of the clinician's time but also a waste of all the administrative processes that surround making and preparing for that appointment. While this varies from specialty to specialty, on average 10% of all patients in the United Kingdom miss their hospital outpatient appointments. The number of missed appointments in 2005 to 2006 was calculated to be 6.8 million [7] and while the estimated costs vary from £300 million to over £600 million a year, any reduction in the percentage of failure to show up for an appointment would save significant resources.

The main reason that patients miss appointments is simply that they forget. Applications are now being used to send a timely reminder SMS message to the patient's mobile phone, stating the time and date of their appointment. Advocates of these systems report great success in reducing the failure to attend rate and resulting efficiencies in throughput in busy clinics and surgeries.

Another reason for failure to attend is difficulty in canceling or changing an appointment time. Again, SMS can help here since it is technically possible to cancel or confirm an appointment using reply text.

15.2.3.2 Sexual Health Service

One medical area where the characteristics of the mobile phone can be usefully exploited is in the treatment and prevention of sexual diseases, since young people are an important risk group and they are most likely to have a mobile phone constantly available. Diseases such as chlamydia are a high disease burden and the United Kingdom has initiated a National Chlamydia Screening Programme to control the disease by early detection and treatment. One of the challenges to overcome is the long waiting times in genitourinary medicine clinics.

Newham General Hospital in London has been using an SMS messaging service for patients waiting for sexual health test results to reduce unnecessary hospital visits and waits. Instead of returning for a second lengthy visit to the clinic, eligible patients are given a negotiated time and date for receiving their results by an SMS message on their mobile phone. By preventing delays and inconvenience for patients, it also frees up appointment time in the clinic, allowing services to be offered to more patients. The hospital reports that news of the user-friendly service has spread among the community and among young people and that as a result more people are coming forward for screening [8].

At the same hospital, a text messaging service was piloted, allowing patients to send blood sugar readings via SMS. These were read by a diabetes specialist nurse who texted or called to give appropriate advice. The project was aimed at improving care for women with antenatal diabetes by reducing the frequency of outpatient attendances and providing aggressive management of blood sugar levels by the diabetes nurse. The service was very popular with the patients.

15.2.3.3 Medication Noncompliance

Failure by patients to take medication as prescribed by clinicians is a significant problem. While there are many reasons for this, undoubtedly forgetfulness is one cause. The results of not taking a medication can be very serious for some patients. Companies offer to help patients and caregivers by sending SMS reminders directly to a mobile phone. Developments in this area can include the ability for a message to be sent to a central system recording when a pill container is opened as the best indication that the patient has taken the medication. The central server could be programed to send an alert to a caregiver or health worker if no compliance message was received.

15.2.4 Measuring Device Linked by PHA to Expert Systems

These applications can be used for a wide range of conditions. One important group is those patients suffering from long-term chronic diseases such as diabetes, asthma, and arthritis. To this list we need to add obesity, which is increasingly recognized as a major health problem in developed countries.

People with long-term conditions (LTC) are the most intensive users of health services. In England, patients with an LTC account for 80% of all GP consultations and they use approximately a third of all sick days, while 78% of expenditure [9] in the United States is spent on patients with chronic diseases. Technology that automates chronic care processes can improve the life of the individual and release funds to be reallocated within healthcare.

An example of such automation is the use of measuring devices linked by a handheld PHA to monitor chronically ill patients as they go about their normal daily routines. Daily monitoring can help patients with chronic diseases maintain appropriate drug, diet, and exercise regimes.

Figure 15.3 shows an application of a PHA device that can give instant feedback to the patient or send the recorded data to be stored and then analyzed by either a software decision support system or by a clinician treating the patient. The software decision support could be useful for either a diagnosed patient actively managing their condition or for a nondiagnosed individual who wants to prevent illness from occurring.

Figure 15.3 shows a blood pressure device, but equally this could be a blood glucose monitor, bathroom scales fitted with a wireless module, a wireless ECG heart monitor, lung function spirometer, or pulse oximeter. All of these and more can be used in conjunction with a PHA operating as a mobile health monitor that

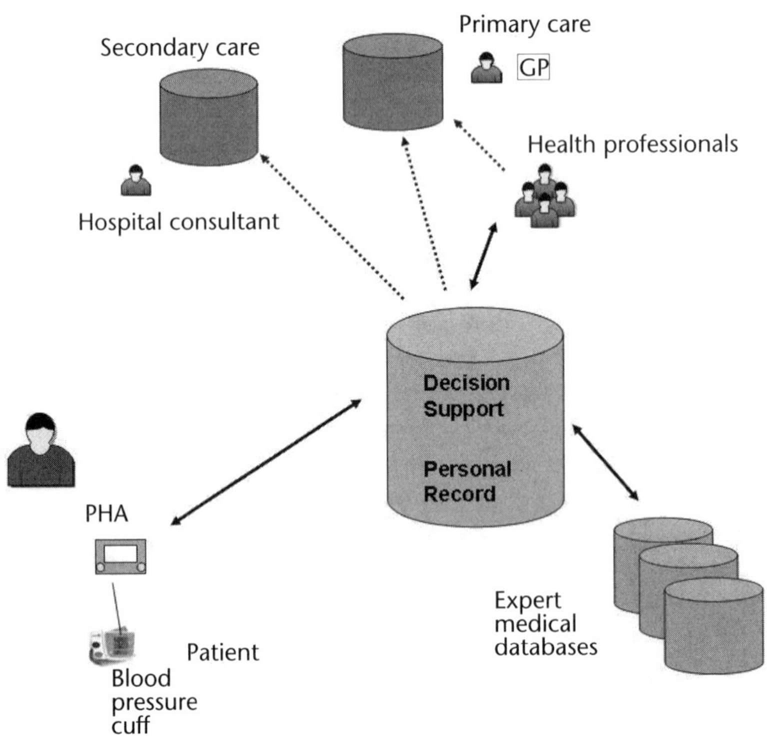

Figure 15.3 A PHA device is a stand-alone feedback system and an interface of an individual to the chronic disease management system.

transmits data to a central server where the results are analyzed and the individual given expert advice from decision support software or from connected clinicians.

An increasing number of people are opting for self-diagnosis. In Germany, for example, 1.2 million self-measurement devices are sold annually [10].

In Figure 15.4, a scenario is depicted of the possible automated interaction between an electronic patient record and a decision support system accessing directly or indirectly trusted evidence based medical databases such as EBM Guidelines and The Cochrane Collaboration. The red arrows depict input to the decision support and the green arrow the output of recommendations. It is now possible to fit this decision support system onto handheld devices like the PHA with the necessary security protection [11].

15.2.5 Medical Data on a SIM Card

Issues such as privacy and cost are delaying the introduction of electronic ID cards. However, mobile phone users already carry a form of ID card called a subscriber identification module (SIM) that is embedded in their handset. Modern SIM cards have enough spare capacity to hold basic medical data applications including patient identification, and the storage of basic medical data such as drug allergies

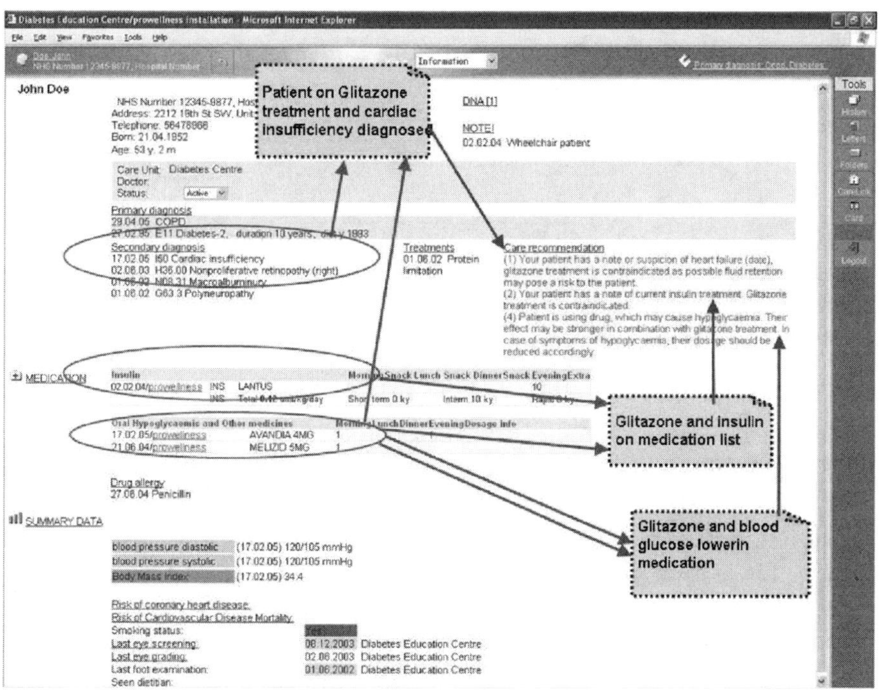

Figure 15.4 In this diagram below a scenario is depicted of the possible automated interaction between an electronic patient record and a decision support system accessing directly or indirectly trusted evidence based medical databases such as EBM Guidelines and The Cochrane Collaboration. The red arrows depict input to the decision support and the green arrow the output of recommendations. It is now possible to fit this decision support system onto hand-held devices like the PHA with the necessary security protection. [11]

and blood group. This would also provide an important step towards the patient taking ownership of their own electronic medical records.

While governments are concentrating on building up national databases within the EU, there are significant numbers of people traveling both for pleasure and for business. There are also large numbers of people in Europe, especially among the retired, who live in more than one country each year for extended periods of time. There is a need for cross-border healthcare for when someone falls sick in another country. A mobile phone with a SIM card containing basic medical data would certainly be a step forward although protocols for accessing that data in emergencies would obviously need to be worked through.

A plan started in the United Kingdom following the London bombings in July 2005 to standardize an emergency contact name in the mobile phone's address list continues to gain support. The entry ICE, which stands for "in case of emergency," is stored and against it is the number of the person you would want to be contacted in an emergency situation. Using this system, police and medical staff can then quickly find out who they need to phone.

15.2.6 Exercise and Rehabilitation

To manage a condition or to prevent illness, individuals need to exercise. Resource levels mean that patients often have to make do with only an occasional visit to a physical therapist. Wireless and mobile devices can be used to record, monitor, and provide feedback to motivate an individual to keep to a planned exercise routine. Once again, a healthcare worker can be informed if the exercises are not being carried out so that the reason for noncompliance is explored and the plan adjusted or further advice given if required. The use of such devices to encourage exercise could alleviate depression and reduce lethargy.

15.2.7 Links to Nutrition Information Expert Knowledge Systems

The importance of diet to health is well known but the information in recent years about eating has often left people confused. Some of the blame for this confusion lies in the hands of the media, who will often pick up issues based partly on their entertainment value. Good expert advice is available but it needs to be packaged and delivered in a meaningful manner for the public.

Existing technologies can be applied in a number of ways. Smart phones could capture the bar code and be send it to an central database that could provide individual-based advice, for example, if the person has an allergy that may be aggravated by a food product or requires a low-salt diet.

The individual could subscribe to a service that allows them to record their food consumption and receive personalized coaching on whether it is matching a target plan.

Just as in Japan, where mobile phone companies provide immediate notification of earthquakes and other natural disasters to their customers, in cases when there has been a contamination of a food product, an alert message could be sent by a central service to mobile phones.

15.2.8 Epilepsy Alert

In the Western world, approximately 5 million individuals are affected by epilepsy and many of these are children. Of the approximately 400,000 new cases diagnosed each year, nearly one-third are children. Of great concern to parents is the thought that the child will have a seizure when they are on their own, for example outside the house or in bed asleep. A prolonged seizure can lead to brain damage and even result in death.

An Israeli company has developed a watchlike sensor unit worn by the patient on the hand or the foot that detects and processes the specific vibrations of epileptic seizures and transmits an alarm to caregivers.

15.2.9 Body-Worn Sensors

Wired sensor devices restrict the movement of the patient being monitored. As a result, some suppliers have developed the ability to capture information wirelessly from sensors embedded in clothes or within a "bandage." This enables the patient to live a normal life, moving about their home and work environment freely. These sensors are able to provide continuous monitoring and, for example, detect early life-threatening signs of imminent cardiac problems. With appropriate software, data can be transmitted to a central remote server using GSM or GPRS and an alert sent to the patient and clinical support.

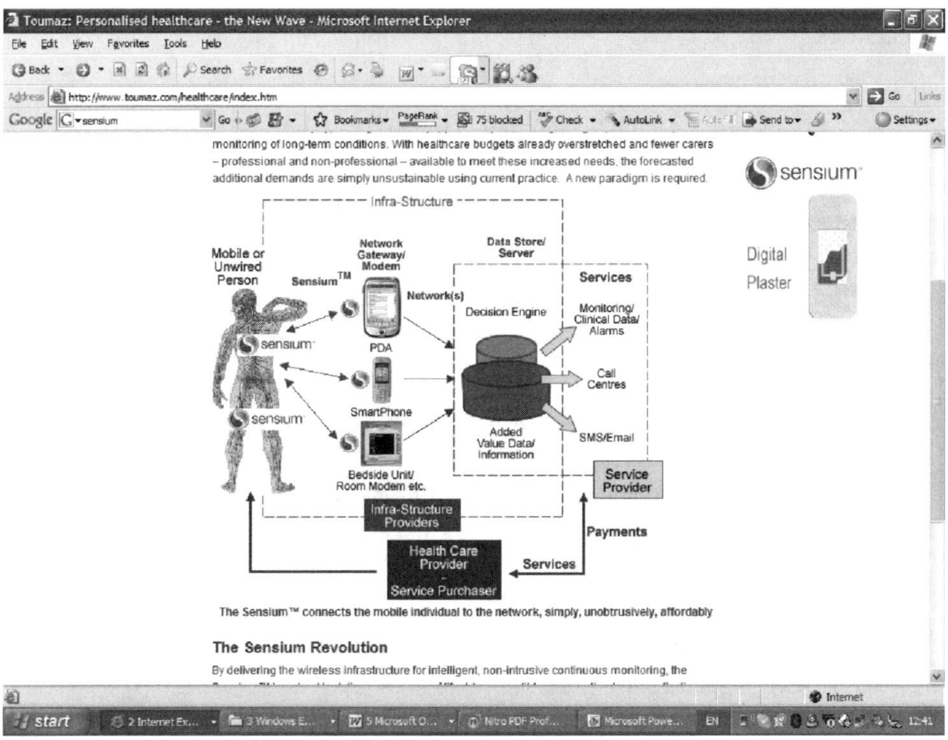

Figure 15.5 Body-worn sensors have been tested widely [12].

15.3 Conclusion

David Colin-Thomé, National Director for Primary Care of NHS wrote,"When the NHS was conceived it was logical to bring the patient to the specialist; now, advances in technology and medicine present us with the opportunity to bring the specialist to the patient. Bringing care closer to home. Making it personal rather than impersonal."

Over the last 50 years, we in the West have concentrated on providing excellent services for the diagnosed patient. We have done this very well, but the resulting increase in lifespan now needs to trigger a new phase that will see strategies developed that will provide both the diagnosed patient and the healthy individual with the information and tools to self-monitor and manage their lifestyles so they can stay as healthy as possible.

These strategies can build on health innovations such at those described in this chapter. The traditional boundaries of healthcare will need to be expanded to take in the currently healthy individual, and employ the Internet and personal handheld devices to provide information and feedback to enable people to make the right lifestyle choices to prevent or delay the onset of chronic diseases. By employing familiar consumer technologies and actively encouraging self-care, the largest collaborative resource available, the public, can be motivated and mobilized to help meet the challenges we face and improve health outcomes and patient satisfaction within current funding levels.

References

[1] European Commission ICT, "Information and Communication Technologies," *European Commission* C (2007) 2460, June 11, 2007.

[2] Wikipedia, www.wikipedia.org/wiki/Mobile_phone.

[3] "Handheld Supercomputers On Way," BBC News, October 22, 2007, www.news.bbc.co.uk.

[4] NHS Choices site, About Us, www.nhs.uk/Pages/homepage.aspx.

[5] Map of Medicine, www.healthguides.mapofmedicine.com.

[6] Lorig, K. R., D. S. Sobel, and A. L. Stewart, et al., "Evidence Suggesting That a Chronic Disease Self-Management Program Can Improve Health Status While Reducing Hospitalization: A Randomized Trial, *Med. Care*, Vol. 37, No. 1, January 1999, pp. 5–14.

[7] "Missed Hospital Appointments Up," BBC News, May 8, 2006, www.news.bbc.co.uk.

[8] *Newham University Hospital Annual Report*, 2004/5.

[9] Horvath, J., " Reining in the Cost of Chronic Illness" 1998.

[10] Eckert, S., U. Gleichmann, O. Zagorski, et al., "Validation of the Omron R3 Blood Pressure Self-Measuring Device through Simultaneous Comparative Invasive Measurements According to Protocol 58130 of the German Institute for Validation," *Blood Press. Monit.*, Vol. 2, 1997, pp. 189–192.

[11] The Finnish Doctors Association, Duodecim, 2007.

[12] www.toumaz.com.

CHAPTER 16

Information Technology Networks, Data Management, and Electronic Medical Records

Thomas Jepsen

16.1 Introduction

As health information technology evolves from stand-alone data processing applications to integrated knowledge systems, the information technology network becomes an increasingly important component of the health IT infrastructure. In the past, the IT network in most healthcare facilities did little more than connect patient management applications to billing systems; now the IT network is viewed as a critical infrastructure that integrates patient admission systems, online medical records, billing and accounting systems, pharmacy systems, and radiology and imaging systems. In the future, as telemedicine, home healthcare, interoperable electronic record sharing, and biosurveillance applications proliferate, the demands placed on the healthcare IT network can be expected to increase dramatically.

As the amount of data managed by the typical healthcare provider increases both in size and in complexity, data management becomes an increasingly important function. Patients must be identified unambiguously, and access to their records limited to those with a need for their medical information. Billing information must be accurate and secure, to ensure that all coded procedures are reimbursed correctly. Image data, such as x-rays, CT scans, PET scans, and ultrasound, has unique storage requirements and must be readily accessible for diagnosis and treatment. Backup and retrieval functions must be implemented to ensure data reliability and survivability, and archival functions must exist to efficiently manage older and less frequently accessed records.

Adoption of online electronic medical records (EMRs) has been one of the primary driving forces for the growth and expansion of health IT networks[1]. While EMRs were originally largely text-based, they now frequently include imaging data, which creates a requirement for increased bandwidth and quality of service. The development of interoperability standards for EMRs and the increased sharing of medical records among healthcare providers in arrangements such as Regional

1. The terms "electronic medical record" and "electronic health record" are frequently used interchangeably. HIMSS Analytics has recommended that the usage be standardized to use EMR to refer to the electronic record within a single healthcare facility, and EHR to refer to the electronic record exchanged between facilities. This terminology distinction will be used in this chapter. See Dimick, C.,"Health IT, Piece by Piece," *Journal of AHIMA (American Health Information Management Association)*, Vol. 78, No. 6, June 2007 p. 37.

Health Information Organizations (RHIOs) have created a requirement for health IT local area networks (LANs) in individual institutions to be interconnected via the wide area network (WAN).

16.2 IT Networks for Medical Information Management

16.2.1 Online Medical Records Systems

Healthcare information technology applications began as standalone computer applications which performed functions such as patient admissions and billing, and reduced the amount of paperwork required for these functions. Soon it became obvious that efficiencies could result from information-sharing among disparate systems. With the widespread adoption of Ethernet technology in the 1980s, LANs were used to interconnect connect workstations and computers to form rudimentary healthcare information networks (Figure 16.1).

As the networks evolved, more applications were integrated into the network, and the network itself came to be seen as a separate entity that required full-time support in the form of network administration. Some of the functions that are commonly integrated into a modern healthcare IT network include

- *Access management.* This includes patient admissions, scheduling of procedures, benefits eligibility checking, and patient discharge.
- *Biosurveillance.* This refers to the gathering of information for public health and epidemiological purposes, and to report any incidence of contagious dis-

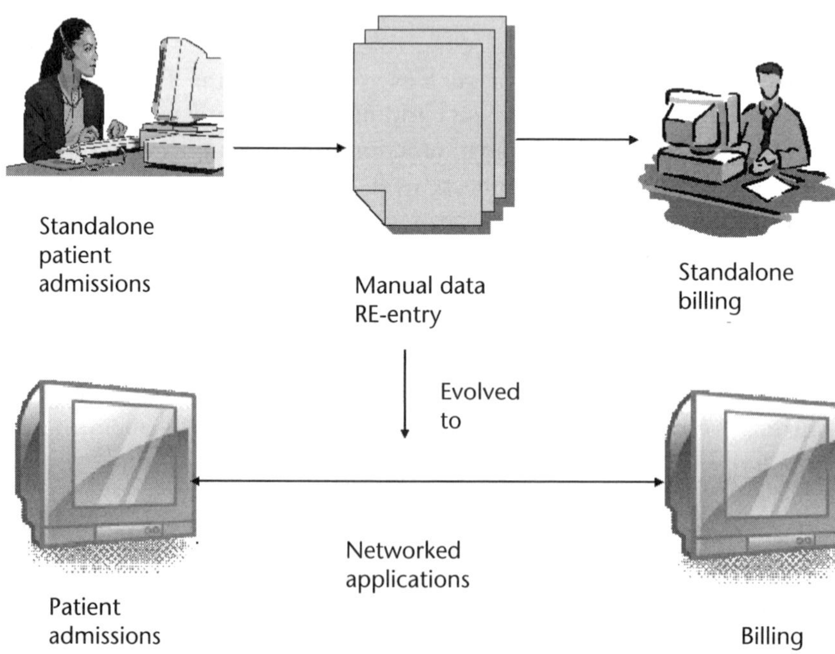

Figure 16.1 Evolution of early health IT networks.

ease in the patient population. Typically this requires that the information be "deidentified" to protect patient privacy.

- *Cardiovascular data.* Data from electrocardiograms (ECGs) can be placed online for quicker access and integration into the electronic medical record.
- *Computerized provider order entry (CPOE) systems.* CPOE systems allow for the online entry of medical information.
- *Electronic medical record (EMR).* A record of a patients encounters with healthcare providers, including diagnostic and treatment information. An EMR is said to be "longitudinal" if it covers a patient's entire healthcare history over the patient's lifetime.
- *Emergency room.* Emergency room healthcare providers need to track admissions, triage, and diagnosis, and quickly locate information on admitted patients.
- *Imaging.* Healthcare IT networks increasingly provide access to image data as well as text-based data. Image data includes diagnostic images such as x-rays, PET scans, CAT scans, MRI images, and ultrasound, as well as nondiagnostic images such as patient photos and scanned documents.
- *Healthcare information management.* Diagnostic and procedural information obtained in patient encounters must be captured using standard classification systems such as ICD9/10 and CPT for statistical and billing purposes. Online EMRs and coding applications increase the efficiency of this process.
- *Laboratory.* Results of laboratory tests can be requested and integrated into online medical records.
- *Pharmacy.* Physicians can submit prescriptions and drug requests directly to pharmacies using online systems.
- *Healthcare devices.* Instrumentation used to monitor vital signs (pulse, blood pressure, oximeter) may be connected to the network to permit real-time monitoring of patient condition at the nursing station.
- *Home healthcare.* Devices used to monitor vital signs or administer drug dosage may be placed in the home to reduce the number of healthcare facility visits by patients with noncritical chronic conditions.
- *Accounting and billing.* Accounting and billing functions use the network to capture charge information from patient encounters, medical procedures, and pharmacy orders.
- *Supply chain.* Standard supply chain applications enable automated ordering and stocking of supplies as they are used.

16.2.2 Health IT Networks for the Private Practice

Health IT networks in small- to medium-sized private practice clinics perform many of the same functions as the corresponding networks in large clinics, but on a smaller scale. A recent study by the American Academy of Family Physicians Center for Health Information Technology indicated that 90% of private practice physicians used computerized billing systems and had access to the Internet; 70% of private practices used personal digital assistants (PDAs) or similar handheld computers. However, a much smaller percentage of AAFP members (13% to 15%) have successfully implemented EMR/EHR systems. Advantages of EHR for small practices include online access to laboratories for lab results, and access to pharmacies for e-prescription [1, 2].

16.2.3 Picture Archiving Communications Systems

Picture archiving communications systems (PACS), are a specialized form of health IT network used for the acquisition, interpretation, and storage of medical images in digital format. Formerly, medical images were captured on photographic film; while photographic film provided good resolution for image interpretation, it created problems with image storage, archival, and retrieval, especially when images had to be transported from one facility to another. The use of electronic imaging technology enables medical images to be readily and easily stored, retrieved, and transmitted from one site to another.

PACS systems are generally based on the Digital Imaging and Communication in Medicine (DICOM) standard for communication of medical images between computers. The DICOM standard allows for interoperability among imaging devices such as CT scanners, MR imagers, and film digitizers. Some of the image modalities supported by PACS systems include

- *Magnetic resonance imaging (MRI).* MRI scanners use powerful magnetic fields to create very detailed images of soft tissue in various parts of the body.
- *Computerized axial tomography (CAT or CT).* CT scanners use multiple x-ray emitters and detectors to image cross sections of the human body in great detail. Each image (e.g., abdomen, chest, or head) is called a slice.
- *Computed radiography (CR).* CR represents the modern evolution of the x-ray. CR uses digital images instead of photographic film to record the x-ray image. CR is used mainly for the chest, skeletal, and abdominal areas, to check for fractures, deterioration of the joints, lung congestion, and disease.
- *Ultrasound (US).* US employs sound waves and the interpretation of their echoes to create images of soft tissue. US is used primarily in pediatrics and studies of the liver, eye, and abdominal organs. Since it uses sound waves instead of x-rays, US is safe for imaging pregnant women.
- *Nuclear medicine (NM).* In NM imaging, a patient is injected with a radiopharmaceutical and a gamma ray camera is used to monitor and record the flow of the isotope through the patient's organs. It is commonly used in the diagnosis and treatment of heart disease and cancer.

16.2.4 Assessing the Implementation Success and Maturity of Health IT Networks

According to a recent study, between 17% and 24% of physicians in ambulatory settings make use of electronic medical records. As for adoption by hospitals, the study was only able to determine that, as of 2005, somewhere between 4% and 21% of hospitals used some form of CPOE system [3]. One reason for the wide discrepancy in numbers is that it is difficult to establish exactly what constitutes implementation of an EMR system, and there are wide differences in the maturity of implemented EMR systems. Another is the high failure rate of EMR implementation; the AAFP study cited above for private practice physicians indicated a failure rate of as high as 50% for implementation of electronic medical records systems in small clinics [1] (Figure 16.2).

One approach to quantifying progress toward EMR adoption is to use a maturity model that establishes criteria for different stages of EMR adoption. Health Information Management Systems Society Analytics (HIMSSAnalytics) has developed a seven-stage maturity model to gauge the actual degree of EMR implementation and use:

- *Stage 0:* Some clinical automation may be present, but all three major ancillary department systems—laboratory, pharmacy, and radiology—are not implemented.
- *Stage 1:* All three of the major ancillary clinical systems—laboratory, pharmacy, and radiology—are installed.
- *Stage 2*: Major ancillary clinical systems feed data to a clinical data repository (CDR) that provides physician access for reviewing and retrieving results. The CDR contains a controlled medical vocabulary and a clinical decision/support rules engine for rudimentary conflict checking. Information from document imaging systems may be linked to the CDR at this stage.
- *Stage 3*: Clinical documentation, such as vital signs, flow sheets, nursing notes, care plan charting, and electronic medication administration records (eMARs) is required, as well as general order entry, and they are implemented and integrated with the CDR for at least one service in the hospital. The first level of a Clinical Decision Support System (CDSS) is implemented to conduct

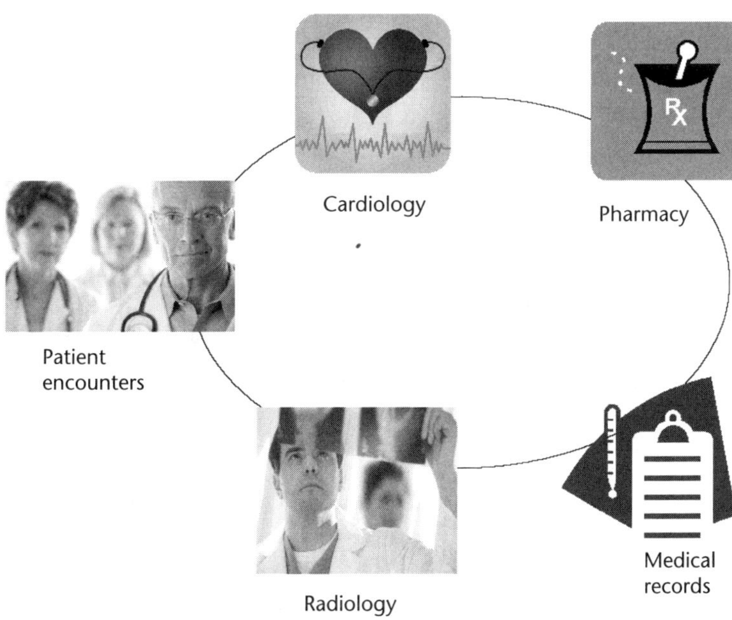

Figure 16.2 The modern health IT network.

error checking with order entry (i.e., drug-drug, drug-food, and drug-lab conflict checking normally found in the pharmacy). Some level of medical image access from PACS is available via the organization's intranet or secure networks outside radiology.

- *Stage 4*: CPOE for use by any clinician is added to the nursing and CDR environment along with the second level of clinical decision support capabilities related to evidence-based medicine protocols. If one patient service area has implemented CPOE and completed the previous stages, then this stage has been achieved.
- *Stage 5*: The closed loop medication administration environment is fully implemented in at least one patient care service area. The eMAR and bar coding or other autoidentification technology, such as radio frequency identification (RFID) are implemented and integrated with CPOE and pharmacy to maximize point-of-care patient safety processes for medication administration.
- *Stage 6*: Full physician documentation and charting (structured templates) is implemented for at least one patient care service area. Level three of clinical decision support provides guidance for all clinician activities related to protocols and outcomes in the form of variance and compliance alerts. A full complement of PACS systems displaces film-based images.
- *Stage 7*: This is the ideal. The hospital has a paperless EMR environment. Clinical information can be readily shared via electronic transactions or exchange of electronic records with all entities within a regional health network (i.e., other hospitals, ambulatory clinics, subacute environments, employers, payers, and patients).

Based on a 2006 study of over 4,000 hospitals, the HIMSSAnalytics maturity model showed that most U.S. hospitals are in the early stages of EMR adoption. Table 16.1 shows the results of the HIMSS Analytics study, based on their maturity model [4].

Table 16.1 Implementation of EMR/EHR by Hospitals using HIMSS Maturity Model

Stage	*Capability*	*% of Hospitals*
0	No laboratory, pharmacy, or radiology implemented	20.7
1	All three ancillaries implemented	18.9
2	CDR, controlled medical vocabulary, CDSS inference engine; may have document imaging	38.8
3	Clinical documentation (flow sheets), CDSS (error checking), PACS available outside radiology	18.0
4	CPOE, CDSS (clinical protocols)	3.0
5	Closed-loop medication administration	0.5
6	Physician documentation (structured templates), full CDSS (variance and compliance), full PACS	0.1
7	Medical record fully electronic, the organization is able to exchange data within a regional network	0.0

Source: Dimich, "Health IT Piece by Piece," *Journal of AHIMA*, 78/6 (June 2007), pp. 36–40.

16.2.5 Network Architectures

The earliest computer systems used for healthcare IT in the 1960s and 1970s, typically mainframe systems, employed terminals that were connected directly to the computer (or through a terminal controller). During the 1980s, as the number of users grew and became more physically dispersed, a client/server architecture evolved in which a centrally located application server made applications available to multiple user workstations ("clients") that accessed the server by means of the LAN. Distinct variants of this architecture evolved; if a client performed a significant amount of application processing, it became known as a "fat client"; if the majority of processing took place in the server, the client was called a "thin client." (See Figure 16.3).

With the advent of the World Wide Web and the Internet in the 1990s, many healthcare facilities adopted these protocols for internal use and created an "Intranet" in which user clients accessed server-based applications by using a Web browser. Due to the stringent security requirements for accessing personal health information, a firewall had to be implemented between the Intranet and the external Internet to prevent unauthorized access to protected information.

As more applications became part of the healthcare IT network, the task of integrating them became increasingly complex, different applications required data in differing formats, and used different protocols for communications. This led to the development of an architecture in which applications are accessed as "services," using standard protocols based on the Extensible Markup Language (XML) standards developed by the World Wide Web Consortium. XML provides a common data format that enables any application to exchange data with another application. This architecture is referred to generically as the Service Oriented Architecture

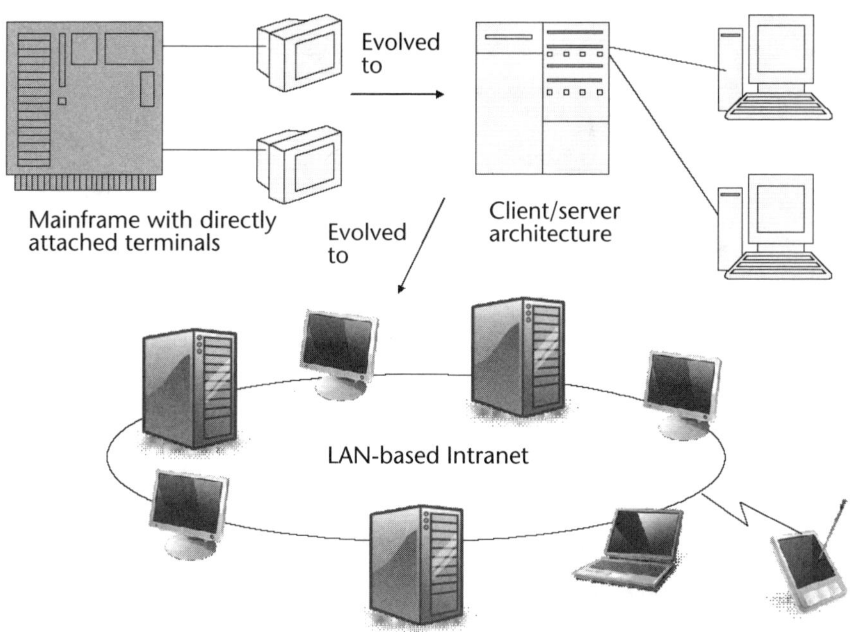

Figure 16.3 Evolution of IT network architecture.

(SOA) (Figure 16.4); a subset of SOA using Web http protocol is referred to as Web Services. A fundamental notion of SOA is that service providers (e.g., applications) and service consumers (e.g., system users) are loosely coupled; this means that it is possible to add new services or change the implementation of services without major impact to the system.

16.2.5.1 Network Topologies

Wired Networks

Almost all wired LANs use the Ethernet protocol, as defined by the IEEE 802.3 group of standards. The IEEE standards support a variety of data rates, including 10-Mb/s Ethernet (IEEE 802.3), 100-Mb/s Fast Ethernet (IEEE 802.3), 1,000 Mb/s-Gigabit Ethernet (IEEE 802.3z), and 10-Gigabit Ethernet (IEEE 802.3ae). Ethernet supports both a LAN topology using carrier sense multiple access/collision detect (CSMA/CD), and a switched point-to-point topology. Each is described below.

LAN Topology

Most Ethernet LANs today use a star or hub configuration, as shown in Figure 16.5. Each node is connected to a central hub or switch by means of twisted pair wiring or optical fiber.

Ethernet operation in a LAN environment uses the CSMA/CD protocol. CSMA/CD enables multiple stations to share a common transport medium without requiring a central arbitration function to determine which station is allowed to send at any time. *Carrier sense* refers to the ability of each station to listen to the medium to determine if anyone else is sending, or if the medium is silent. *Multiple access* means that multiple stations may attempt to transmit, if the medium is silent. *Collision detect* means that if two stations begin transmitting simultaneously, they

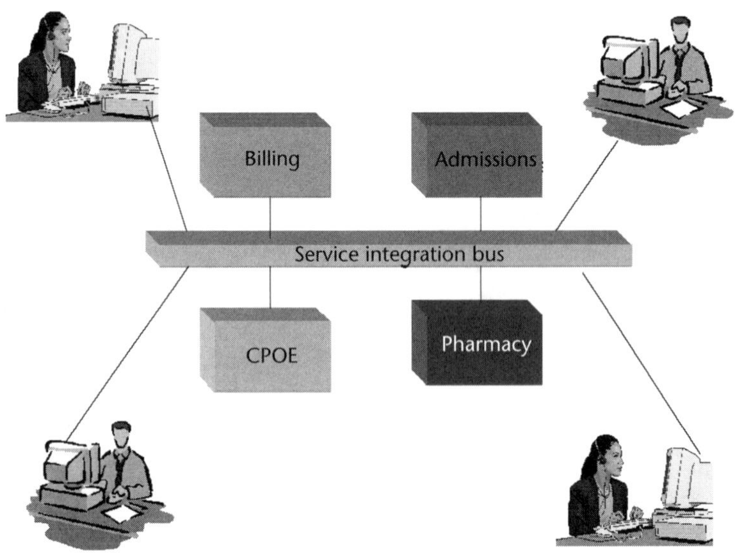

Figure 16.4 A Service Oriented Architecture for healthcare IT.

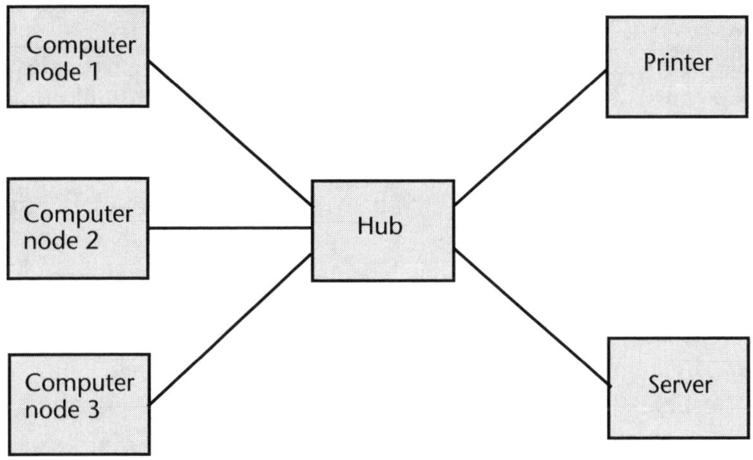

Figure 16.5 Conventional Ethernet LAN topology.

will interfere or collide with each other. Each station must be able to detect collisions and stop transmission if they occur. Each station will then wait an algorithmically determined back-off time before trying to transmit again.

CSMA/CD operation allows half-duplex mode operation, in which a given station may receive or transmit at any time, but not both at the same time. Figure 16.6 shows operation of the CSMA/CD protocol for Ethernet [5].

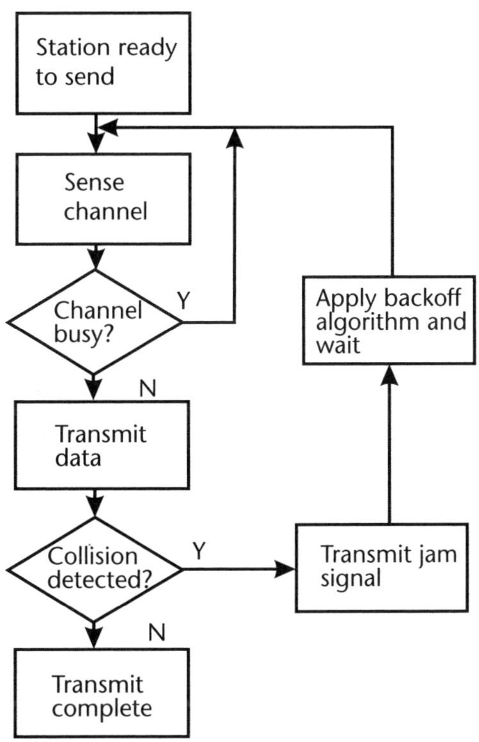

Figure 16.6 CSMA/CD protocol.

Point-to-Point Switched Topology

Full-duplex operation is an optional capability that allows simultaneous bidirectional communication over point-to-point links. Full duplex configurations are typically used for high-bandwidth switch-to-switch links between networks. Figure 16.7 shows a typical network configuration in which a point-to-point Gigabit Ethernet link is used to provide a high-bandwidth path for switched LAN-to-LAN traffic.

In full-duplex operation, each transmitter simply begins sending frames as soon as they are available. A 96-bit interframe gap (IFG) is placed between each frame. No carrier sense or collision detection is employed, and no frame extensions are required.

A flow control mechanism is employed to prevent receiver buffer overloading during congestion. If a receiving node becomes congested, it sends a *pause frame* to the transmitter to tell it to stop sending for a specified period of time. If the congestion is removed before the time period has elapsed, the receiver may send a second pause frame with a wait time of zero to enable the transmitter to begin sending again [6].

Transport Media for Wired Networks

10-Mb/s Ethernet and 100-Mb/s Fast Ethernet are typically carried over Category 5 (CAT5) twisted pair copper cabling for distances of <100m, or 328 feet. 1,000-Mb/s Gigabit Ethernet may be transported over the same distance using Category 5 Enhanced (CAT5e) copper twisted pair, or over optical fiber cable.

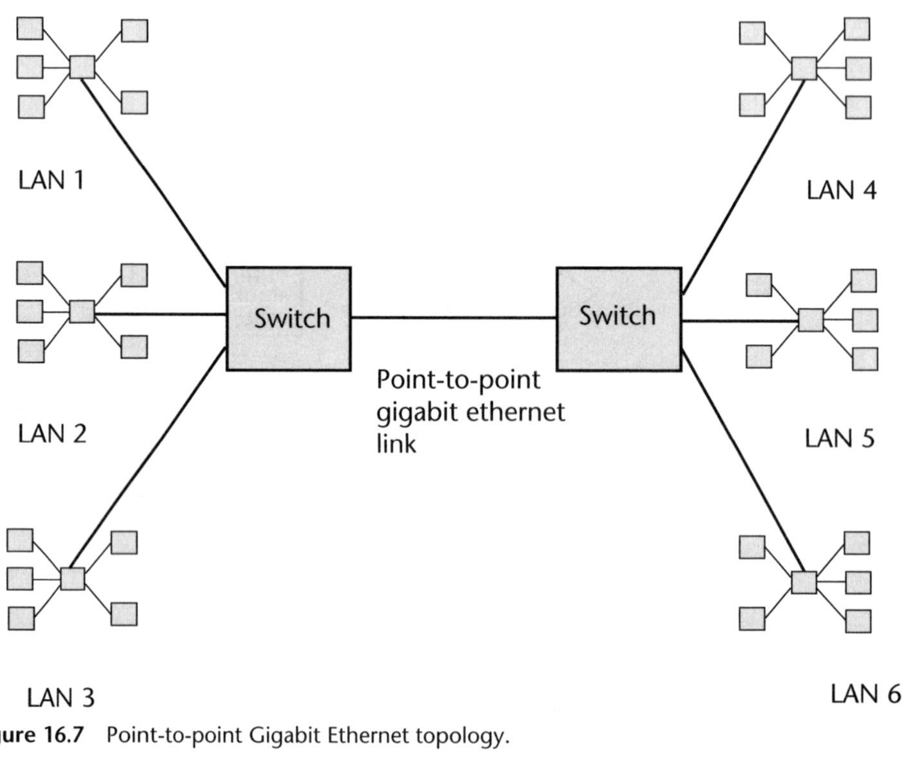

Figure 16.7 Point-to-point Gigabit Ethernet topology.

10-Gigabit Ethernet is carried over CX4 copper cable for distances of up to 50 feet, or optical fiber.

16.2.5.2 Wireless Networks

Wireless technologies are being increasingly used in healthcare settings to eliminate the need for wired connections among medical devices and computers, and to enable the ubiquitous use of portable handheld devices for recording clinical data. Wireless technologies for healthcare applications fall into two general categories: the wireless local area network (WLAN), which is designed to support wireless access over an area of <100m, and the wireless personal area network (WPAN), which supports wireless access over a distance of <10m. WLAN technology is commonly used to replace wire and fiber optics for interconnection of computer systems. Common WLAN standards include IEEE 802.11g (54 Mb/s) and IEEE 802.11b (11 Mb/s). WPAN technology is commonly used to support hand held devices and increasingly also for medical instrumentation. Common WPAN standards include IEEE 802.15.1 (1 Mb/s) and IEEE 802.15.4 (250 Kb/s). All the above technologies make use of the 2,400-MHz Industry, Scientific, Medical (ISM) portion of the frequency spectrum.

Wireless technologies are generally utilized where bandwidth requirements are relatively low, and portability is a primary concern. Care must be taken to avoid interference from nonmedical devices such as cell phones, cordless phones, microwave ovens, and PDAs, which make use of the same frequency spectrum.

Modulation Technique

Modulation type for wireless technologies include direct sequence spread spectrum (DSSS, or just DS) and frequency hopping spread spectrum (FHSS, or just FH). In DS, the carrier signal is phase-modulated with a pseudonoise signal to create "chips" that are then used to modulate the information bits. The sequence of chips transmitted by the sender is known to the receiver to enable demodulation. In FH, the carrier frequency is rapidly switched among many channels in a pseudorandom sequence known to both the sender and receiver.

Access Arbitration

WLAN and WPAN use either carrier sense multiple access/collision avoidance (CSMA/CA) or time division multiple access (TDMA) protocols to arbitrate access to the network by multiple senders. In TDMA, the transmitted signal is divided up into individual time slots, and each user is assigned to a specific slot. CSMA/CA is similar to CSMA/CD as described for wired Ethernet, but employs an additional step to ensure that collisions will not occur. In the CD protocol, the sender simply transmits once the channel is clear, and then listens to see if a collision has occurred. In the CA protocol, once the channel is clear, the sender sends out a message to all other stations in the network to tell them not to transmit before sending a packet. The sender then listens for an acknowledgment from the recipient of the packet to ensure that it was properly received [7].

Table 16.2 lists the characteristics of typical wireless technologies.

Table 16.2 WLAN and WPAN Technologies for Healthcare IT

Standard	*Network*	*Modulation Technique*	*Distance Covered*	*Maximum Data Rate*	*Arbitration Protocol*
802.11g	WLAN	Direct sequence	<100m	54 Mb/s	CSMA/CA
802.11b	WLAN	Direct sequence	<100m	11 Mb/s	CSMA/CA
802.15.1	WPAN	Frequency hopping	<10m	1 Mb/s	TDMA
802.15.4	WPAN	Direct sequence	<10m	250 Kb/s	CSMA/CA

Source: [7].

16.2.6 Bandwidth Requirements

In planning and implementing a health IT network, consideration must be given to the amount of data that will be transported and the performance implications of network loading. In addition to EMRs and other patient records that are transmitted from computer to computer and downloaded from network servers, the network typically must support a variety of data types used for patient monitoring, imaging, and telemedicine. Some, such as the telemetry signals used to monitor patient vital signs, require relatively low bandwidth; others, such as video used for telemedicine interviews and download of functional MRI scans, consume a considerable amount of network bandwidth and may affect overall network performance. Table 16.3 lists typical image size and data rates required for transmission of common medical imaging and monitoring functions.

Table 16.3 Image Sizes and Data Rates for Common Medical Imaging and Monitoring Functions

Digital Device	*Image Size*	*Typical Data Rate (Mode) Required for Transmission*
Digital blood pressure monitor	N/A	<10 kilobits per second (Kb/s)
Digital thermometer	N/A	<10 Kb/s
Digital audio stethoscope and integrated electrocardiogram	N/A	<10 Kb/s
Ultrasound, angiogram	256 KB	256 Kb/s to 1920 Kb/s
Magnetic resonance image	125 KB	10 Mb/s
Full functional MRI scan	30 GB	1 Gb/s (4-minute download) to 10 Gb/s (24-second download)
Scanned x-ray	1.8 MB	10 Mb/s
Digital radiography	6 MB	256 Kb/s to 10 Mb/s (MPEG over ATM)
Computerized tomography (CT) scan	500 KB	10 Mb/s
Mammogram	184 MB (complete exam)	1.6 Mb/s (15 minute download) to 1.5 Gb/s (1 second download)
Compressed and full motion video	N/A	384 Kb/s (ISDN) to 40 Mb/s (MPEG over Asynchronous Transfer Mode (ATM)) connection), depending on required quality
Telesurgery	N/A	10-155 Mb/s (ATM)

Sources: M. Ackerman et al., "Telemedicine Technology," *Telemedicine J. and e-Health*, vol. 8, no. 1, 2002, p. 73; B. Davie et al., "Bringing Healthcare Applications to the Internet," *IEEE Internet Computing*, May/June 2001, p. 44; H. Handels et. al., "Image Transfer and Computer Supported Cooperative Diagnosis," *J. Telemedicine and Telecare*, vol. 3, no. 2, 1997, p. 103.

16.2.7 Protocols

16.2.7.1 Protocol Architecture

The Ethernet standards define the data link and physical layers only (layers 1 and 2 of the Open System Interconnect (OSI) model), and therefore do not address higher-level issues such as transport, routing, and quality of service (QOS). Figure 16.8 shows the relationship of the Ethernet standards to the common application and transport layers it typically supports in a healthcare IT environment. Normally, the TCP protocol is used to supply the layer 4 transport function, and IP is used to support the layer 3 networking function.

16.2.7.2 TCP/IP

The Transmission Control Protocol, or TCP, is one of the fundamental protocols of the Internet and was defined by the Internet Engineering Task Force (IETF) in Request For Comments (RFC) 793 in 1981. TCP is a connection-oriented transport protocol designed to provide reliable communications between processors across a variety of reliable and unreliable networks.

16.2.7.3 IP

The Internet Protocol (IP) is the most commonly used internetworking protocol in the world today. IP normally comprises the networking layer of the TCP/IP protocol suite. It interfaces with the TCP transport layer by means of two service primitives: *Send*, which is used by the transport layer to request delivery of a data unit; and *Deliver*, which is used to notify the transport layer of the arrival of a data unit. The

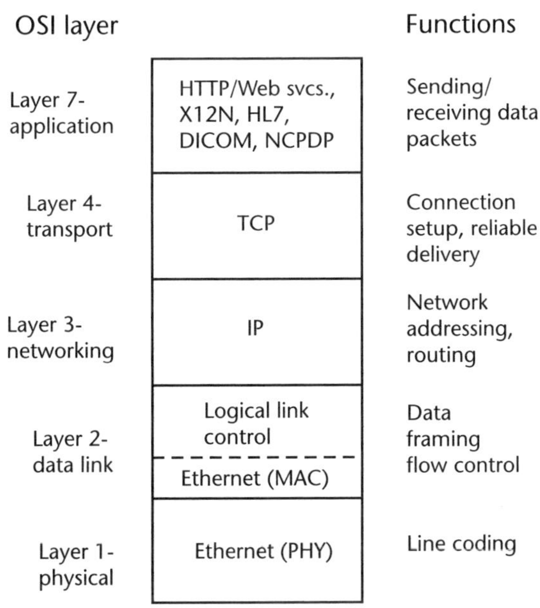

Figure 16.8 Relationship of healthcare/IP/Ethernet standards to OSI Protocol Stack.

IP datagram format for the 32-bit address version of IP, referred to as IP Version 4 (IPv4) is shown in Figure 16.9.

IP datagrams are routed through a network on an individual basis, using destination address. Networks use interior routing protocols, such as Open Shortest Path First (OSPF) to route datagrams among end systems. For exchange of network-to-network traffic, exterior routing protocols, such as Border Gateway Protocol (BGP) are used to exchange routing information among networks.

16.2.7.4 Class of Service

Originally, IP supported only one class of service, referred to as *best effort*. In this class of service, the network will attempt to deliver each packet to its destination, but no guarantees about performance are made, and no resources are dedicated for delivery of the packet. Services that demand higher levels of service, such as real-time voice, video, and storage data, have created requirements for guaranteed quality of service. The Integrated Services Architecture (IntServ) was developed by the IETF in 1994 and standardized as RFC1633. It defines a *flow* as a stream of packets with a common source address, destination address, and port number, and creates a requirement to maintain a state for each flow in the network. IntServ uses *admission control* to check to see if the processor or router is able to support each service request, a *packet classifier* to determine which class each packet belongs to, and a *packet scheduler* to schedule each packet for transmission on the outbound link. IntServ provides two classes of service in addition to best effort: *guaranteed service*, which supports real-time traffic with bounds on delay and is defined in RFC2212; and *controlled load*, which provides an approximation of best-effort service over an uncongested network, and is defined in RFC2211.

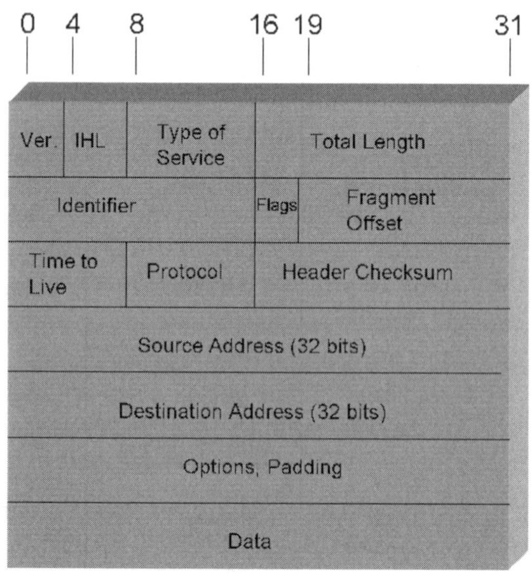

Figure 16.9 IP datagram format (IPv4).

IntServ uses a reservation protocol called the *Resource Reservation Protocol*, or RSVP, to create a path and reserve resources for a flow; RSVP is defined by RFC2205. The data source sends a Path message to the destination. The Path message is used to mark a route and to collect information about the QOS characteristics of each router along the route. When the destination receives the Path message, it returns a RSVP reservation (RESV) message to the source. If sufficient resources are available through the entire path, those resources are reserved specifically for that flow.

One limitation of RSVP is that it is not scalable. State must be maintained for each flow in the network, and each flow state must be periodically updated as conditions change. Another approach to QOS is provided by *Differentiated Services (DiffServ)*, defined in RFC 2474. Instead of maintaining a separate state for each flow, DiffServ classifies service requests at the edge of the network and aggregates individual flows into a few service classes for transport through the network. Packets with similar QOS requirements are aggregated into *behavior aggregates* (BA), and each BA receives a specified *per-hop-behavior* (PHB) as it moves from router to router, specified in terms of queuing and delay. The IETF Working Group has defined two broad service categories: *expedited forwarding* (EF), RFC2598, and *assured forwarding* (AF), RFC2597. EF supports low packet loss, low delay, and low jitter. AF is divided into four relative classes of service, each subdivided into three levels of packet drop precedence. When congestion is encountered, packets with lower drop precedence are dropped before packets with higher drop precedence.

While IP has proven itself as a vital component of the Internet, it has significant limitations that are becoming more apparent as the global use of the Internet increases. One of these is address space. The 32-bit addresses defined by IPv4 provide only 4 billion distinct addresses, which is less than the total population of the earth. Address availability is further limited by the inefficiencies imposed by a two-level addressing scheme in which each address is broken down into variable-length host and network addresses. In response to this, the IETF has developed a next generation addressing scheme, known as *IPv6*. IPv6 provides a 128-bit address, and incorporates some additional functionality, such as integrated multicasting and IPsec security. Initial adoption of IPv6 has been slow, due to the large installed base of IPv4 networks, but is expected to gain momentum as global demand for Internet connectivity increases.

16.2.7.5 IPv6

The IPv6 header uses a simplified 40-byte fixed-length header defined in RFC2460, consisting of a 16-byte source address, a 16-byte destination address, and an 8-byte field for general header information. Figure 16.10 shows the fields of the Ipv6 header. (Bit values, except for bit 31, indicate starting bit of each field.)

16.2.7.6 Ethernet Data Framing

The data link layer specified in the OSI standards is implemented as two sublayers in the IEEE 802 specifications: a media access control (MAC) client layer and a

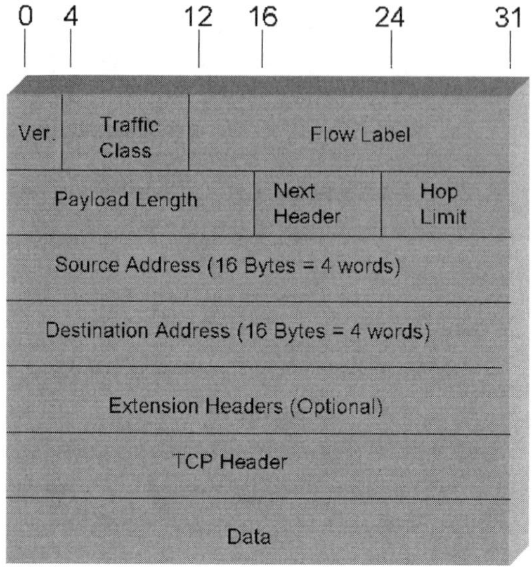

Figure 16.10 Ipv6 header format.

MAC layer. If the node implementing the Ethernet layer is an end station or data terminal equipment (DTE), then the MAC client level is a logical link control (LLC) layer as defined by IEEE 802.2. If the node implementing the Ethernet layer is an intermediate station or data communications equipment (DCE), then the MAC client layer is a bridge entity as defined by IEEE 802.1.

As the application data is passed down the protocol stack, a TCP header, an IP header, and an LLC header are added. This becomes the LLC protocol data unit that is encapsulated in the basic Ethernet frame, or MAC frame, as described below. Figure 16.11 shows the different levels of data encapsulation and their relationship to the MAC frame.

Individual addresses are referred to as *unicast* addresses. They refer to a single MAC and are assigned by the network interface card (NIC) manufacturer from a block assigned by the IEEE. Group addresses, or *multicast* addresses, identify the end stations in a work group and are assigned by the network manager. A special group address of all 1s is referred to as the *broadcast* address and indicates all stations on the network.

16.2.7.7 Virtual LANs and Quality of Service

A virtual LAN (VLAN) is a logical subset of a LAN that is based on geographical location, user type, work group, or similar criteria. VLANs are defined in IEEE 802.1q. Each VLAN consists of a broadcast domain for its users; a broadcast on a specific VLAN will only propagate to the users on that VLAN. Groups of users with similar access requirements can be given access to specific files, and work groups can be extended across geographical areas. VLANs can be used for security purposes as well.

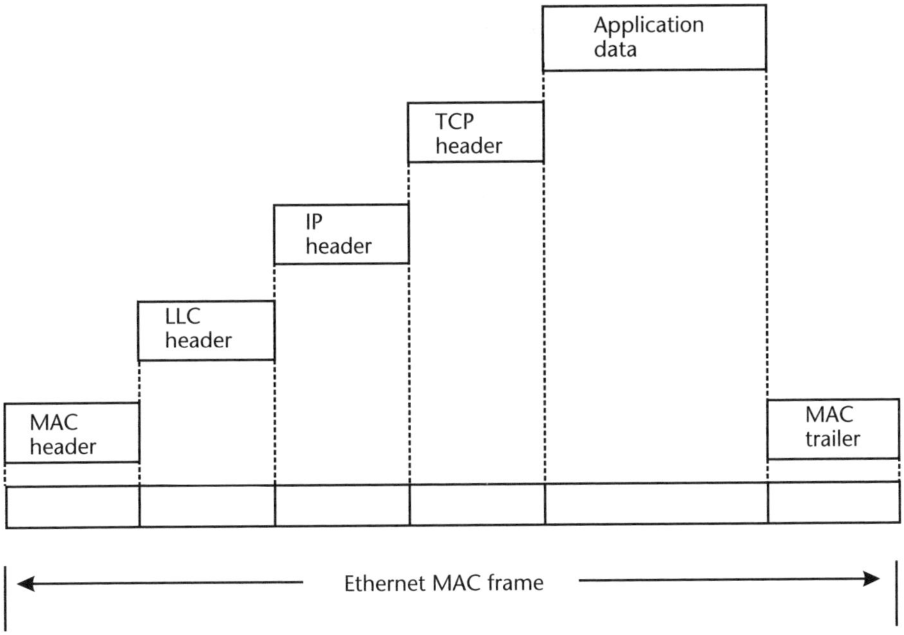

Figure 16.11 MAC frame data encapsulation and protocol layers.

IEEE 802.1p provides quality of service (QoS) for Ethernet packets by enabling user priority values to be encoded in packets that represent different traffic classes. The user priority value is located in the frame header using the extended frame format defined in IEEE 802.1q. RSVP sessions can be mapped to the user priority values specified in IEEE 802.1p.

16.2.7.8 Protocols for Web Services

Small private practice clinics seldom have access to sophisticated IT support systems and high-bandwidth LANs and WANs. The emerging use of Web Services based on Extensible Markup Language (XML) for online business transactions offers a solution that will streamline these processes, particularly in small and rural offices where the use of electronic data interchange (EDI) is neither technically nor economically feasible.

In April 2004, HL7 announced that two Version 3 transport specifications—ebXML and Web Services—were approved as draft standards for trial use (DSTU). To quote the press release:

> Web Services is a way for applications to expose software services using standard interoperability protocols, regardless of the platform on which they are implemented. The development of interoperable standards and the focus on communication and collaboration among people and applications have created an environment where Web Services is becoming the protocol/technology of choice for application integration.

16.2.8 Security Considerations

16.2.8.1 Application Level: HIPAA

Security requirements for healthcare IT in the United States are defined in the Health Insurance Portability and Accountability Act (HIPAA) Security Standards, Final Rule. In general, the HIPAA regulations apply to all "covered entities"—health-care providers, insurers and health plan providers, and claims clearinghouses, which process claims on behalf of healthcare providers. Business associates of healthcare providers and insurers, including medical transcription services, software vendors, and IT consultants, are also affected.

The security standards focus specifically on the EMR/EHR, and cover all healthcare entities that maintain or transmit electronic health information. The HIPAA security standards require covered entities to perform a risk analysis to determine what level of risk is acceptable, and to determine which security and control measures are most cost-effective. Some specific areas to include in a risk analysis are:

- *Computer and peripheral equipment*: Ensure that departments perform routine inventories and manage equipment access.
- *Communications networks*: Ensure that servers are adequately protected by firewalls, and that telephony networks enable secure communication.
- *Software*: Verify that application access is password protected, and limited to users with a need to know. Ensure that common applications such as Web browsers and e-mail programs do not create gaps in the security strategy. Validate the security capabilities of operating systems and databases.
- *Business and clinical procedures*: This includes not only standard functions such as admissions, patient record update, and claims processing, but also activities such as customer relations and visitor information.

Section 1173(d)(2) states that "Each person described in 1172(a) who maintains or transmits health information shall maintain reasonable administrative, technical, and physical safeguards (A) to ensure the integrity and confidentiality of the information; (B) to protect against any reasonably anticipated —(i) threats or hazards to the security or integrity of the information; and (ii) unauthorized uses or disclosures of the information; and (C) otherwise to ensure compliance with this part by the officers and employees of such person."

Security for Home Healthcare/Remote Access

In the response to comments received, the Final Rule states that this language draws no distinction between internal and external data movement; therefore, the same security regulations that cover workers internal to the covered entity must cover workers working at home, or workers accessing the system remotely.

Additionally, paragraph 164.312 states that "A covered entity must, in accordance with 164.306... Implement technical policies and procedures for electronic information systems that maintain electronic protected health information to allow access only to those persons or software programs that have been granted access rights as specified in 164.308."

It is worth noting that while encryption is mentioned in 164.312, it is not required, only "addressable." While most implementers would find it reasonable and prudent to encrypt protected healthcare information for transmission across a network, the presence of clear-text protected healthcare information on a home workstation would pose an obvious risk if another application maintained a nonsecure connection to the Internet.

Another type of risk is posed by the possibility of a virus invading the home workstation through a nonsecure Internet connection, and harming the functionality of the application. Also, operating systems that allow automatic update capability may not bc HIPAA-compliant at this time.

In short, the best way to ensure compliance with HIPAA security regulations is to maintain the same umbrella of security for home systems or remote access as for systems used at the healthcare facility [8].

16.2.8.2 Passwords

Providing secure access to the healthcare IT system is of critical importance. Password protection should be provided for all users, and user access should be restricted to the degree of control required for each function. Public Key Infrastructure (PKI) authentication and digital certificates may also be used to ensure that users possess the proper credentials when logging in. If users are allowed to log in remotely, protection should be provided by means of a firewall.

Validation should be provided at three levels:

1. *User level*: Each user login should be validated using passwords and access control lists. Each user should be assigned to an administrative group with specific access privileges.
2. A*pplication level:* Each time a user attempts to invoke a management application, the attempt should be validated to ensure that the user is entitled to access this function, based on the user's role and permissions.
3. T*ransaction level*: Each transaction should be validated to ensure that the user and application are entitled to perform the transaction.

16.2.8.3 Firewalls

A *firewall* is a set of programs, located at a network gateway server, which is used to protect the resources of a private network located behind the firewall. A firewall examines each packet and decides whether or not to deliver it to its destination, based on security policy algorithms. A standard screening function is to examine each request to ensure that it comes from a known domain name and/or IP address. A firewall may contain a proxy server to process requests on behalf of the protected network users.

A firewall may also allow remote access to the private network through the use of secure logins and authentication procedures. Firewalls are recommended if users are allowed to log in remotely to access the healthcare IT network.

16.2.8.4 Web Services Security

Web Services Security (WS-Security) is an industry standard that addresses security when data is exchanged as part of a Web Service. WS-Security is one of a group of standards developed by the Web Services Interoperability Organization (WS-I). WS-Security specifies enhancements to the Simple Object Access Protocol (SOAP) to protect the integrity and confidentiality of a message, and to authenticate the sender. WS-Security also specifies how to associate a security token with a message.

WS-I standards include the WS-I Basic Security Profile, which defines secure interoperable transactions for Web Services [9].

16.2.8.5 SSL/TLS

Secure Sockets Layer (SSL) is a protocol for managing message transmission security over the Internet, in order to prevent eavesdropping, message tampering, or message forgery. From a protocol architecture standpoint, SSL operates as a layer between TCP and the application level protocol, which is normally HTTP for Internet applications. SSL has gained wide acceptance, and is included in most common Web browsers and Web servers.

SSL was originally developed by Netscape, Inc., and is specified in the Netscape SSL3.0 Specification [10]. SSL3.0 has been superseded by an IETF standard, RFC2246, Transport Layer Security (TLS) 1.0. SSL3.0 and TLS1.0 are functionally similar, but do not interoperate; however, TLS1.0 includes a fallback mechanism that enables SSL3.0 authentication if one of the parties involved does not support TLS1.0. The remainder of this description focuses on TLS1.0.

The TLS1.0 protocol consists of two sublayers: the TLS Record Protocol, and the TLS Handshake Protocol. The TLS Record Protocol provides privacy and reliability, as well as encapsulating the higher-level Handshake Protocol. Symmetric cryptography, such as the Data Encryption Standard (DES) defined in ANSI X3.92, is used for data encryption. Keys are generated uniquely for each connection, and are based on a secret negotiated by the Handshake Protocol.

The TLS Handshake Protocol allows the server and client to authenticate each other, to negotiate an encryption algorithm, and to exchange cryptographic keys before data transfer begins. Peers authenticate each other using asymmetric public key cryptography such as RSA (named after its developers, Rivest, Shamir, and Adleman) or Digital Signature Standard (DSS) [11, 12]. The Handshake Protocol also provides secure negotiation of a shared secret that is unavailable to eavesdroppers and man-in-the-middle attacks.

One of the advantages of TLS (and SSL) is that it is protocol-independent. It sits on top of the TCP layer, and therefore requires no operating system or kernel changes to implement. SSL/TLS may provide adequate security for Web browser access to management systems, if other security precautions are implemented. Figure 16.12 shows the protocol architecture of TLS.

16.2.8.6 PKI

A public key infrastructure (PKI) enables users of an insecure network (such as the Internet) to securely exchange data through the use of a public and a private crypto-

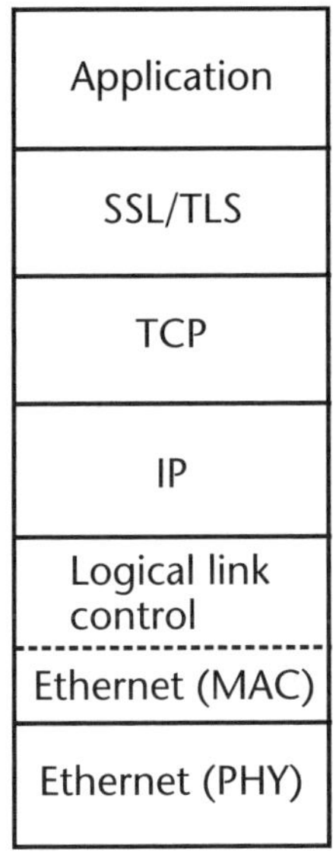

Figure 16.12 Protocol architecture of TLS1.0.

graphic key pair that are obtained from a trusted authority. The PKI allows for the use of a *digital certificate* that can uniquely identify an individual or organization, and a directory service that can store (and, when necessary, revoke) the certificates.

PKI assumes the use of public key cryptography. Message encryption may be based on a *private key* (referred to as *symmetric cryptography*), or on a *public key* (referred to as *asymmetric cryptography*). For private key cryptography, the same key is known to the sender and the receiver. The sender uses the key to encrypt the message, and the receiver uses it to decrypt the message. The disadvantage of the private key approach is that if the private key is discovered, then all encrypted messages may easily be decoded.

For public key cryptography, a user, whom we shall call User A, requests a certificate authority to create a public key and a private key. The two keys are created simultaneously, using an algorithm such as RSA. The private key is given only to User A, but the public key is made available as part of a digital certificate that can be obtained from a public directory. A second user, User B, may then encrypt a message using the public key obtained from the directory, and send it to User A. User A then uses his or her private key to decrypt the message that was encrypted using the public key. User A may then authenticate himself/herself to User B by using the private key to encrypt a signature, and sending it to User B. User B may decrypt the signature by using the public key.

PKI is recommended for authentication of users, especially users who may be accessing the system remotely.

16.2.9 Network Management

As the size and complexity of health IT networks increases, network management becomes an increasingly important function. While network management traditionally was performed on a per-application basis, a more holistic approach, in which the entire health IT network is the focus of management activities, is now being implemented by many healthcare providers. Network management functions include:

- User administration;
- Security administration;
- System monitoring;
- Adding new services;
- Optimizing network performance.

Each is discussed below.

16.2.9.1 User Administration

User administration is a critical part of network management. New users must be given appropriate access to specific network functions, and users that no longer have a need for network access must have their network privileges removed. Procedures must be put in place to automate as much as possible of the user administration process, such as automatic reminders to users to change their passwords.

16.2.9.2 Security Administration

In addition to password administration and network access administration, network managers must establish procedures to prevent, detect, and deal with security breaches. This includes not only technology-based approaches, such as firewalls, encryption, and VPNs, but also procedural approaches, such as equipment inventory, restricted physical access to system resources, and routine security audits.

16.2.9.3 System Monitoring

System monitoring functions include monitoring system health, detecting faults and errors, and responding to system problems in a timely fashion. Network management software that can detect and respond to errors is particularly useful in this context. An important aspect of fault detection is the ability to perform event correlation and detect the root cause of a network problem.

16.2.9.4 Adding New Services

One of the goals of network planning and architecture is to design a network that is scalable, both in terms of adding additional users and also in terms of adding new services. Care should be taken to ensure that adding a new network component or application will not impact negatively on other network functions.

16.2.9.5 Optimizing Network Performance

The ability to monitor network performance is a critical requirement of network management. The system should provide the ability to measure parameters such as system throughput, application response time, and bandwidth usage. Having this kind of information makes it possible to plan for future growth and eliminate network bottlenecks caused by inadequate resource provisioning.

16.2.10 Storage Management

Placing image data online requires not only a tremendous increase in storage capacity, but also new techniques for efficient management and retrieval. Many healthcare facilities are investigating the use of storage networks and content-addressable storage to efficiently manage the storage and retrieval of these vast amounts of data. *Storage networks* are special-purpose networks that transport data between storage devices, such as tape or disk, and computing devices, such as mainframes or servers.

16.2.10.1 Storage Area Networks

A storage area network, or SAN, is a storage network that provides shared access to storage by multiple computers and servers. SANs improve the efficiency of information retrieval by allowing block-oriented data to be distributed across multiple storage devices, providing improved reliability and availability.

SANs enable storage devices from different vendors and using different access protocols to be pooled for efficient management; they permit standard data management functions such as backup, recovery, mirroring, and data migration to be performed reliably and efficiently. SANs primarily use the Fibre Channel protocol to transfer data; however, SANs are able to include devices that use other protocols by employing bridges and gateway devices to perform protocol translation.

In its purest sense, the SAN topology is based on a switched architecture; a fabric switch is used to provide connectivity between a computing node and a storage node for the duration of a data transfer operation. However, the SAN architecture has evolved over time from simple point-to-point topologies to arbitrated loops to the present switched topology, so many SANs contain elements of these earlier topologies as well (see Figure 16.13).

16.2.10.2 Network Attached Storage

Network attached storage (NAS) is another technique for providing managed storage. NAS consists of a *file manager* (or filer) attached to a LAN that manages and provides access to stored data. The NAS concept emerged from the informal practice of designating one computer node on a LAN as a pure file server, with no resident applications. The file management function was optimized to provide better performance and throughput for NAS.

NAS transfers file-structured data between a processor and LAN-attached storage, while a SAN transfers I/O blocks between processors and storage. NAS uses LAN protocols, such as Ethernet or Gigabit Ethernet, to transfer data [13, 14].

16.2.10.3 Comparison of SAN and NAS

There is no simple answer as to whether SAN or NAS is the preferred architecture for networked storage. Generally speaking, SAN is the preferred architecture for

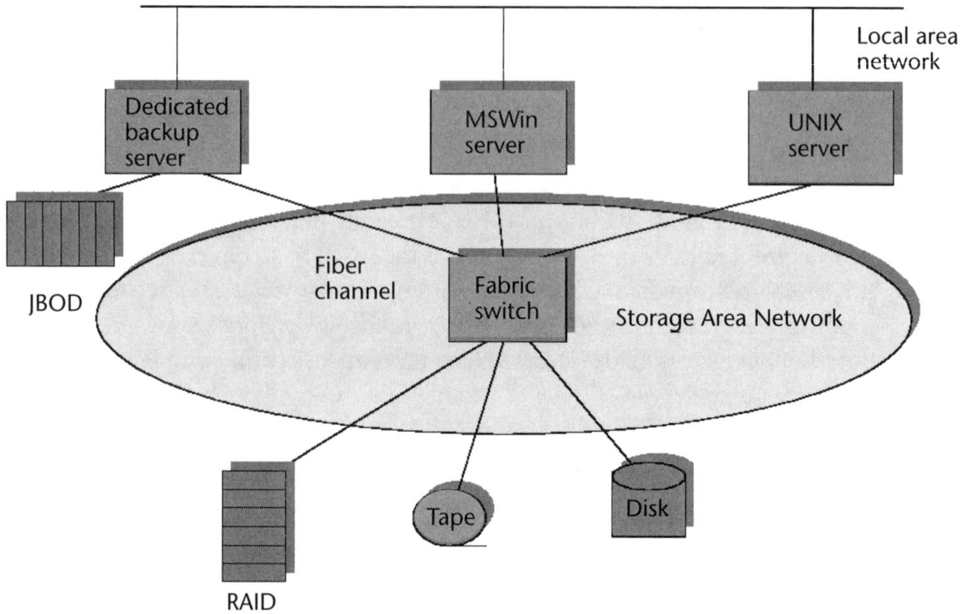

Figure 16.13 Storage area network (SAN) architecture.

large institutions that maintain large amounts of data, and where performance is a critical requirement. NAS is generally adequate for smaller entities, where cost and ease of implementation are most important (see Figure 16.14).

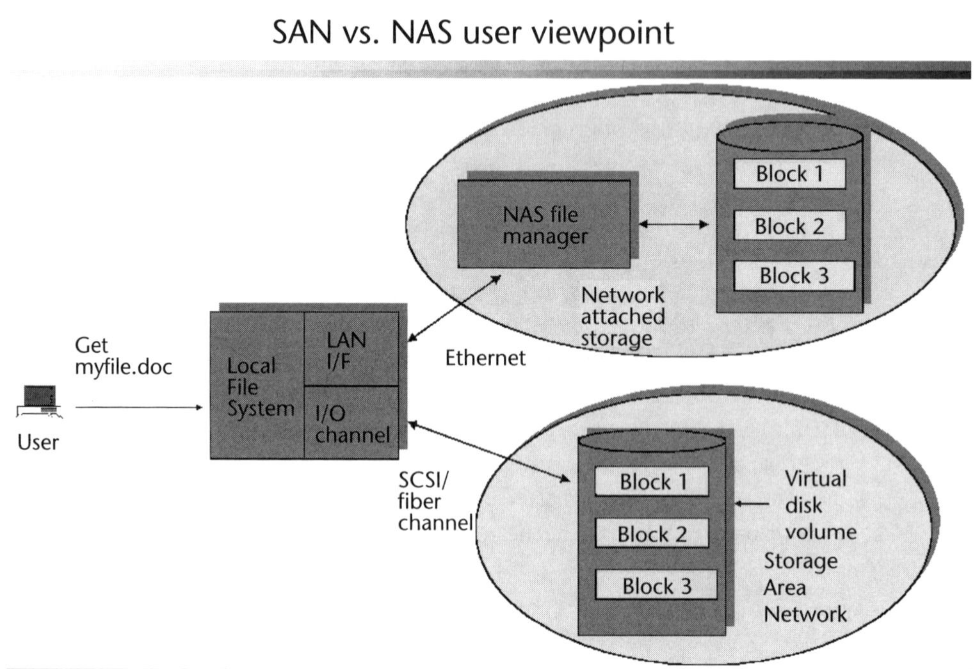

Figure 16.14 SAN and NAS—a comparison of architectures.

16.3 Data Management

16.3.1 Introduction

As the amount of data in the health IT network increases in size and complexity, management becomes a critical factor in maintaining data integrity and availability. In a healthcare setting, data typically passes through a number of well-defined stages:

- *Creation.* Data is created when a new patient record is created, or when a patient is admitted.
- *Modification.* As tests and procedures are performed, and as the patient is transferred from one ward or clinic to another, the original data is modified.
- *Integration.* Data from many sources such as admissions, CPOE, cardiology, radiology, and other departments must be integrated into a single patient-identifiable record.
- *Storage and retrieval.* Stored data must be available, reliable, and secure. Provision must be made to back up critical data on a regular basis to guard against loss due to system malfunction.
- *Archival.* Once a patient has been discharged or is no longer under medical care, the encounter and treatment data must be maintained for future reference.

Implementation of data management techniques enables many of these functions to be performed efficiently, with a minimum of human interaction required.

16.3.2 Architecture for Data Management

16.3.2.1 Databases

Simply stated, a *database* is a collection of information that is structured in such a way as to provide ease of retrieval by an application. In a healthcare environment, databases might contain patient encounter or admissions data, laboratory data, pharmacy data, or radiological image data. The most common form of database currently in use is the relational database. In a relational database, data is stored in a number of formally described tables, which enable data to be accessed in a number of ways without having to reorganize the tables. Tables in turn consist of columns, which define the categories of data, and rows, which contain unique instances of the data defined by each column. The formal description of the tables, columns and rows in a database is referred to as *metadata*.

Data in relational databases is accessed by means of structured query language, or SQL. With the increasing use of XML in medical applications, databases are becoming available that store data in XML format and permit direct query using XML syntax. Most databases used in medical applications are commercially available products from the major software vendors.

16.3.2.2 Data Warehouses

A *data warehouse* is a central repository for many types of data, generally assembled from a number of disparate sources and databases. Unlike a database, the data is not organized in an application-specific or user-specific manner. Data warehouses are generally used for research purposes, such as data mining. For example, a data warehouse might contain information on all patients with a particular condition over a period of years; a researcher might do "data mining" to see if any patterns emerge linking the condition to some characteristic shared by the patients. Another common use of data warehouses is *text mining*, in which textually based data, such as is found in journal articles, is searched to find connections between disease, genomics, and pharmacology.

A *data mart* is a specific type of data warehouse in which the data organization has been optimized for a specific user or purpose (Figure 16.15).

16.3.2.3 The Clinical Data Repository

A clinical data repository (CDR) is a database (actually, a data mart) that consolidates data from a variety of sources to provide a unified view of a single patient. Typical data types that may be found in a CDR include laboratory test results, patient demographics, pharmacy information, radiology reports and images, pathology reports, ICD-9 codes, and admission and discharge information.

A critical component of the CDR is the patient identification information. Typically this is in the form of an enterprise identifier that is taken from the master patient index (MPI), which performs the function of unambiguously identifying the patient, based on patient demographics and other information (see below). The CDR must also have electronic linkage to ancillary sources of data, and a semantic mapping to enable compatibility among the disparate sources of data. Typically, a CDR is created by means of database federation.

16.3.2.4 Federated Databases

As noted above, databases are typically maintained on a per-application basis; thus patient admissions, laboratory data, and radiology imaging for a single patient may be maintained in three separate databases. To create the single-patient view needed for the CDR, a *federated database*, in which the data from multiple sources is pre-

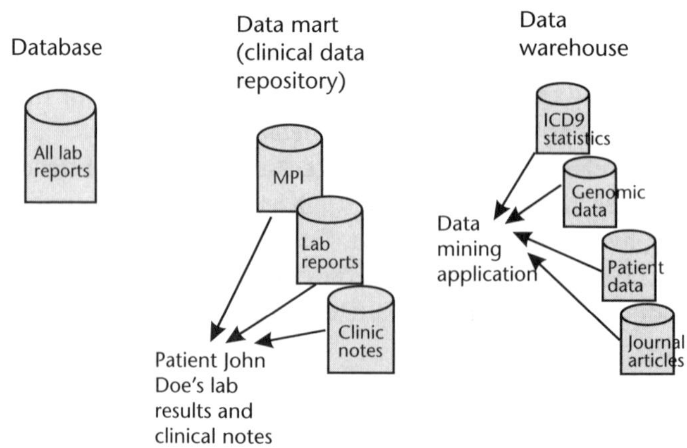

Figure 16.15 Databases, data marts, and data warehouses for healthcare IT.

sented to an application user in a single view, may be employed. Thus patient identification data would reside in only one location (the MPI), and would appear to be part of the admissions data to the admissions user, and would appear to be part of the billing data to the billing user.

Care must be taken when federating databases and eliminating redundant data to ensure that the consolidated data is correct and accurate. Small differences in the duplicated data can lead to errors and lost productivity. This is often accomplished by a process known as *data cleansing*. Data cleansing involves removal or repair of data that is incorrect, incomplete, or improperly formatted. Tools are available to automate the process of data cleansing.

16.3.3 The Master Patient Index

Many healthcare facilities maintain a special-purpose database referred to as the master patient index or master person index (MPI). The purpose of the MPI is to ensure that patients are correctly identified in their admissions, diagnosis and treatment, billing, and discharge records. An MPI may be clinic-, facility-, or system-specific. Some of the identifiers commonly in use in MPIs include:

- Enterprise identifier: An identifying number assigned by the facility or system providing treatment;
- Facility identifier: A code representing the facility or system providing treatment;
- Internal identifier: A medical record number;
- Patient's name;
- Patient's date of birth;
- Patient's gender;
- Patient's race and/or ethnicity;
- Patient's address;
- Patient's alias, previous name, or maiden name;
- Patient's Social Security Number;
- Patient's telephone number;
- Patient's insurance information: Group and individual identifier.

16.3.3.1 Progress Toward the Universal Patient Identifier in the United States

While the MPI can provide accurate patient identification within a single healthcare organization, it is not adequate for uniquely identifying a patient within a regional or national healthcare network. The task of unambiguously identifying a patient within a regional/national healthcare system would be greatly simplified if a standard existed for a universal patient identifier (UPI). While the original HIPAA legislation of 1996 called for the establishment of a UPI, work on such a standard in the United States has been halted due to concerns about privacy and the implications of a national repository of identification information [15].

According to "Analysis of Unique Patient Identifier Options," a report issued by the U.S. Department of Health and Human Services in 1997, a universal patient identifier must consist of six components:

1. Identifier (numeric, alphanumeric, etc.) scheme;
2. Identifying information;
3. Index;
4. Mechanism to hide, or the tool to encrypt the identifier;
5. Technology infrastructure including the software, hardware, and communication technologies to search, identify, match, encrypt, etc.;
6. Administrative infrastructure including the central governing authority.

These components must work together to effectively fulfill the objectives of the unique patient identifier [16].

In the absence of a standardized UPI, two alternative approaches to patient identification have been proposed by the Integrating the Healthcare Enterprise (IHE) Initiative, sponsored by the American College of Cardiology (ACC), the Healthcare Information and Management Systems Society (HIMSS), and the Radiological Society of North America (RSNA). Patient identifier cross-referencing (PIX) provides cross-referencing of patient identifiers from multiple patient identifier domains. Patient demographics query (PDQ) provides ways for multiple distributed applications to query a central patient information server for a list of patients, based on user-defined search criteria, and retrieve a patient's demographic information directly into the application [17, 18].

16.3.3.2 Universal Patient Identifiers in Other Countries

France

In the French national healthcare system, patients are identified by a healthcare identification number, used primarily for reimbursement purposes. Children are identified by the healthcare identification number of one of the parents. France is in the process of implementing a unique identifier for each person in the healthcare system, the Numéro d'Inscription au Registre INSEE, or NIR [15].

Australia

Australia uses a unique patient identifier similar to the U.S. Medicare number. By law, it may be used for reimbursement purposes only and may not be used for the transfer of medical information [15].

16.3.4 Data Management Functions

16.3.4.1 Data Access and Retrieval

Data for real-time access is normally stored in disk storage devices. Dramatic increases in disk storage capacity and decreases in the price-per-bit cost of disk storage have made it possible to store increasing amounts of image data and other data types that consume large amounts of storage capacity; healthcare facilities routinely store hundreds of gigabytes, or even terabytes of data. However, this increase in the amount of stored data has also created a requirement for more efficient means of storage access, data retrieval, and storage management.

One such technique is the use of a redundant array of independent (formerly "inexpensive") disks, or *RAID*. RAID is a technology for managing a group of disk units to provide increased performance, reliability, or both (Figure 16.16). A RAID employs a form of storage virtualization; the application accessing data sees a single virtual disk, while data is actually written to or read from a group of physical disk devices managed by a RAID controller using a technique called *striping*. In striping, the RAID controller divides up the blocks of data that constitute an I/O write request and a portion of the total number of blocks (i.e., a stripe) is written to each physical disk. Conversely, the RAID controller satisfies a read I/O request by reading blocks of data from multiple physical disk devices simultaneously. This technique improves performance by performing multiple disk read/write operations concurrently rather than serially.

A stripe may be as small as a single sector (512 bytes) or as much as several megabytes. For accessing large records or datasets for single users, a small stripe gives better performance, since I/O throughput is maximized if the data is spread across all available physical devices. In multiuser environments, performance is improved if the stripe size is set equal to a typical record size, since this permits overlapping of I/O operations.

In addition to striping, RAID controllers may also employ *mirroring*. Mirroring improves reliability by writing each block redundantly to two different locations in physical disk storage.

Many different combinations of striping and mirroring are possible to provide different levels of performance and reliability. Some of the common RAID types are as follows:

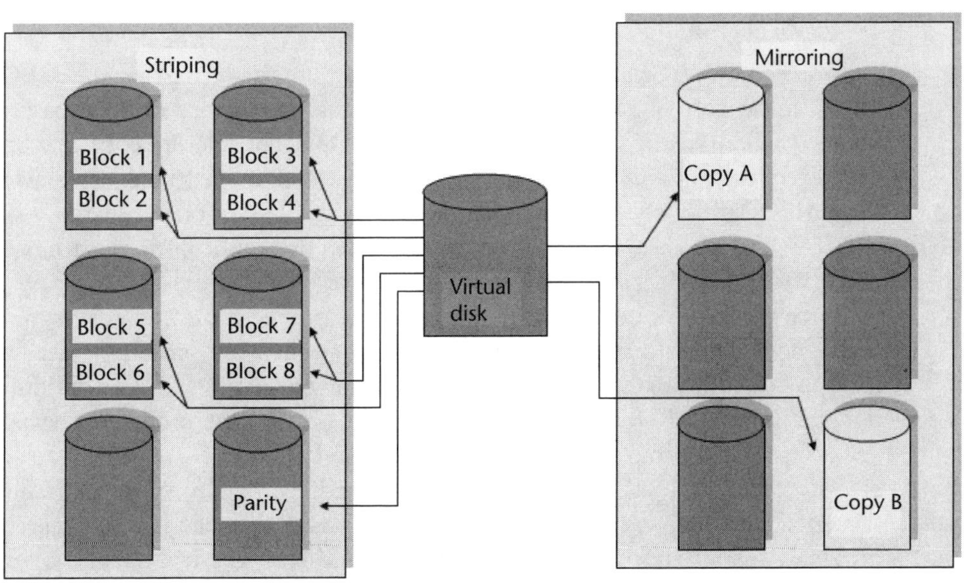

Figure 16.16 RAID architecture,

- RAID 0: Striping but no mirroring. This provides the best performance, but is not fault-tolerant.
- RAID 1: Mirroring but no striping. This provides redundancy for fault tolerance, but no performance improvement over a single disk.
- RAID 2: Striping with error checking and correction (ECC) information stored on some disks.
- RAID 3: Striping with ECC. One drive is dedicated to storing parity information (fixed parity). RAID 3 does not permit overlapped I/O. RAID 3 provides high availability at a lower cost than RAID 1.
- RAID 4: Striping with large stripes and fixed parity. This allows a complete record to be written to a single drive; therefore the complete record can be read in a single-read operation. This permits overlapping of read operations. However, write operations may not be overlapped, since the parity drive must be written to.
- RAID 5: Striping with parity distributed across multiple disks (floating parity). This allows overlapping of both read and write operations. Best for multiuser systems.
- RAID 10: Striping and mirroring. RAID 10 is an array of stripes in which each stripe is a RAID 1 array of drives. Offers higher performance than RAID 1.

Another technique that is sometimes employed is just a bunch of disks, or *JBOD*. JBOD is an acronym used to describe a management technique in which a group of hard disk units are managed as a single logical drive. Unlike RAID, JBOD does not provide performance or reliability improvements. JBOD is generally used to organize multiple disk drives of different capacities into a larger virtual disk. JBOD is also referred to as *spanning* [14].

16.3.4.2 Backup

Data stored on disk must be periodically backed up to ensure that the data will not be lost in the case of equipment failure. Data may be backed up to streaming tape, another disk drive, or simply written to CD-ROM. The backup copy may be stored offsite to ensure availability in the event of physical damage to the primary facility.

SANs provide the ability to automate the backup process. In the "islands of storage" architecture, each individual computer or server node must have a dedicated backup device, leading to inefficient use of resources. To improve backup resource utilization, network designers first employed a *backup server*, a separate server node attached to the same LAN as the application servers, which managed backup for all application nodes in a centralized fashion. This configuration is referred to as *LAN-based backup* (Figure 16.17). In this mode of operation, the backup server is programed to perform backup on data belonging to each application server at specified intervals. When the scheduled backup time for an individual server occurs, the backup server sends commands to the specified server, requesting it to back up all or some portion of its backend storage. The backup operation is performed by transferring the data through the application server, across the LAN, and through the backup server to the dedicated backup storage. While this approach improves resource utilization by centralizing the backup operation, it

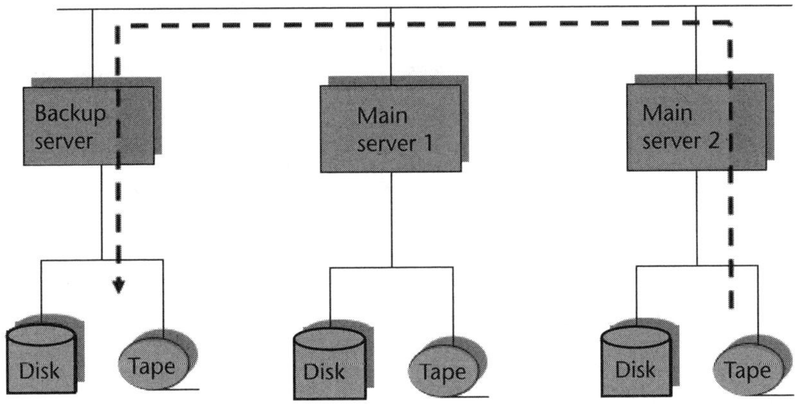

LAN-based backup uses server processor cycles and LAN bandwidth

Figure 16.17 LAN-based backup.

impacts system performance through its use of application server processor cycles and LAN bandwidth to perform backup. As a result, LAN-based backup is generally performed during off-peak hours when system performance will not be affected by the operation.

16.3.4.3 LAN-Free Backup

Implementation of a SAN allows backup to be performed without utilizing significant amounts of either LAN bandwidth or server processing cycles. In *LAN-free backup*, the LAN is used only for the purpose of sending backup commands to the individual servers. The servers then use the SAN fabric switch to connect the application storage to the backup storage in order to back the data up directly, eliminating the need to send data through the servers and over the LAN.

16.3.4.4 Server-Free Backup

A further refinement to the backup process is provided by *server-free backup*. In this configuration, scheduled backups are performed by moving data directly from the application backend storage to the backup storage without the involvement of the application server. Server-free backup may be performed by a separate SAN backup appliance, or by software resident in the storage director. It is sometimes referred to as *third-party copy*. This approach provides full backup capability with a minimum of impact on the network and application servers (Figure 16.18) [14].

16.3.4.5 Archival

Archival consists of relocating data that is infrequently accessed to a storage location where it may be maintained at a lower cost, and with less performance impact, than the frequently accessed data maintained for real-time access. Archival data may consist of EMRs for patients who are no longer being treated, or general medical data being kept for research or statistical purposes.

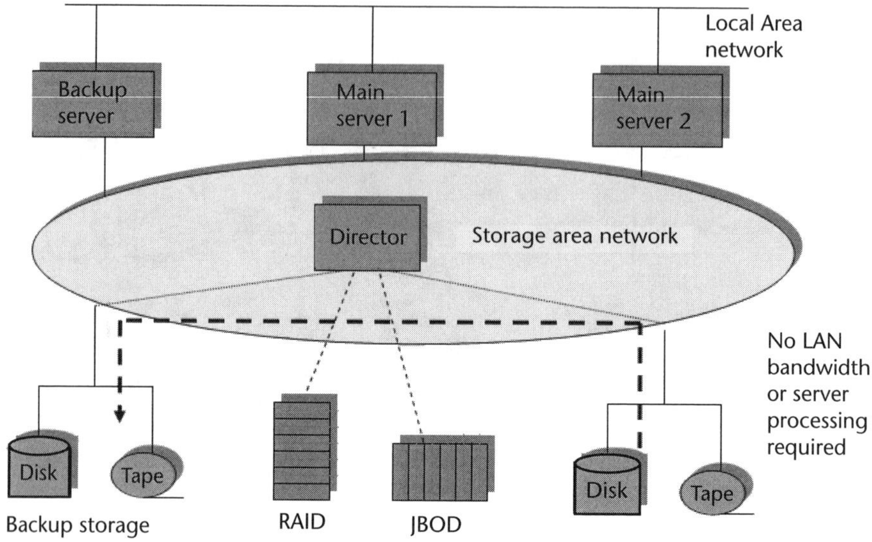

Figure 16.18 SAN-based backup.

The archival process is similar to the backup process in that the data is written to backup media, such as streaming tape, another disk drive, or CD-ROM diskettes, and generally taken to an offsite location for storage. However, while backup makes a copy of frequently accessed data to guard against data loss in the event of equipment failure, archival permanently removes the data from the real-time system in order to free up storage and improve performance.

Since archival data is intended for long-term storage, and may become valuable for research purposes in the future, care should be taken to ensure that the media it is stored on will be readable in the future. Technology changes may make older media and data formats obsolete; aging may make certain media types unreadable. Therefore it may be wise to migrate data from older media (e.g., older tape formats, removable disk packs, diskettes) to newer media as part of the archival process.

In addition to EMR, certain other types of administrative information may need to be archived for legal purposes. For example, the HIPAA Final Privacy Rule requires that changes to policies or procedures related to privacy must be documented, and the documentation maintained for at least 6 years; this includes consent forms and authorizations for release of PHI. Similarly, the HIPAA Final Security Rule requires that policies and procedures related to security must be maintained for at least 6 years [19, 20].

16.3.4.6 Disaster Recovery

Disaster recovery, sometimes referred to as business continuity, is a strategy for recovery of not only data, but also processing capability, in the event of damage to a physical facility due to natural disaster, acts of war, or similar destruction. Disaster recovery planning involves not only backing critical operational data up to an offsite location, but also creating a backup processing capability at an offsite location. For the healthcare facility, this would enable continued access to and processing of patient records in the event of a disaster. Providing disaster recovery capability is a critical requirement for healthcare facilities that would be expected to provide healthcare services in the event of a disaster.

16.3.5 Design Considerations

16.3.5.1 Media Selection

Selection of media for data storage should be based on the operational requirements for a specific task. For real-time access, the primary consideration is access latency, and therefore disk storage and RAID arc normally chosen. For backup and archival purposes, cost and storage capacity are the main considerations, and streaming tape and CD-ROM are common options.

16.3.5.2 Storage Capacity

Online storage of image data greatly increases the need for storage capacity, and this should be taken into account when planning and implementing a PACS system. PACS systems typically consume hundreds of gigabytes of storage, and may require terabytes of storage in a large institution.

16.3.5.3 Access Latency

As mentioned above, disk storage and RAID are generally chosen for applications requiring real-time access due to their superior access latency. Access latency for RAID configurations can be optimized by selecting the best striping algorithm for the type of user access that is anticipated.

16.3.5.4 Scalability

Scalability is an important consideration when planning and implementing a health IT network. The network should not only be planned to include sufficient capacity for future needs, but should also be designed for ease of expansion and update. As a specific example, one of the benefits of the SAN architecture is that it enables additional storage to be added to the network without taking the network offline.

16.3.5.5 Security

Storage media create additional security issues for network administrators. Care should be taken that archive and backup media are stored in secure locations with controlled access. One area that is frequently overlooked is secure disposal of obsolete storage media. When disk drives, diskettes, and magnetic tapes are disposed of, they should first be erased to remove any data they may retain in a way that does not allow recovery of erased data; this entails writing streams of random data onto the erased drives/ tapes. Similarly, CD-ROM diskettes containing sensitive data should be physically destroyed before disposal.

16.4 Electronic Medical Records

16.4.1 Introduction

According to the *HL7 Functional Model*, an electronic medical record or EMR is "a comprehensive, structured set of clinical, demographic, environmental, social, and financial data and information in electronic form, documenting the healthcare given to a single individual."[2] (See Figure 16.19).

16.4.2 Interoperability

If EMRs/EHRs are to be exchanged among healthcare institutions and insurers, they must be interoperable. According to the *IEEE Standard Computer Dictionary*, interoperability is the "ability of two or more systems or components to exchange information and to use the information that has been exchanged."[3][21].

IEEE 1073, *Point of Care Medical Device Communications*, defines two different types of interoperability—"functional" and "semantic" interoperability. Both are defined and discussed below [22].

16.4.2.1 Functional Interoperability

Functional interoperability is defined as "the capability to reliably exchange information without error." The IEEE 1073 definition of functional interoperability lists the following characteristics:

- Shared architectures;
- Shared methods;
- Shared frameworks.

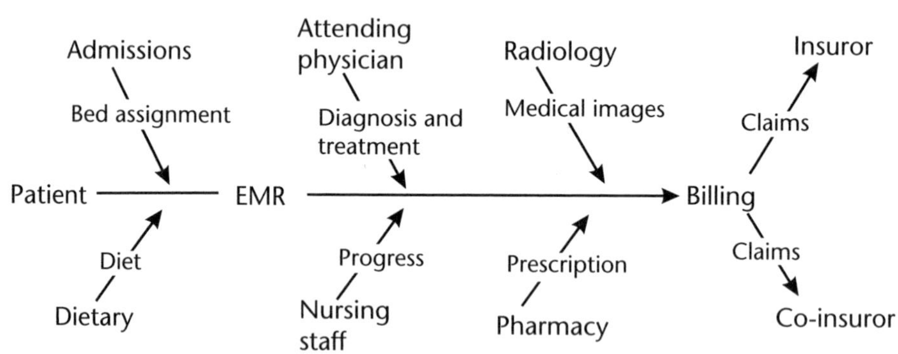

Figure 16.19 Electronic medical record functional flow.

2. As noted above, when this information is shared with other healthcare institutions or payers, it becomes an electronic health record, or EHR.
3. For an extended discussion of interoperability and interoperability definitions, see "Coming to Terms: Scoping Interoperability for Healthcare," which can be downloaded from HL7 at http://www.hl7.org/ehr/downloads/index_2007.asp.

An *architecture* is the conceptual design of the system. Systems interoperate if their architectures are similar enough so that functions that execute on one system execute identically (or nearly identically) on another system. (For example, accessing a Web site or sending an e-mail using the Internet should function the same whether I am an aol.com subscriber or an earthlink.net subscriber.)

Shared methods refers to the processes and procedures that a system performs. These processes and procedures would include operations such as accessing EHRs, submitting insurance claims, and broadcasting alerts. To ensure interoperability, these operations must be capable of being performed identically at any point in the network, regardless of implementation. An example of a definition of shared methods is the EHR Functional Model developed by Health Level 7.

A *shared framework* is a shared set of goals and strategies. Effective EHR implementation requires that all stakeholders share a common set of goals and principles.

16.4.2.2 Semantic Interoperability

Semantic interoperability is "the ability to interpret, and, therefore, to make effective use of the information so exchanged." The IEEE 1073 definition of semantic interoperability consists of the following:

- Shared datatypes;
- Shared terminologies;
- Shared codings [22].

Shared datatypes refers to the types of data exchanged by systems. Interoperability requires that systems share datatypes on many different levels, including messaging formats (e.g., XML, ASCII), and programming languages (e.g., integer, string).

Shared terminologies refers to the establishment of a common vocabulary for the interchange of information. Standardized terminology is a critical requirement for healthcare applications in order to ensure accurate diagnosis and treatment, and has led to the development of standards such as SNOMED-CT (see below).

Shared codings refers to the establishment of standard encodings to be shared among systems. In the context of EHR this refers not only to the encoding of software functions, but also to the encoding of medical diagnoses and procedures for purposes of claims processing, research, and statistics gathering (e.g., ICD9/10, CPT).

Full semantic interoperability is a desirable goal for EMR/EHR because it permits any arbitrary function involving exchange, analysis, or modification of patient information to be automated in a seamless fashion.

16.4.2.3 Mapping Functions

Mapping functions play an important role in interoperability. Mappings are sets of relationships established between two vocabularies in order to allow automated translation or connection between them. Mappings are typically implemented as software functions that automate the translation process between two encoding standards, such as a mapping between laboratory procedure codes and the codes used to bill for the procedures. Automated maps create efficiency by minimizing duplicative data entry and patient data integration across a wide variety of applications.

16.4.2.4 Standards Introduction

Standards are a necessary component of interoperability. They define the components of interoperability (shared architectures, methods, frameworks, terminologies, datatypes, codings) and enable these types of interoperability to be tested for compliance and certification.

Many technical standards, such as the SNOMED standards for clinical terminology, the HL7 Electronic Health Record standards, and the Internet Engineering Task Force (IETF) standards for the Internet, are international in scope. Security and privacy standards, however, such as HIPAA in the United States and the European Union's Directive 95/46/EC, tend to be national or regional in scope, reflecting political and cultural influences.

The following sections propose a taxonomy of standards for the EMR/EHR (Figure 16.20).

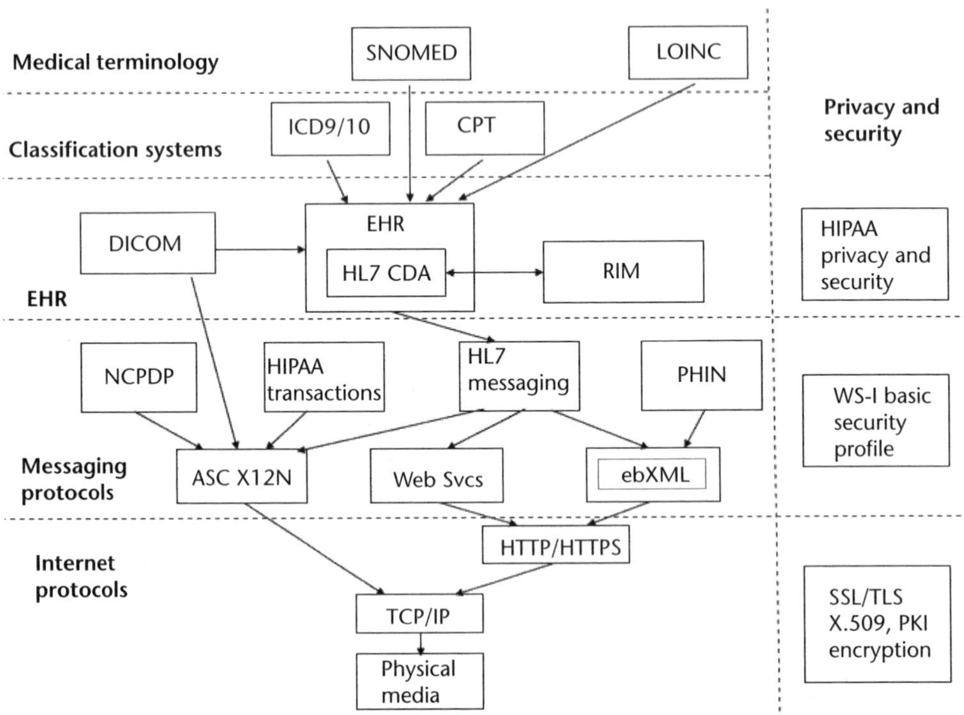

Figure 16.20 Taxonomy of standards for healthcare IT.

16.4.3 Medical Terminology

Medical terminology standards define the standard terminology to be used in clinical and laboratory reporting and electronic health records.

16.4.3.1 SNOMED

Standardized Nomenclature for Medicine–Clinical Terminology (SNOMED-CT) is the world's most comprehensive clinical terminology database, comprising 350,000+ terms. It was originally developed and maintained by SNOMED International, a division of the College of American Pathologists. The United States Department of Health and Human Services (HHS) acquired a license to make SNOMED-CT available to U.S. healthcare entities via the Unified Medical Language System (UMLS) at the National Library of Medicine.

On June 23, 2005, SNOMED International signed an Associate Charter Agreement with Health Level Seven, Inc. This agreement formalized the relationship between SNOMED and HL7, and provided the basis for harmonization between SNOMED-CT and the HL7 Reference Information Model (RIM) [23].

In April 2007, SNOMED-CT became an international standard when ownership was transferred from the College of American Pathologists to the International Health Terminology Standards Development Organization (IHTSDO) [24].

16.4.3.2 LOINC

Logical Observation Identifier Names and Codes (LOINC) provides universal identifiers for laboratory and clinical observations. The laboratory portion of the LOINC database contains the categories of chemistry, hematology, serology, microbiology (including parasitology and virology), and toxicology, as well as categories for drugs and the cell counts you would find reported on a complete blood count or a cerebrospinal fluid cell count. Antibiotic susceptibilities are a separate category. The clinical portion of the LOINC database includes entries for vital signs, hemodynamics, intake/output, EKG, obstetric ultrasound, cardiac echo, urologic imaging, gastroendoscopic procedures, pulmonary ventilator management, selected survey instruments, and other clinical observations.

The Regenstrief Institute (www.regenstrief.org) maintains the LOINC database and its supporting documentation. The U.S. National Library of Medicine has developed a mapping between the LOINC codes used for ordering laboratory procedures, and the Current Procedural Terminology (CPT) codes used for billing for these procedures [25].

16.4.4 Classification Systems

Classification systems provide a standard encoding for diagnoses and clinical procedures, for statistical purposes as well as insurance claims processing.

16.4.4.1 ICD9/10

The International Classification of Diseases, 9th/10th Revision (ICD9/10), provides classification of morbidity and mortality information for statistical purposes, and for the indexing of hospital records by disease and operations, for data storage and retrieval. It is maintained by the World Health Organization. The Health Insurance Portability and Accountability Act (HIPAA) Final Rule for transactions and code sets named ICD-9-CM volumes 1 and 2 as the code set for diagnosis codes, and ICD-9-CM volume 3 for inpatient hospital services.

While ICD-9-CM remains the standard classification system for use in the United States, it has been superseded in much of the rest of the world by newer versions, ICD-10-CM and ICD10-PCS. While ICD9 consists of approximately 20,000 classification codes, ICD10 supports approximately ten times as many codes, allowing for much finer granularity of classification [26].

16.4.4.2 CPT

Current Procedural Terminology (CPT) is a listing of descriptive terms and identifying codes for reporting medical services and procedures performed by physicians. It was developed by and is maintained by the American Medical Association (AMA). CPT is designated by HHS as the national coding standard for physician and other healthcare professional services under HIPAA. The CPT code set must be used for all financial and administrative healthcare transactions sent electronically [27].

The *HIPAA Final Rule for transactions and code sets* names CPT (including codes and modifiers) and Healthcare Common Procedure Coding System (HCPCS) as the procedure code set for:

- Physician services;
- Physical and occupational therapy services;
- Radiological procedures;
- Clinical laboratory tests;
- Other medical diagnostic procedures;
- Hearing and vision services;
- Transportation services including ambulance.

CPT codes are updated annually and become effective for use on January 1 of each year.

16.4.5 Health Level 7 (HL7) Standards for EMR/EHR

Health Level 7 (HL7) is a not-for-profit, ANSI-accredited standards development organization with a focus on clinical and administrative data, founded in 1987. HL7 is currently engaged in providing a comprehensive framework and related standards for the exchange, integration, sharing, and retrieval of electronic health information that supports clinical practice and the management, delivery, and evaluation of health services. HL7's more than 2,000 members represent approximately 500 corporate members, including 90% of the largest information systems vendors serving healthcare.

HL7 established an EHR special interest group in 2001 to begin efforts focused on EHR systems. HL7 was chosen to organize the accelerated EHR initiative led by the Center for Medicare and Medicaid Services (CMS).

The standards developed by Health Level 7 (HL7) for Electronic Health Records (EHR), including the HL7 EHR Functional Model, the HL7 Clinical Document Architecture (CDA), and the HL7 Reference Information Model (RIM).

Electronic Health Record (EHR) standards define the content and format of an EHR. EHR standards include the format of the messages used to transmit EHR across a network.

16.4.5.1 HL7 EHR Functional Model

The HL7 EHR System Functional Model provides a reference list of functions that may be present in an electronic health record system (EHR-S). The function list is described from a user perspective with the intent to enable consistent expression of system functionality. This EHR-S model, through the creation of functional profiles, enables a standardized description and common understanding of functions sought or available in a given setting (e.g., intensive care, cardiology, or office practice in one country or primary care in another country) [28].

Some of the profiles currently under development include

- The *legal EHR profile* identifies a subset of functionality that organizations require to maintain a legally sound health record within an EHR system;
- The *long-term care profile* addresses the needs of the long-term care setting;
- The *emergency care and disaster planning profile* for emergency department information systems identifies appropriate functionality for handling disasters;
- The *behavioral health profile* applies to settings within the behavioral health environment;
- The *child health profile* identifies the functionality for general child healthcare;
- The *regulated clinical research profile* contains a subset of functions to support clinical research [29].

In February 2007, the HL7 EHR System Functional Model became an approved standard of the American National Standards Institute (ANSI). As an approved standard, the EHR System Functional Model will be used in the development of EHR product certification criteria by the U.S. Certification Commission for Healthcare Information Technology (CCHIT) [30].

16.4.5.2 HL7 Clinical Document Architecture

The Clinical Document Architecture (CDA) is a document markup standard that specifies the structure and semantics of clinical documents for the purpose of exchange. A clinical document has the following characteristics:

- *Persistence:* A clinical document continues to exist in an unaltered state, for a time period defined by local and regulatory requirements;
- *Stewardship:* A clinical document is maintained by an organization entrusted with its care;

- *Potential for authentication*: A clinical document is an assemblage of information that is intended to be legally authenticated;
- *Context:* A clinical document establishes the default context for its contents;
- *Wholeness:* Authentication of a clinical document applies to the whole and does not apply to portions of the document without the full context of the document;
- *Human readability:* A clinical document is human-readable.

CDA was approved by the American National Standards Institute (ANSI), in May 2005. CDA Release 2 (January 2005) defines an XML architecture for exchange of clinical documents. Because CDA Release 2 utilizes XML, the HL7 Reference Information Model, and coded vocabularies, CDA documents are both machine-readable and human-readable (e.g., in an XML-aware browser) [31].

Some clinical documents that have been developed using the CDA standard include anesthetic records, claims attachments, genetic testing reports, dental reports, and imaging reports, as well as the Continuity of Care Document (CCD). The CCD is a joint effort of HL7 and the American Society for Testing Materials (ASTM) to define interoperable medical summaries. CCD is a CDA implementation of the ASTM Continuity of Care Record (CCR), which is described below [32].

16.4.5.3 HL7 Reference Information Model (RIM)

According to the HL7 Web site, "The Reference Information Model (RIM) is the cornerstone of the HL7 Version 3 development process. An object model created as part of the Version 3 methodology, the RIM is a large pictorial representation of the clinical data (domains) and identifies the life cycle of events that a message or groups of related messages will carry. It is a shared model between all the domains and as such is the model from which all domains create their messages. Explicitly representing the connections that exist between the information carried in the fields of HL7 messages, the RIM is essential to our ongoing mission of increasing precision and reducing implementation costs."[4]

The HL7 Version 3 Messaging Standard includes RIM.

16.4.6 Longitudinal EHR

The *longitudinal EHR* refers to the capability of the EHR to provide information on a patient's medical history throughout the patient's life, to be ubiquitously accessible, and to be storable in a persistent format. *ISO/TR 21089, Trusted End to End Information Flows*, and *ASTM E31 Continuity of Care Record*, are two standards in wide use that define aspects of the longitudinal EHR. Both are discussed below.

4. "Health Level Seven, Inc. (HL7) Releases Two Version 3 Transport Specifications as Draft Standards for Trial Use (DSTU) : Web Services and ebXML," HL7 Web site, http://www.hl7.org/Press/20040427b.asp.

16.4.6.1 ISO/TR 21089 Trusted End-to-End Information Flows

ISO/TR 21089, Trusted End to End Information Flows, is a published standard issued by the International Standards Organization (ISO). It lists guidelines for trusted end-to-end information flow for healthcare records, key trace points, and audit events in the electronic entity/act record lifecycle, from point of record origination to each point of record access/use. ISO/TR 21089 provides recommendations for:

- Interchange content;
- Accountability;
- Auditability;
- Traceability;
- Audit trails;
- Authentication;
- Data integrity;
- Persistence of health record;
- Persistent health event/act contexts [33].

16.4.6.2 ASTM E31 Continuity of Care Record

The *CCR*, or *Continuity of Care Record*, is a standard specification developed jointly by the American Society for Testing Materials (ASTM) International, the Massachusetts Medical Society (MMS), the Health Information Management and Systems Society (HIMSS), and the American Academy of Family Physicians (AAFP). It is intended to foster and improve continuity of patient care, to reduce medical errors, and to assure at least a minimum standard of health information transportability when a patient is referred or transferred to, or is otherwise seen by, another provider.

The CCR is completed by physicians, nurses, and ancillary providers (e.g., social work, physical therapy, occupational therapy) upon referral or transfer or other transition of a patient from one caregiver to another, whether it is outpatient, inpatient, or community based. Because the CCR is an XML standard document, it is both machine- and human-readable, and the data content may be displayed or printed in a variety of formats, including by Web browser, PDF reader, and word processor [34].

16.4.7 Web Services Standards

16.4.7.1 HL7 Draft Standards for Trial Use for Web Services[5]

In April 2004, HL7 announced that two Version 3 transport specifications—ebXML and Web Services—were approved as draft standards for trial use (DSTU). To quote the press release:

5. "Health Level Seven, Inc. (HL7) Releases Two Version 3 Transport Specifications as Draft Standards for Trial Use (DSTU) : Web Services and ebXML," HL7 Web site, http://www.hl7.org/Press/20040427b.asp.

"Web Services is a way for applications to expose software services using standard interoperability protocols, regardless of the platform on which they are implemented. The development of interoperable standards and the focus on communication and collaboration among people and applications have created an environment where Web Services is becoming the protocol/technology of choice for application integration."

To further quote the HL7 press release on Web Services standards:

"There are many definitions of Web Service, but almost all definitions have these things in common:

• Web Services expose useful functionality to users through the Web Services protocols

• Web services provide a way to describe their interfaces in enough detail to allow a user to build a client application to talk to them. This description is usually provided in an XML document called a Web Services Description Language (WSDL) document

• Web services are registered so that potential users can find them easily. This is done with Universal Discovery Description and Integration (UDDI)

The Web Services Profile for HL7 promotes the use of Web Services to exchange HL7 messages and to ease interoperability between implementations. The profile focuses on basic Web services protocols and technologies like SOAP (Simple Object Access Protocol) and WSDL (Web Services Description Language), which lay the groundwork for more complex interactions based on higher-level Web services specifications."[6]

16.4.7.2 Web Services Interoperability Organization

The Web Services Interoperability Organization (WS-I) is an open industry organization chartered to promote Web Services interoperability across platforms, operating systems, and programming languages. WS-I deliverables include the following:

- *Profiles* provide implementation guidelines for how related Web services specifications should be used together for best interoperability.
- *Sample applications* demonstrate Web services applications that are compliant with WS-I guidelines. These implementations are developed using multiple platforms, languages, and programming tools, demonstrating interoperability in action, and providing readily usable resources for the Web services developer.

Testing tools are used to determine whether the messages exchanged with a Web service conform to WS-I guidelines. These tools monitor the messages and analyze the resulting log to identify any known interoperability issues [9].

6. What is HL7? HL7 Web site, http://www.hl7.org/about/hl7about.htm#WhatisHL7.

16.4.8 EMR/EHR Privacy and Security

16.4.8.1 Introduction

Privacy and security regulations for protection of personal health data vary widely from one nation to another, based on differing legal environments, cultural differences, and degree of maturity of healthcare IT systems.[7]

16.4.8.2 European Union

In an effort to harmonize the laws of European member states, the European Union adopted the *Directive 95/46/EC of the European Parliament and of the Council of 24 October 1995 on the Protection of Individuals with Regard to the Processing of Personal Data and on the Movement of Such Data* in October 1998. While this document does not specifically address issues of privacy and security related to healthcare information technology, it covers the more general area of automatic and manual processing of personal data of an identifiable person. Directive 95/46/EC establishes that a person has the right to give or withhold consent for the processing of his or her data, and that those responsible for the processing must ensure that the data is protected against alteration or unauthorized access. The individual nations of the European Union are responsible for implementing Directive 95/46/EC in national law [35].

To more specifically address issues related to the automatic processing of medical data, and the transborder flow of such data, the Council of Europe issued *Recommendation No. R (97) 5 on the Protection of Medical Data* in 1997 [36]. This recommendation states that medical data should only be collected and processed by healthcare professionals with appropriate safeguards provided by domestic law. It further states that transborder flow of medical data may only take place if the states involved have ratified the Council of Europe's *Convention for the Protection of Individuals with regard to Automatic Processing of Personal Data* [37].

16.4.8.3 United States

The Health Insurance Portability and Accountability Act (HIPAA), first passed in 1996 and amended by the Administrative Simplification Compliance Act of 2001, provides a uniform national standard for electronic healthcare transactions and code sets, and regulations for healthcare information privacy and security. In general, HIPAA regulations apply to all "covered entities"—healthcare providers, insurers, health plan providers, and claims clearinghouses that process claims on behalf of healthcare providers. HIPAA also affects business associates of healthcare providers and insurers, including medical transcription services, software vendors, and IT consultants.

7. For an overview of healthcare-related privacy regulations from an international perspective, see R. J. Rodrigues, P. Wilson, and S. J. Schanz, "The Regulation of Privacy and Data Protection in the Use of Electronic Health Information: An International Perspective and Reference Source on Regulatory and Legal Issues Related to Person-Identifiable Health Databases," Pan American Health Organization, World Health Organization Division of Health Systems and Services Development, 2001.

Title II of HIPAA, Administrative Simplification, provides seven sets of rules that affect online healthcare:

- Electronic healthcare transactions;
- Health information privacy;
- Unique identifiers for employers;
- Security requirements;
- Unique identifiers for providers;
- Unique identifiers for health plans;
- Enforcement procedures.

HIPAA Standards for Privacy of Individually Identifiable Health Information

The HIPAA Standards for Privacy of Individually Identifiable Health Information became effective in April 2003 [38]. The Privacy Rule provides general rules for use and disclosure of protected health information for healthcare facilities and other covered entities.

The HIPAA privacy rules are intended to ensure that protected health information is used only for healthcare purposes, and that access to this information is limited to those with a need to see it. All medical records, including EHRs, are covered by the HIPAA privacy rules.

HIPAA regulations require that all covered entities appoint a privacy officer who is responsible for implementing the rules, and for developing a set of privacy procedures that ensure the privacy of protected health information. Since the health record is stored electronically, an important part of ensuring privacy is providing proper access control through the use of passwords, encryption, firewalls, and physical security of equipment. The IT staff must work closely with the privacy officer to develop and implement HIPAA-compliant privacy procedures.

HIPAA Security Standards

The HIPAA Security Standards, Final Rule, became effective for large covered entities in April 2005, and for small health plans in April 2006 [39].

While the privacy rules cover all forms of healthcare information, whether written, spoken or electronic, the security standards focus specifically on the EHR, and cover all healthcare entities that maintain or transmit electronic health information. The HIPAA security standards require covered entities to perform a risk analysis to determine what level of risk is acceptable, and to determine which security and control measures are most cost-effective.

16.4.9 The Personal Health Record

In order to give healthcare consumers more access to and control of their personal health information, several healthcare organizations have proposed the adoption of a standard for a personal health record (PHR). The PHR would provide the information needed to make healthcare decisions over a person's lifetime, and would be ubiquitously available online. While the information in the PHR would come from healthcare professionals, it would be owned and managed by the healthcare consumer. The healthcare consumer would reserve the right to limit access to the PHR. Since the PHR would contain personally identifiable health information, it would be subject to privacy and security considerations similar to the EHR.

The PHR would not replace the EHR, but would contain a subset of the information in the EHR. Some of the information in the PHR might include

- Information on allergies or adverse drug reactions;
- Current and past medication history;
- A record of treatments and hospitalizations;
- Vaccination records;
- Lab tests and results;
- Family history.

Individuals would have the ability to create their own PHR, or to use a PHR created by a healthcare provider, insurer, employer, or healthcare product vendor. A PHR might be paper-based, personal computer based, available on the Internet, or on a storage medium such as a DVD or USB drive [40].

16.5 The Nationwide Health Information Network in the United States

In the United States, considerable progress has been made towards the development of the Nationwide Health Information Network (NHIN). Conceptually, the NHIN is a "network of networks" that will securely connect healthcare consumers, healthcare providers, and others who require access to health-related data and services. From an architectural standpoint, the NHIN utilizes a shared architecture and therefore does not require a central database of patient-related data. Participants in the NHIN would include:

- Healthcare providers that use EHRs;
- Consumer organizations that provide and maintain PHRs;
- Health information exchanges (HIEs) that enable the transfer of healthcare-related data among local, state, or regional organizations;
- Secondary users such as public health departments, research organizations, and healthcare quality assurance [41].

16.5.1 Advantages of the NHIN

Benefits (see Figure 16.21) could include

- *Reducing the number of medical errors.* The fast and ubiquitous access to patient records and other medical information provided by the NHIN could reduce the number of medical errors due to inadequate information regarding a patient's history, prescribed medication, and current condition.
- *Restraining the rising cost of providing healthcare.* Use of an electronic network for transmittal of healthcare-related information could reduce the over-

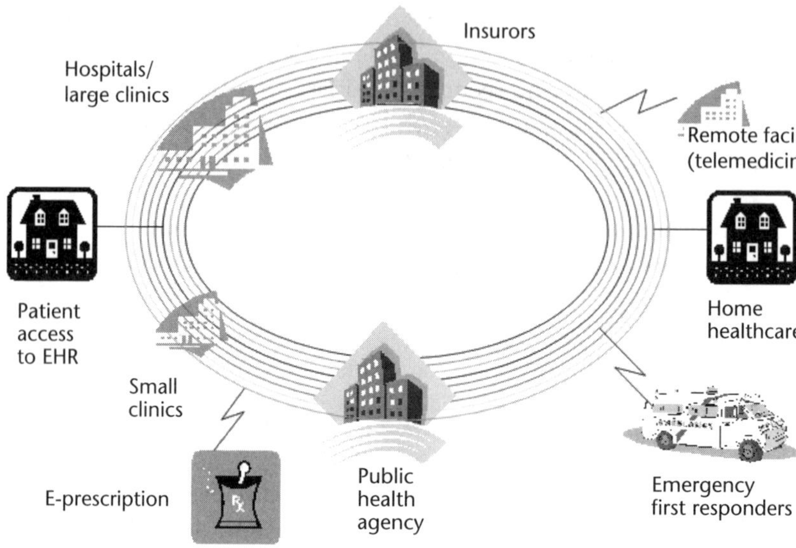

Figure 16.21 NHIN conceptual view.

all cost of providing healthcare by eliminating much of the current paper-based processing of patient records and healthcare insurance claims.

- *Providing fast access to healthcare information in emergency situations.* Implementation of an NHIN could enable the sharing of confidential medical record information with enough speed and accuracy to be of value to a physician examining an emergency patient at a remote site.
- *Coordination of federal health information systems.* Use of the NHIN could expand access to healthcare information by healthcare providers and consumers. The NHIN must interoperate with existing health information systems to make sharing of public health information, healthcare delivery, reimbursement, and oversight efficient and cost-effective.
- *Use of telehealth systems to provide access to healthcare services in remote, rural, or otherwise underserved areas.* A NHIN could make use of information technology to expand healthcare coverage to underserved areas.
- *Accelerate adoption and standardization of electronic health records (EHR).* Medical records in electronic format could be more easily transmitted across the NIHN than current paper records—which would require laborious processing before transmission. EHRs that could be opened by any remote healthcare provider is the only practical means of making such records available to emergency room physicians at remote sites. If such records were also fully standardized, they could also be used for research, quality assurance, and other purposes at state and national levels.

16.5.2 Implementing the NHIN: Implications for Healthcare IT

Participating in the NHIN presents some challenges for the healthcare IT networks maintained by healthcare providers. First and foremost, healthcare IT networks must be developed with interoperability in mind. Interoperability can be implemented through the use of standardized vocabularies, technologies, and protocols. Some specific issues to address include

- *Implementation of IT standards.* By implementing the healthcare IT standards discussed in Section 16.4 above, the healthcare provider ensures that the exchange of healthcare information with other healthcare providers will be seamless and transparent.
- *Planning for external access.* Design of standard healthcare databases such as the CDR and MPI should include capability for secure external access, so that functions such as patient identification and EHR transfer can be made available to local, state, and national network entities. One example would be the use of Web Services interfaces for these functions to enable external access.
- *Availability.* While many health IT networks function on a 24-hour "around the clock" basis, availability of healthcare-related information may be limited to business hours at small clinics and private practices. In these cases, provision should be made to periodically back data up to local or regional repositories, so that it may be accessed by other healthcare providers if necessary.

References

[1] *Brief Report of the AAFP's EHR Pilot Project: Key Learnings from Six Small Family Practices.* American Academy of Family Physicians Center for Health Information Technology, March 8, 2005, p. 4.

[2] Thompson, T., and Brailer, D., *The Decade of Health Information Technology: Delivering Consumer-Centric and Information-Rich Healthcare—Framework for Strategic Action,* U.S. Department of Health and Human Services, July 21, 2004.

[3] *Health Information Technology in the United States: The Information Base for Progress.* Robert Wood Johnson Foundation, 2006, p. 1:3.

[4] *EMR Sophistication Correlates to Hospital Quality Data: Comparing EMR Adoption to Care Outcomes at UHC Hospitals, Including Davies Award Winners, Using HIMSS Analytics' EMR Adoption Model Scores,* HIMSS Analytics, 2006, http://www.himss.org/content/files/UHCresearch.pdf.

[5] *IEEE 802.3-2002, Carrier Sense Multiple Access with Collision Detection (CSMA/CD) Access Method and Physical Layer Specifications,* Institute of Electrical and Electronics Engineers, 2002.

[6] Stallings, W., *Data and Computer Communications,* Upper Saddle River, NJ: Prentice-Hall, Inc., 1997.

[7] Cypher, D., N. Chevrollier, and N. Montavont, et al.,"Prevailing Over Wires in Healthcare Environments: Benefits and Challenges," *IEEE Communications Magazine,* Vol. 44, No. 4, April 2006, pp. 56–63.

[8] *HIPAA Security Guidance,* U.S. Department of Health and Human Services, http://www.cms.hhs.gov/SecurityStandard/Downloads/SecurityGuidanceforRemoteUseFinal122806.pdf, September 2007.

[9] WS-I Web site, http://www.ws-i.org/, September 2007.

[10] Freier, A., P. Karleton, and P. Kocher, "The SSL Protocol, Version 3.0," http://wp.netscape.com/eng/ssl3/draft302.txt, September 2007.

[11] RSA Web site, http://www.rsasecurity.com, September 2007.

[12] "Digital Signature Standard (DSS)," *Federal Information Processing Standards Publication 186*, National Institute of Standards and Technology, http://www.itl.nist.gov/fipspubs/fip186.htm, September 2007.

[13] Barker, R., and P. Massiglia, *Storage Area Network Essentials: A Complete Guide to Understanding and Implementing SANs*, New York: John Wiley & Sons, Inc., 2002.

[14] Jepsen, T., *Distributed Storage Networks: Architecture, Protocols and Management*, Chichester, UK: John Wiley & Sons, Ltd., 2003.

[15] St. George, J., "Analysis of Approaches to Uniquely Identifying Patients in Massachusetts," Massachusetts Health Data Consortium, 2004, p. 7.

[16] Appavu, S. I., "Analysis of Unique Patient Identifier Options," Department of Health and Human Services, http://www.ncvhs.hhs.gov/app0.htm, September 2007.

[17] "Patient Identifier Cross-Reference (PIX) and Patient Demographic Query (PDQ) HL7 V3," *IHE IT Infrastructure Technical Framework Supplement 2007-2008 (Version for Public Comment)*, Integrating the Healthcare Enterprise, http://www.ihe.net/Technical_Framework/upload/IHE_ITI_TF_Suppl_PIX_PDQ_HL7v3_PC_2007_06_25.pdf, September 2007.

[18] Henderson, M., and Y. Bao, "Patient Identifier Cross-Referencing for MPI (PIX)," IHE IT Infrastructure Technical Committee, http://www.ihe.net/Presentations/upload/PatientIdXref_webinar_2005-08-18.ppt, September 2007.

[19] Paragraph 164.530, Administrative Requirements, Subsection (j)(2), Implementation Specification: Retention Period, *HIPAA Final Privacy Rule*, http://www.hhs.gov/ocr/hipaa/privrulepd.pdf , September 2007.

[20] Paragraph 164.316, Policies and Procedures and Documentation Requirements, Subsection (b)(2)(i), Time Limit, *HIPAA Final Security Rule*, http://www.cms.hhs.gov/SecurityStandard/Downloads/securityfinalrule.pdf , September 2007.

[21] *IEEE Standard Computer Dictionary: A Compilation of IEEE Standard Computer Glossaries*, IEEE, 1990.

[22] IEEE 1073/ISO TC215 WG2.1/CEN TC251 WGIV, "Point of Care Medical Device Communications," http://ieeexplore.ieee.org/xpl/standardstoc.jsp?isnumber=11758 , September 2007.

[23] "SNOMED, HL7 Sign Associate Charter Agreement," *SNOMED International News Release*, June 23, 2005, http://www.snomed.org/news/documents/062305_E_HL7_Assoc_Char_FINAL_000.pdf , September 2007.

[24] International Health Terminology Standards Development Organization, http://www.ihtsdo.org/.

[25] "LOINC to CPT Mapping," http://www.nlm.nih.gov/research/umls/mapping_projects/loinc_to_cpt_map.html, September 2007.

[26] "International Classification of Diseases (ICD)," World Health Organization, http://www.who.int/classifications/icd/en/, September 2007.

[27] Beebe, M., et. al., *Current Procedural Terminology (CPT) 2004*, Chicago, IL: AMA Press, 2004.

[28] HL7 Web site, http://www.hl7.org/ehr/, September 2007.

[29] Dougherty, M., "It's Official: HL7's EHR Model Becomes Approved Standard," *Journal of AHIMA*, Vol. 78, No. 4, May 2007, pp. 56–57.

[30] "HL7 Announces Industry's First Electronic Health Record System (EHR-S) Functional Requirements Standard," http://www.hl7.org/documentcenter/public/pressreleases/20070221.pdf , September 2007.

[31] "HL7 Receives ANSI Approval of Three Version 3 Specifications Including CDA, Release 2," HL7 Web site, http://www.hl7.org/Press/20050505.pdf , September 2007.

[32] "Health Level Seven's Clinical Document Architecture Gains Momentum in Healthcare Industry," http://www.hl7.org/documentcenter/public/pressreleases/20070227.pdf , September 2007.

[33] "ISO/TR 21089, Health Informatics—Trusted End-to-End Information Flows," *International Standards Organization*, 2004, Geneva, Switzerland. An abstract can be viewed at http://www.iso.ch/iso/en/CatalogueDetailPage.CatalogueDetail?CSNUMBER=35645, September 2007.

[34] *Continuity of Care Record: The Concept Paper of the CCR*, http://www.astm.org/COMMIT/E31_ConceptPaper.doc, September 2007.

[35] *Directive 95/46/EC of the European Parliament and of the Council of 24 October 1995 on the Protection of Individuals with Regard to the Processing of Personal Data and on the Movement of Such Data*, http://ec.europa.eu/justice_home/fsj/privacy/docs/95-46-ce/dir1995-46_part1_en.pdf and http://ec.europa.eu/justice_home/fsj/privacy/docs/95-46-ce/dir1995-46_part2_en.pdf, September 2007.

[36] *Recommendation No. R (97) 5 on the Protection of Medical Data*, http://www1.umn.edu/humanrts/instree/coerecr97-5.html, September 2007.

[37] *Convention for the Protection of Individuals with Regard to Automatic Processing of Personal Data*, http://conventions.coe.int/Treaty/en/Treaties/Html/108.htm, September 2007.

[38] *Standards for Privacy of Individually Identifiable Health Information (45 CFR Parts 160 through 164)*, Department of Health and Human Services, http://www.hhs.gov/ocr/hipaa/privrulepd.pdf , September 2007.

[39] *Health Insurance Reform: Security Standards; Final Rule (45 CFR Parts 160, 162, and 164)*, Department of Health and Human Services, http://www.cms.hhs.gov/SecurityStandard/Downloads/securityfinalrule.pdf , September 2007.

[40] Heubusch, K., "Piecing Together the PHR," *Journal of AHIMA*, Vol. 78, No. 4, April 2007.

[41] *Summary of the NHIN Prototype Architecture Contracts: A Report for the Office of the National Coordinator for Health IT*, Department of Health and Human Services, http://www.hhs.gov/healthit/healthnetwork/resources/ summary_report_on_nhin_Prototype_architectures.pdf , September 2007.

About the Editors

Robin Felder, Ph.D. is a professor of pathology and associate director of clinical chemistry at the University of Virginia, Charlottesville, VA. He received his Ph.D. at Georgetown University (Washington, DC), and completed a postdoctoral fellowship at the National Institutes of Health in Bethesda, MD. He has been at UVA for 24 years, where he served as the director of the Medical Automation Research Center from 1986–2008. He was founder and president of the Association for Laboratory Automation (http://labautomation.org), and Medical Automation (http://medicalautomation.org).

Majd Alwan, Ph.D., is the director of the Center for Aging Services Technologies (CAST), at the American Association of Homes and Services for the Aging (AAHSA), Washington, DC. He received his masters of science with distinction from Bradford University (Bradford, UK), and his Ph.D. from Imperial College of Science, Technology, and Medicine, University of London (London, UK). Prior to joining CAST in 2007, he was assistant professor and director of the robotics and eldercare technologies at the University of Virginia's Medical Automation Research Center (MARC), in Charlottesville, VA. Dr. Alwan is one of the founding members of CAST. He served as a commissioner for CAST since its inception in 2003 where he chaired the Funding and Pilots committees.

Mingjun Zhang received his doctor of science degree from Washington University in St. Louis, and his Ph.D. from Zhejiang University, P. R. China. He also holds an M.S. in bioengineering and an M.S. in electrical engineering from Stanford University. Since January 2008, he has been an associate professor of biomedical engineering program at the Department of Mechanical, Aerospace and Biomedical Engineering at the University of Tennessee, Knoxville. From 2001–2007, he worked on research and development for bioinstrumentation and DNA/protein microarray technology with the Life Science and Chemical Analysis Division of Agilent Technologies, Santa Clara, California. He is an associate editor of *IEEE Transactions on Automated Science and Engineering*, and a member of the Nanomedicine Editorial Board. He was awarded the Early Government/Industrial Career Award by the IEEE Robotics and Automation Society in 2003.

List of Contributors

Majd Alwan, Ph.D.
Director
Center for Aging Services Technologies (CAST)
2519 Connecticut Ave. NW
Washington, DC 20008-1520
Email: Malwan@aahsa.org

Dr. Hua Cao
Research Assistant
Computer Science Department
Louisiana State University
Baton Rouge, LA 70803
Email: hcao@csc.lsu.edu

David Deng
Director of Assay Development
Artemis Health, Inc.
480 Arsenal Street Suite 130
Watertown, MA 02472
Email: ddeng@artmishealthinc.com

Robin A. Felder
Professor of Pathology
Department of Pathology
The University of Virginia
P.O. Box 800403, Charlottesville, VA 22903
Email: rfelder@virginia.edu

Elizabeth A. Flynn
Center for Pharmacy Operations and Designs
Harrison School of Pharmacy
128 Miller Hall
Auburn University, AL 36849-5506
Email: flynnel@auburn.edu

Thomas Jepsen
Consultant—IT Solutions Online
Chair, IEEE-USA Medical Technology Policy Committee
515 Morgan Creek Road
Chapel Hill, North Carolina 27517-4931
Email: tjepsen@ieee.org

Ou Ma
Associate Professor of Mechanical Engineering
P.O. Box 3001, MSC 3450
New Mexico State University
Email: oma@nmsu.edu

Dr. Habib Mohamadian
Professor and Dean of College of Engineering
Southern University and A & M College
P.B.S. Pinchback Room 206
Phone: (225) 771-5290
E-mail: mohamad@engr.subr.edu

Timo R. Nyberg
Professor and Dr.
GoodIT Ltd.
Tekniikantie 12, 02150 Espoo Finland
Email: timo.nyberg@goodit.f
timornyberg@yahoo.com

Karen S. Rheuban
Department of Pediatrics
P.O. Box 800386
University of Virginia Health System
Charlottesville, VA 22908
Email: krheuban@virginia.edu

Jonathan M. Sackier
Department of Pathology
The University of Virginia
P.O. Box 800403, Charlottesville, VA 22903
Email: sackier@aol.com

Mark W. Scerbo
Department of Psychology
Room 346A, Mills Godwin Building
Old Dominion University
Norfolk, VA 23529-0267
Email: mscerbo@odu.edu

Norbert Stoll
Center for Life Science Automation—celisca
Friedrich-Barnewitz-Strasse 8
Email: norbert.stoll@uni-rostock.de

Weimin Tao
Senior Engineer
Brooks Automation, Inc.
401 Charcot Ave.
San Jose, CA 95131-1192
Email: david.tao@sbcglobal.net

Yue Xie
Administrative Director, Finance
St. Luke's Sugar Land Hospital
St. Luke's Episcopal Health System
1317 Lake Pointe Pkwy
Sugar Land, TX, 77478
Phone: (281) 277-1214
Email: yxie@sleh.com

Zhengmao Ye
Assistant Professor
Department of Electrical Engineering
Southern University and A&M College
408 PBS Pinchback Hall
Baton Rouge LA 70813
Email: zhengmaoye@engr.subr.edu

Mingjun Zhang
Associate Professor of Biomedical Engineering
414 Dougherty Engineering Building
1512 Middle Drive
The University of Tennessee
Knoxville, TN 37996-2210
Email: mjzhang@utk.edu

Index

D

E

N